ENVIRONMENTAL FLUID DYNAMICS

ENVIRONMENTAL FLUID DYNAMICS

Flow Processes, Scaling, Equations of Motion, and Solutions to Environmental Flows

JÖRG IMBERGER

Centre for Water Research, University of Western Australia, Australia

ELSEVIER

Amsterdam • Boston • Heidelberg • London
New York • Oxford • Paris • San Diego
San Francisco • Sydney • Tokyo
Academic Press is an imprint of Elsevier

Academic Press is an imprint of Elsevier
The Boulevard, Langford Lane, Kidlington, Oxford OX5 1GB, UK
Radarweg 29, PO Box 211, 1000 AE Amsterdam, The Netherlands
225 Wyman street, Waltham, MA 02451, USA
525 B Street, Suite 1900, San Diego, CA 92101-4495, USA

First edition 2013

Library of Congress Cataloging-in-Publication Data
Imberger, Jorg.
Environmental fluid dynamics: fluid processes, flow scales and processes, and equations of motion / Jorg Imberger.
p. cm.
Includes bibliographical references and index.
ISBN 978-0-12-088571-8 (alk. paper)
1. Aquatic ecology. 2. Fluid dynamics. 3. Marine eutrophication. I. Title.
QH541.5.W3I43 2012
577.6--dc23

201207724

British Library Cataloguing in Publication Data
A catalogue record for this book is available from the British Library

Printed and bound in USA
Transferred to Digital Printing in 2014

ISBN: 978-0-12-088571-8

To

Barbara, Nicole and Rohin

Thank you for your support, guidance and love.

CONTENTS

PREFACE

Nature uses the properties of water and air to deliver nutrients and dispose of all its waste products. Nature's strategy is based on three underlying principles. First, water and air move, so waste is removed from the immediate vicinity of the disposal site and so is "out of sight, out of mind". Second, as the fluid moves the waste, turbulence disperses the concentrated waste quickly and effectively and dilutes the waste, to a small fraction of the original concentration, rendering it accessible for biological recycling. Third, once diluted the waste is transformed to food for other organisms by microbes, billions of workers transforming waste from one organism to food for another. This is the basis of all evolution, organisms filling niches where others have created a habitat with appropriate temperature, chemical composition, food and protection from predators.

Human traditional engineering designs adopted this strategy most effectively, but never closed the loop. Designs were and still are, based on the misconception that the aquatic and atmospheric domains are infinite and that caretaker microbes would always be available to take render harmless the waste being emitted into the domains. As we now know this is no longer the case, our chimneys, our sewage outfalls, our chemical disposal sites, both solid and liquid and most recently the increasing number of large-scale accidental spills are no longer being dispersed to low enough concentrations to be transformed to food in small contained domains; the initial concentrations are increasingly effecting very large areas and are, as a consequence, killing a huge number of living plants and animals, depleting our biodiversity at an alarming rate. Even more importantly, this depletion is breaking the very engine that sustains the transformation from waste to food and so the whole strategy is beginning to show signs of leading to a major ecological break down.

The transport and mixing that a fluid habitat services the biogeochemical world with are thus more important than ever. Historically, effluent removal by fluid transport was the key, assuming that somehow nature would do the rest. Then in the 1960s, it was observed that our wastes were too toxic and had to be diluted in the near field. Thus, an emphasis was placed on diffuser designs, engineering structures that would quickly and effectively dilute the waste with local ambient fluid. In the last 10 years, engineers have started to think in terms of "assimilative capacity" were the biota can effectively carry out the transformation from waste to food, before being impacted on by a neighboring effluent. However, at present this is only a vague concept, with the pressures of development usually outweighing the domain constraints of assimilative capacity.

This book is an attempt to provide some of the tools for waste cycling, rather than disposal, where all three factors are given equal importance. First, removal of the waste to a domain size where sufficient dilution may take place to allow for the biological transforms, second ensure appropriate dispersal to levels conducive for the microbial world and third, to provide a far field fluid environment that has mixing properties, from large domain scales to microbial rates of strain scales that foster biota that perform the transformations of waste to food.

Clearly, this is an enormous task, so I have focused on providing the basic overview, sufficient for the general environmental engineer and scientist, to a level that will allow the specialist to then turn to advanced texts in the particular special area of interest.

On and off I have taught this material over the last 40 years, at Berkeley, Caltech, Karlsruhe, Padova and Western Australia. Chapter 1 is material for a first one-semester course, bringing together students from different undergraduate backgrounds and giving them a common foundation in ideas and notation. Chapters 2 and 3 provide material for a second course after which the student may claim to have some knowledge of fluid dynamics. Chapters 4, 5, 6 and 7 present material from which an instructor may wish to put together a third one-semester course designed for second year graduate students in environmental engineering and/or science. Chapters 8 and 9 are designed to provide the foundation for a serious student in Physical Limnology, Estuarine Dynamics or Coastal Oceanography. A student having completed a one-semester course using these two chapters should be able to move onto specialist text in Limnology, Oceanography and even Meteorology.

ACKNOWLEDGMENTS

This book arose from the simple need to find a text that provided the foundations of environmental fluid dynamics; the study of fluid transport, dispersion, mixing and small-scale straining for courses I was teaching in this subject over the last 30 years or so. Then when I started writing an advanced book on Physical Limnology about 20 years ago, the need became acute. There is simply no text available that either has the right mix of material or includes recent results from the field. On realizing this, I put writing the book on the Physical Limnology on hold and started to put my Environmental Fluid Dynamics lecture notes in order and here is the result. Any project, such as this, taking close to 20 years to complete will, by default, include explicitly and implicitly, influence from many people.

The confidence that I can make a contribution was given to me by Harry Levey when I came, in 1963, to Western Australia to take up a tutorship in Mathematics at UWA. Harry simply assumed that I would do something useful with my life and in the process also taught me a great deal of mathematics. I shall always be grateful to him.

Most of the fluid dynamics I know, I absorbed while I was a graduate student at Berkeley. Frederick Sherman taught me to order my thoughts, Gilles Corcos provided an antidote to too much order and John Wehausen opened the world of rational mechanics to me, this formed the basis of Chapter 2. Hugo Fischer consolidate the ability to act independently and the Berkeley campus atmosphere is responsible for my inability to cope with bureaucratic authority. Berkeley in the late 1960s and early 1970 was simply wonderful!

The year at Caltech made me aware of the whole field of environmental fluid dynamics and I was very fortunate to form friendships with Norman Brooks, Frederich Raichlan and Don Cohen. Norman has been a long-term mentor, thank you Norman, you may take credit for this book as it was you who showed me the sincerity inherent in nature.

On returning to Western Australia, I was very fortunate to have the support of Stewart Turner in his position as Chair of the Special Research Centre Programme. Without the funding from this programme, that lasted for 18 years and the understanding protection by the then Vice Chancellor, Robert Street, I would never had the research freedom that I have enjoyed. My brief study leave at Stanford also made a great impression on me in that

I gained a much deeper knowledge of turbulence during that stay. Robert Street thank you!

Then there were numerous mentors that visited CWR over time, most numerous times. Retaining focus in a far-off place requires visits by friends and mentors who are willing to provide guidance. In the general area of environmental fluid dynamics, I am indebted to George Veronis, Owen Phillips, David Halpern, Stephen Thorpe, Ernie Tuck, Tony Maxworthy, Tom Berman, David Farmer, Robert Street, Ian Wood, Bill Wiebe, Larry Armi, Sally MacIntyre, John Melack, Colin Reynolds, George Batchelor, Walter Debbler, Angus McEwan, Philip Saffman, Peter Rhines, Larry Redekopp, Chris Garrett and Peter Baines. All left behind from their visit some new ideas for me to build on and good advice on how to pursue new fields of endeavor.

The field research team, that has supported me over the last 20 years must also take credit for what is in this book. Without the sensor eyes that Roger Head provided me with, fluid dynamics would never have come alive for me. Carol Lam makes real-time data stream come onto our computer screens from far-flung places around the globe; Greg Attwater's organizational skills makes every field trip a exhilarating experience where science is always at the fore; logistic problem never cloud the experience, Lee Goodyear is the real-time web guru and Caroline Wood provided support for all, in this 20 year adventure.

In terms of research collaboration, I have enjoyed the support, in the last 8 years, of Clelia Marti, whose passion for environmental fluid dynamics surpasses even mine. I would also like to thank all the 90 or so graduate students who I have had the privilege to supervise and who have contributed to my education. Where ever you are, thank you!

In closing, I would also sincerely thank Mohanambal Natarajan from Elsevier for her fantastic support in bringing this book to print.

CHAPTER 1

Physical Quantities, Dimensional Analysis, Scaling and Bulk Conservation Equations

Contents

In this chapter, we present an overall introduction to the remainder of this book, from the point of view of a student who has had an undergraduate education in applied science or engineering or a researcher in the chemical or biological sciences who wishes to gain an introduction to fluid dynamics.

Environmental Fluid Dynamics
ISBN 978-0-12-088571-8, DOI: 10.1016/B978-0-12-088571-8.00001-2

The calculus that is required to understand the material is summarized in the Appendix.

1.1. PHYSICAL QUANTITIES

Physical quantities are entities that quantify and describe some particular physical state. Everybody is familiar with quantities such as the volume of a bottle of milk, the weight of a suitcase, the speed of a motorcar and so on. As we shall see later such physical quantities may differ widely, but all possess two main attributes, a number to describe the magnitude or size of the quantity and the unit of measure used in assessing the magnitude or size. The particular unit chosen for a particular physical quantity is not uniquely determined, but depends on a number of factors. Tradition is certainly the main determinant. In Europe, the unit to measure the distance between two places is called the kilometer (equivalent to 1000 m) whereas in the US, people use miles to describe the same physical quantity. How did this come about? To see this, consider what determines the choice of the unit. It is very convenient to match the unit to the magnitude so that the mind is confronted with manageable numbers as magnitudes. Again consider a simple example. The size of the old small format photographic negative was specified as 35 mm, whereas the medium format negative was specified as 6 cm × 6 cm. We see immediately, from this example, that the units, mm and cm, were chosen in each case to keep the magnitude 35 and 6 manageable.

On the other hand, if we speak of the width of a road we would quite naturally speak of a certain number of meters, but if we speak about the length of the road between say two cities we would again quite naturally use kilometers or miles. All of this came about as it is convenient for our mind to visualize the physical quantity when we communicate, and this is best done in manageable units for which we have a feel. However, there is a second very important underlying reason for matching the choice of units to the magnitude of a physical quantity and that is the inherent accuracy when communicating the information. In most common experiences, it is sufficient to specify a physical quantity to an accuracy of, say, one part between 10 and 1000. Obviously, this is not universal, but for common usage this would usually suffice. Hence, we speak of the distance between two cities is say 876 km, the width of a road is say 20.5 m, the diagonal of a computer screen is say 40.5 cm and the size of a pixel is 0.25 mm. The mind associates the unit with the quantity being specified and through every day use is able to quickly visualize the "size" of the quantity. Accompanying this visualization is the accuracy of the specification. The necessary accuracy is achieved with a reasonable and manageable number of digits when specifying the magnitude. It is therefore not surprising that different units developed in different countries such as the foot in England and the US and the meter in the European countries. The foot being a measure of a "standard" foot and the meter being, in the first instance, a measure of a step, both derived from human experience. The third influence in describing physical units was the choice of the ease of subdivision. For instance, there are 1000 m in a kilometer, 100 cm in a meter, 10 mm in a cm and so on. On the other hand, there are 12 in. in a foot and 3 feet in a yard or 36 in. in a yard. The number 36 may be divided by 36, 18, 12, 6, 4, 2, and 1, a total of seven possible divisions. By contrast a meter has 100 cm and so has the following simple divisions, 100, 50, 25, 10, 5, 1 a total of only six divisions; in general the decimal system has the advantage of ease of computation and the 12 system has the advantage of having slightly more simple divisions, making mental calculations easier. These differences have receded with the increasing use of electronic aids to do calculations.

The photographic image illustrates a further transition brought about by the digital world, making it convenient to use the building block, or the pixel, as the unit. This is convenient as it measures the resolution of the image, but it really implies nothing about the linear size or quality of the image, only the storage size. Digital building blocks have their own magnitudes when related to the original physical quantity.

The choice of units is, however, not arbitrary. As we know, the world is governed by a series of physical laws. The most relevant one to our study of fluid mechanics is Newton's law of motion that states, for a single particle, simply:

$$\text{Force} = \text{Mass} \times \text{Acceleration} \tag{1.1.1}$$

Hence, if we fix units for mass (kg) and acceleration ($\mathrm{m\,s^{-2}}$) then we cannot arbitrarily also fix the unit for force, as then the magnitude derived by multiplying the magnitudes of the mass and acceleration would not match our force magnitude. Hence, each country has now adopted a consistent set of units such as (meters, kilograms, seconds, Newtons) in Europe or (feet, slug, seconds, pounds) in the US. This means that the chosen units are consistent with the physical laws in which they occur.

Scalars: A scalar physical quantity is a quantity that is uniquely defined by specifying a unit of measurement and the quantity of these units as a number.

Simple examples are area (m^2), volume (m^3), time (s), speed ($m\,s^{-1}$) and power ($kg\,m^2\,s^{-3}$ or W). For a scalar, it is sufficient to specify a single number indicating the amount and a set of units indicating the size of the measure.

Consider further the concept of a simple area S_0,

$$[S_0] = m^2, \tag{1.1.2}$$

where [ ] denotes the unit of measure or the dimensions of the physical quantity.

Suppose we are dealing with a simple rectangle of width B and length L, then

$$S_0 = LB \tag{1.1.3}$$

Both sides of (1.1.3) have the same unit (m^2) and we may note that:

$$\left[\frac{S_0}{LB}\right] = 1 \tag{1.1.4}$$

indicating that the quotient $\frac{S_0}{LB}$ is dimensionless. It is often convenient to write our equations in a dimensionless form as in (1.1.4) so that consistency of units is easily recognized. Suppose we are dealing with the volume, V, under a surface as shown in Fig. 1.1.1 and the surface, S, is described by:

$$x_3 = f(x_1, x_2). \tag{1.1.5}$$

The volume V, defined by (1.1.5), is given by:

$$V = \int_a^b \int_c^d f(x_1, x_2)\mathrm{d}x_2\mathrm{d}x_1. \tag{1.1.6}$$

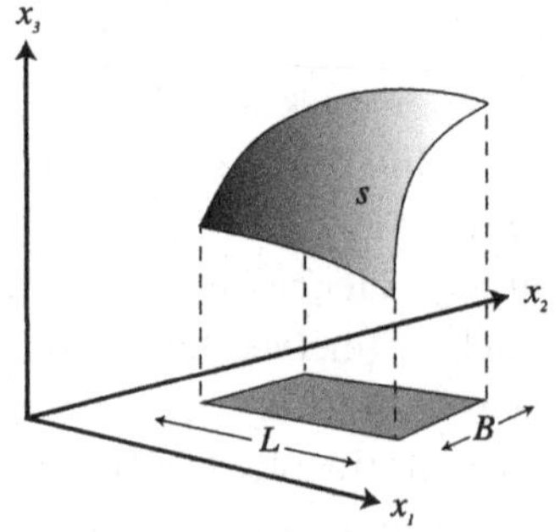

Figure 1.1.1 Surface subtended by a rectangle in the base plane.

Suppose we define the mean height of $f(x_1, x_2)$ by $\bar{f}$ then we may rewrite (1.1.6) in the form:

$$V^* = \int_{a^*}^{a^*+1} \int_{b^*}^{b^*+1} f^*(x_1^*, x_2^*)\mathrm{d}x_2^*\mathrm{d}x_1^*, \tag{1.1.7}$$

where we have introduced the new dimensionless variables:

$$\begin{aligned} x_1^* &= \frac{x_1}{L}, \\ x_2^* &= \frac{x_2}{B}, \\ f^* &= \frac{f}{\bar{f}}, \\ a^* &= \frac{a}{L}, \\ c^* &= \frac{c}{B}, \\ V^* &= \frac{V}{BL\bar{f}}. \end{aligned} \tag{1.1.8}$$

The right hand side of (1.1.7) is now a simple dimensionless number. Since we have chosen $\bar{f}$ to be the mean height, the quantity V^* is not only dimensionless, but is also about unity. Rendering an equation dimensionless with magnitudes that make the leading term in the equation about one is called scaling; it is a convenient tool to highlight which terms in an equation are important and which are small and may thus be neglected.

Vectors: A vector is a physical quantity that is made up of three scalar components. Alternatively, a vector is a quantity that requires specification of a magnitude and a direction. Many examples are common experience; velocity ($\mathrm{m\,s^{-1}}$), momentum ($\mathrm{kg\,m\,s^{-1}}$), force ($\mathrm{kg\,m\,s^{-2}}$) and acceleration ($\mathrm{m\,s^{-2}}$).

Implied in the above definition, however, is the concept of direction and the three scalar components are the magnitudes of the physical quantity resolved in the three directions. Now the nature of a physical quantity is such that it is independent of the frame of reference and remains unchanged when the axes (x_1, x_2, x_3) are changed (rotated). This invariance to the orientation may be likened to the invariance of a scalar to a change in units; the magnitude changes so as to keep the physical quantity unchanged.

Thus, if the reference axes are changed from (x_1, x_2, x_3) to (y_1, y_2, y_3) via a rotation transformation, then if a vector v given in (x_1, x_2, x_3) by:

$$\boldsymbol{v} = v_1\hat{\boldsymbol{i}}_1 + v_2\hat{\boldsymbol{i}}_2 + v_3\hat{\boldsymbol{i}}_3, \tag{1.1.9}$$

then this changes to:

$$\tilde{\boldsymbol{v}} = \tilde{v}_1\hat{\boldsymbol{j}}_1 + \tilde{v}_2\hat{\boldsymbol{j}}_2 + \tilde{v}_3\hat{\boldsymbol{j}}_3, \tag{1.1.10}$$

where $(\tilde{v}_1, \tilde{v}_2, \tilde{v}_3)$ are the components of the same vector in the new coordinate system and $(\hat{\boldsymbol{j}}_1, \hat{\boldsymbol{j}}_2, \hat{\boldsymbol{j}}_3)$ are the new unit vectors.

Mathematically, this may be expressed as follows. Suppose that the (y_1, y_2, y_3) and (x_1, x_2, x_3) coordinate systems are connected by the rotation transformation:

$$y_1 = \ell_{11}x_1 + \ell_{21}x_2 + \ell_{31}x_3, \tag{1.1.11}$$

$$y_2 = \ell_{12}x_1 + \ell_{22}x_2 + \ell_{32}x_3, \tag{1.1.12}$$

$$y_3 = \ell_{13}x_1 + \ell_{23}x_2 + \ell_{33}x_3, \tag{1.1.13}$$

that may be shortened to:

$$y_j = \ell_{ij}x_i, \tag{1.1.14}$$

where the repeated subscript i indicates there is a sum over i (see Appendix 1) and $j = (1, 2, 3)$. The linear transformation (1.1.14) is a rotation of the axis so that the length of any vector is preserved. This constraint implies

$$\ell_{ik}\ell_{kj} = \delta_{ij}, \tag{1.1.15}$$

where the symbol

$$\delta_{ij} = \begin{cases} 1 & i = j \\ 0 & \textit{otherwise} \end{cases}. \tag{1.1.16}$$

Multiplying (1.1.14) by $\ell_{j\alpha}$ and summing over j, leads to the inverse of (1.1.14):

$$x_\alpha = \ell_{j\alpha}y_j. \tag{1.1.17}$$

Now the triplet (x_1, x_2, x_3) may be viewed as the position vector of a point in space and (y_1, y_2, y_3) are the components in the transformed coordinate system. Thus, the general definition of a vector $\boldsymbol{a}$ is a quantity

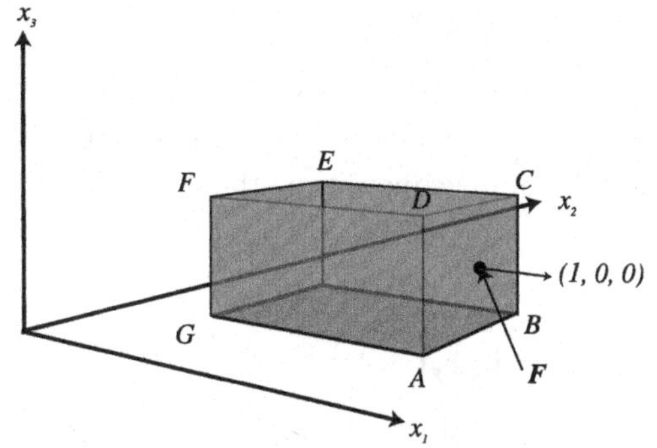

Figure 1.1.2 Infinitesimal volume cube.

that has three components (a_1, a_2, a_3) that transform into components (b_1, b_2, b_3) under a rotation transformation (1.1.14) such that:

$$b_j = \ell_{ij} a_i. \tag{1.1.18}$$

Tensor of second order: Many physical quantities in environmental fluid mechanics are vectors that act on a surface. A simple example of this is the pressure, or more generally the stress $\boldsymbol{\tau}$ on a particular surface. A little reflection reveals that if we fix the surface then the stress $\boldsymbol{\tau}$ (force divided by area) is a vector because force is a vector. On the other hand, the area on which the stress acts has its own orientation and is itself a vector; an area is usually defined by specifying the magnitude of the area (or scalar) and the direction of the unit normal $(\hat{\boldsymbol{n}}_1, \hat{\boldsymbol{n}}_2, \hat{\boldsymbol{n}}_3)$. It is common that we have a particular force acting on a particular area. For example, consider the infinitesimal cube shown in Fig. 1.1.2.

The face (A B C D) in Fig. 1.1.2 has a unit normal $\hat{\boldsymbol{n}}_1 = (1, 0, 0)$ and in general a force $\boldsymbol{F}$ acting on (A B C D) will have components (F_1, F_2, F_3) leading to stresses $(\tau_{11}, \tau_{12}, \tau_{13})$, where τ_{ij} is the component of force in the jth direction on a surface with unit normal in the ith direction; in general such physical quantities have nine components. This is called a second order tensor if all the components are transformed under a rotation so that the net physical quantity remains unchanged. Given that such a transformation acts on two vectors (the force and the surface) the transformation rule for tensors becomes:

$$\tau_{pq} = \ell_{ip} \ell_{jq} \tau_{ij}. \tag{1.1.19}$$

A tensor is thus, a quantity, with nine components, that transform according to (1.1.19) under a rotation (1.1.14) and each component has a magnitude and a unit of measurement.

1.2. DIMENSIONAL ANALYSIS

Physical laws, expressing one physical quantity in terms of other physical quantities, should have the property that the dimensions of the terms on the left and right hand sides of the equation are the same. For instance Newton's second law:

$$\boldsymbol{F} = m\boldsymbol{a}, \tag{1.2.1}$$

defines the units for F such that,

$$[\boldsymbol{F}] = \text{kg ms}^{-2}, \tag{1.2.2}$$

where the square bracket is used to denote units of the physical quantities. In other words, physical laws are dimensionally homogeneous. Another way to express this is to say that, physical laws that model the behavior of physical processes should be independent of the system of units used to define the individual variables in the equation. This property may be used to rewrite any physical law in a non-dimensional form. For example, given (1.2.1) we may write:

$$\frac{\boldsymbol{F}}{m\boldsymbol{a}} = 1, \tag{1.2.3}$$

where now both sides are now non-dimensional. In general, if a particular physical law depends on say n state variables x_i, $i = 1, 2, \ldots n$, then the relationship between the state variables x_i,

$$f(x_1, x_2, x_3, \ldots, x_n) = 0, \tag{1.2.4}$$

must be independent of the set of units used. Here $f(\ldots)$ represents the functional relationship between the independent variables x_i, $i = 2, \ldots, n$. We say that the function $f(\ldots)$ is invariant to a transformation of the units. This observation has far reaching implications as may be seen from the following examples.

Example: Suppose a car is traveling with a speed U m s^{-1} along a straight road and the car and the passengers have a total mass M kg. If the driver suddenly brakes so that a constant force F kg m s^{-2} (Newton's) is applied in the direction opposite to the motion, how far will the car take to stop?

In order to illustrate the power of dimensional reasoning, we shall look at this problem in three distinct ways.

Informal Reasoning: Let L be the distance the car takes to stop. Then we may write (1.2.4) in the form:

$$f(L, U, M, F) = 0. \tag{1.2.5}$$

The dimensions of the state variables are:

$$[L] = \text{m}; \quad [U] = \text{m s}^{-1}; \quad [M] = \text{kg}; \quad \text{and} \quad [F] = \text{kg m s}^{-2}.$$

As noted above, the function f must be independent of the units used to specify L, U, M and F. We notice that only M and F have the unit kg, so that these two variables must occur as the quotient $\frac{F}{M}$ in order to remove the dependence on mass:

$$\frac{F}{M} = \frac{\text{kg m s}^{-2}}{\text{kg}} = \text{m s}^{-2}, \tag{1.2.6}$$

If we further divide this by U^2 then

$$\left[\frac{F}{MU^2}\right] = \frac{\text{m s}^{-2}}{\text{m}^2\ \text{s}^{-2}} = \text{m}^{-1}, \tag{1.2.7}$$

becomes independent of time and has only units of length. The only variable not included yet is the distance L with dimensions m so that:

$$f\left(\frac{LF}{MU^2}\right) = 0, \tag{1.2.8}$$

since,

$$\left[\frac{LF}{MU^2}\right] = 1.$$

Thus, in order for the function f to be non-dimensional the dependent variables must collapse into one group, that we shall call π where

$$\pi = \frac{LF}{MU^2}, \tag{1.2.9}$$

so that (1.2.5) reduces to:

$$f(\pi) = 0. \tag{1.2.10}$$

However, $f(\pi)$ is a simple function of π so the above implies that there exists a solution or root at the equation π_0 such that:

$$\pi = \pi_0. \tag{1.2.11}$$

Hence, a simple expression for the stopping distance is given by:

$$L = \pi_0 \frac{MU^2}{F}. \tag{1.2.12}$$

The value of the constant π_0 cannot be determined by such dimensional analysis; this must be determined through theoretical consideration, numerical modeling or experimentation.

So suppose we carry out a braking experiment and find that:

$$\begin{aligned} M &= 1000\ \text{kg} \\ U &= 60\ \text{km h}^{-1} = 16.66\ \text{m s}^{-1} \\ F &= 10,000\ \text{kg m s}^{-2}(\text{N}) \\ L &= 13.88\ \text{m} \end{aligned}$$

Substituting this into the above formula yields a value:

$$\pi_0 = 5,$$

so that:

$$L = 0.5\frac{MU^2}{F}. \tag{1.2.13}$$

This formula must now hold general and may be used to calculate the stopping distance for other circumstances. For instance, if $U = 80\ \text{km h}^{-1}$, then $L = 24.69\ \text{m} \gg 13.88\ \text{m}$. This is the reason for the $60\ \text{km h}^{-1}$ speed limit.

Two important consequences result from the above discussion. First, simple dimensional invariance provides an expression for the results in term of all the relevant state variables. Second, (1.2.12) provides a powerful aid for the experimentalist. Instead of needing to carry out numerous experiments using different car sizes and different brake forces to learn the characteristics of the car, (1.2.12) shows that it is sufficient to carry out one experiment that will allow the determination of the dimensionless coefficient π_0 and then we have a universal law for all cars under all braking scenarios; such dimensional reasoning finds wide application among experimentalists. In both interpretations, it is assumed that the behavior of the car is governed by (1.2.1).

Formal application of dimensional reasoning: Once again assume that (1.2.5) applies that we repeat here for convenience:

$$f(L, U, F, M) = 0.$$

The assumption that the solution is independent of the system of units used implies that (1.2.5) will be invariant to the transformation:

$$\begin{aligned} \ell &= \alpha\tilde{\ell} \\ m &= \beta\tilde{m} \\ t &= \gamma\tilde{t}, \end{aligned} \tag{1.2.14}$$

where ℓ is the unit of length, m is the unit of mass and t is the unit of time and the tilde superscript indicates the transformed set of units.

Introducing (1.2.14) into (1.2.5) leads to the equation:

$$f\left(\alpha\tilde{L}, \frac{\alpha}{\gamma}\tilde{U}, \frac{\beta\alpha}{\gamma^2}\tilde{F}, \beta\tilde{M}\right) = 0. \tag{1.2.15}$$

Now

$$\alpha = \frac{L}{\tilde{L}}, \beta = \frac{M}{\tilde{M}}, \frac{\alpha}{\gamma} = \frac{U}{\tilde{U}}, \frac{\beta\alpha}{\gamma^2} = \frac{F}{\tilde{F}}, \tag{1.2.16}$$

leading to four equations for the three unknowns α, β and γ. Eliminating α, β and γ we are left with one equation that may, for convenience, be written:

$$\frac{MU^2}{FL} = \frac{\tilde{M}\tilde{U}^2}{\tilde{F}\tilde{L}}, \tag{1.2.17}$$

implying that $\frac{MU^2}{FL}$ is independent of α, β and γ so that:

$$f(\pi) = 0; \quad \textit{where } \pi = \frac{MU^2}{FL}. \tag{1.2.18}$$

This is the same result as (1.2.11).

Power Transformation: This method involves noting that in (1.2.18) the state variables occur to some power and so we may write (1.2.5) in the form:

$$f(L^{\alpha_1}U^{\alpha_2}F^{\alpha_3}M^{\alpha_4}) = 0, \tag{1.2.19}$$

where (α_1, α_2, α_3, α_4) are chosen so that the product is dimensionless. Inserting the units of each state variable leads to:

$$f(\ell^{\alpha_1}\ell^{\alpha_2}t^{-\alpha_2}m^{\alpha_3}\ell^{\alpha_3}t^{-2\alpha_3}m^{\alpha_4}) = 0. \tag{1.2.20}$$

Now as we wish (1.2.20) to be dimensionless, we equate the power of each unit to zero, leading to three equations:

$$\alpha_1 + \alpha_2 + \alpha_3 = 0, \tag{1.2.21}$$

$$-\alpha_2 + 2\alpha_3 = 0, \tag{1.2.22}$$

$$\alpha_3 + \alpha_4 = 0. \tag{1.2.23}$$

These three equations allow us to solve for all three unknowns in terms of the value of α_4:

$$\alpha_3 = -\alpha_4, \tag{1.2.24}$$

$$\alpha_2 = -2\alpha_3 = 2\alpha_4, \tag{1.2.25}$$

$$\alpha_1 = -\alpha_2 - \alpha_3 = -2\alpha_4 + \alpha_4 = -\alpha_4. \tag{1.2.26}$$

Substituting these into the original equation leads to:

$$f(L^{-\alpha_4} U^{2\alpha_4} F^{-\alpha_4} M^{\alpha_4}) = 0, \tag{1.2.27}$$

or

$$f\left(\left(\frac{U^2 M}{LF}\right)^{\alpha_4}\right) = 0. \tag{1.2.28}$$

This may be rewritten in the form:

$$f^*\left(\frac{U^2 M}{LF}\right) = 0. \tag{1.2.29}$$

Once again we have arrived at the same result.

The result from the above three different methods may be generalized into a single theorem.

Buckingham π Theorem: If n dependent state variables x_1, x_2,...,x_n are related by a physical law:

$$f(x_1, x_2, ..., x_n) = 0, \tag{1.2.30}$$

and the variables are themselves depend on m independent units, then the invariance to a change in units requires that there are $(n - m)$ dimensional groups or π's, such that:

$$f(\pi_1, \pi_2, ..., \pi_{n-m}) = 0. \tag{1.2.31}$$

However, while this theorem is true, considerable care must be taken when applying this theorem to ensure that the units used in the state variables are independent. We illustrate this with an example.

Example: A sound wave is propagating through a fluid medium that has a density ρ kg m^{-3} and a bulk modulus of compressibility E_v kg m^{-1} s^{-2}. How does the speed of sound depend on ρ and E_v?

Answer: Given that sound travels in the form of compression-expansion waves we may write, without loss of generality:

$$f(C, \rho, E_v) = 0, \tag{1.2.32}$$

where $[C] = \mathrm{m\,s^{-1}}$, $[\rho] = \mathrm{kg\,m^{-3}}$ and $[E_v] = \mathrm{kg\,m^{-1}\,s^{-2}}$.

Superficially, it would appear, from (1.2.32) that $n = 3$ and $m = 3$, so Buckingham's π Theorem implies that there are no dimensionless groups, yet it is easy to see that:

$$\pi = \frac{C}{\left(\frac{E_v}{\rho}\right)^{\frac{1}{2}}}, \tag{1.2.33}$$

is a dimensionless group. This apparent contradiction may be resolved by noting:

$$\begin{aligned} [C] &= \mathrm{m\,s^{-1}} \\ [\gamma] &= \mathrm{kg\,m^{-3}} \\ [E] &= (\mathrm{kg\,m^{-3}})(\mathrm{m\,s^{-1}})^2 \end{aligned} \tag{1.2.34}$$

So we see that we have only two independent units $\mathrm{m\,s^{-1}}$ and $\mathrm{kg\,m^{-3}}$, hence $m = 2$.

Once the right independent units have been identified, the theorem and the informal method for obtaining the non-dimensional groups can provide great insight into most problems. As a last example, consider:

Example: What is the pressure in a fluid at depth h?

Answer: Suppose the density of the fluid is $\rho\ \mathrm{kg\,m^{-3}}$, the depth underneath the surface is $h\ m$ and the acceleration due to gravity is $g\ \mathrm{m\,s^{-2}}$, then

$$\begin{aligned} f(p, h, g, \rho) &= 0 \\ [p] &= \mathrm{kg\,m\,s^{-2}\,m^{-2}} = \mathrm{kg\,m^{-1}\,s^{-2}} \\ [h] &= \mathrm{m} \\ [g] &= \mathrm{m\,s^{-2}} \\ [\rho] &= \mathrm{kg\,m^{-3}}, \end{aligned} \tag{1.2.35}$$

so that $n = 4$, $m = 3$, resulting in one π. Simple inspection implies:

$$f\left(\frac{p}{\rho g h}\right) = 0, \tag{1.2.36}$$

so that:

$$p = \pi_0(\rho g h). \tag{1.2.37}$$

This result has a simple interpretation. Suppose we think of a one square meter horizontal surface at depth h, then $\rho g h$ is equal to the weight of the column of fluid that this 1 m^2 supports. Hence the pressure is just equal to the weight of the fluid above the point. Since pressure acts perpendicularly to the vertical surfaces of the column, all the weight of the column is taken by the lower square meter:

$$\pi_0 = 1, \tag{1.2.38}$$

and

$$p = \rho g h + p_0, \tag{1.2.39}$$

where p_0 is the pressure acting on the free surface.

1.3. FLUID PROPERTIES

1.3.1. Nature of Fluids: Gases and Liquids

The basic characteristic that distinguishes a fluid from a solid is that a fluid cannot resist a shearing stress without continuous deformation. Whenever a shearing stress is applied, as shown, for example in Fig. 1.3.1, the fluid will accelerate until the associated internal rate of strain is high enough for the internal resistance in the fluid to equal the applied external load and the upper plate velocity reaches a steady state velocity of V.

In this sense, both gases and liquids are similar. By contrast a solid will deform under an applied shear stress, but once deformed will come to rest provided, of course, that the solid can resist the force without yielding. The difference between a gas and a liquid is best understood by looking at the molecular structure of the two. From molecular theory, it is known that the force on a molecule of a common substance, such as water, in the

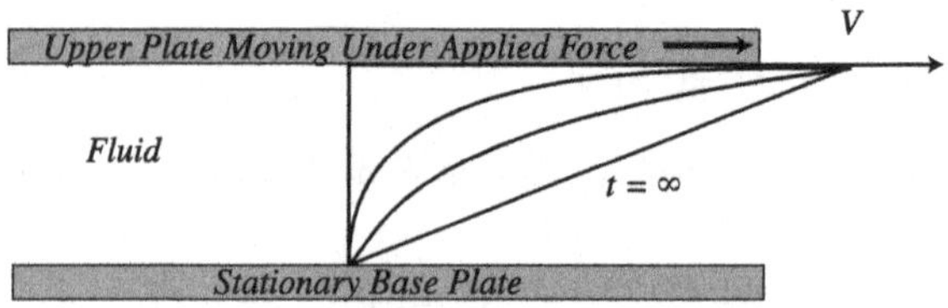

Figure 1.3.1 Flow between two plates.

vicinity of another molecule changes from one of weak attraction at large separation distances to one of repulsion at very small distances; this is shown schematically in Fig. 1.3.2.

For the case of water, when two molecules are closer than about 3.5×10^{-8} m they repel each other; for separation distances larger than this value, the molecules attract each other. By contrast, gases are at equilibrium under atmospheric pressure at distances nearer to 3.5×10^{-7} m or about 10 times greater than the spacing in liquids. The implications of this, put simply, is that molecules have more room to move in a gas than in a liquid. From this it follows that gases are more compressible than liquids and the ability to resist stress is much less in gases, so that they need to experience a much greater rate of distortion in order to balance a particular applied stress than liquids; this ability to resist an applied stress is called the viscosity of fluid.

Because gases have very much larger molecular mean separation, the bonds between the molecules tend to be much weaker that in a liquid and in gases the molecules have a tendency to exchange momentum via inter-molecular collisions rather than inter-molecular forces. The mean free path, or the average distance between molecular collisions, changes from about 3.5×10^{-7} m at atmospheric pressure to infinity in outer space.

During the early part of the nineteenth century, a number of gases, such as carbon dioxide, sulfur dioxide, hydrogen sulfide, ammonia, etc., were

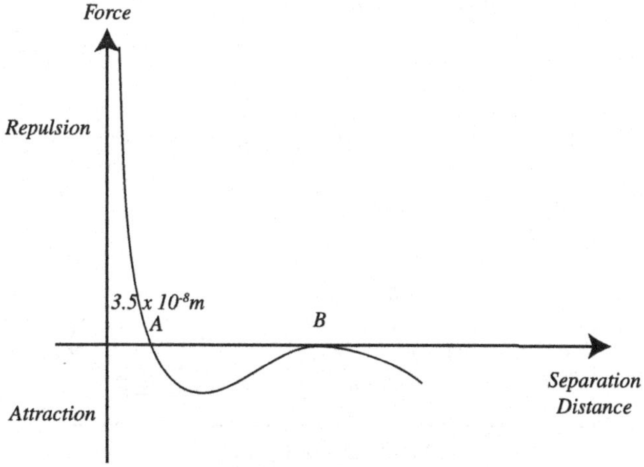

Figure 1.3.2 Schematic of inter-molecular forces.

liquefied by the simultaneous use of high pressure and low temperature. Further, by allowing compressed liquid carbon dioxide to evaporate, the temperature was lowered sufficiently for a solid to form. The same principle is used at the present time in the manufacture of "dry ice", as solid carbon dioxide is called. By mixing solid carbon dioxide with ether, M. Thilorier (1835) was able to obtain temperatures as low as −110 °C, and this permitted the liquefaction of gases such as ethylene, phosphine and silicon tetrafluoride. In spite of numerous attempts, however, involving the use of pressures up to 3000 atm, the gases hydrogen, oxygen, nitrogen and carbon monoxide could not been liquefied. The general opinion in the middle of the last century was, therefore, that certain gases, called "permanent gases", could not be converted into liquids under any circumstances.

The essential conditions for the liquefaction of gases were discovered by T. Andrews (1869) as the result of a study of the pressure–volume–temperature relationships of carbon dioxide. A definite amount of the gas was enclosed in a glass tube kept at a constant temperature, and the volumes at different pressures were measured; the results for a series of constant temperatures were plotted in the form shown in Fig. 1.3.3.

At the lowest temperature employed by Andrews, 13.1 °C, carbon dioxide is entirely gaseous at low pressures, as at A; upon increasing the pressure the volume decreases, as indicated by the curve AB, approximately in accordance with Boyle's law. At the pressure B, however, liquefaction

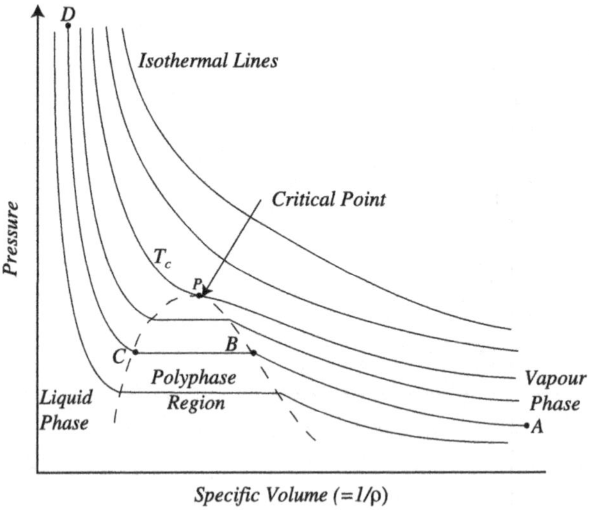

Figure 1.3.3 Isotherms for a typical liquid vapor system.

commences, and the volume decreases rapidly as the gas is converted to liquid with a much higher density; the molecules come within their attraction spacing. At C carbon dioxide has been completely liquefied, and the steepness of the curve CD is evidence of the fact that the liquid is not easily compressed. It should be noted that the portion AB of the isothermal represents gas only, and CD represents liquid only; along BC, however, gas and liquid can co-exist. Since BC is parallel to the volume axis, it follows that the pressure remains constant while gas and liquid are present together, irrespective of the relative amounts of the two forms. The constant pressure represented by BC is called the *vapor pressure* of the liquid at the temperature of the isotherm.

The pressure–volume curve at 21.5 °C is similar to that for the lower temperature, except that the horizontal portion, above which liquefaction occurs, is shorter. As the temperature is raised, this section of the isotherm becomes shorter until, at 31.1 °C and a pressure of 75 atm, the line becomes a point; for temperatures higher than this, the gas cannot be liquefied no matter how much pressure is applied. This result is true in general, except that the limiting temperature varies from gas to gas.

The maximum temperature at which a gas can be liquefied, that is, the temperature above which liquid cannot exist, is referred to as the critical temperature, and the pressure required to cause liquefaction at this temperature is the critical pressure. The pressure–volume curve for the critical temperature is called the critical isotherm. The point P in Fig. 1.3.3 represents carbon dioxide in its critical state, the temperature, pressure and volume being the critical values. It may be remarked that the term vapor is used to describe a gaseous substance when its temperature is below the critical value; a vapor can, therefore, be condensed to a liquid by pressure alone.

Conversely, reducing the pressure will form more vapor. This requires the molecules to move from C to B in Fig. 1.3.2, which requires energy; this energy is called the latent heat of vaporization.

In the limit of larger temperatures the curves in Fig. 1.3.3 approach a hyperbola and are given by the relationship:

$$pv = RT, \tag{1.3.1}$$

where p is the pressure, v is the volume per unit mass, R ($= 8.314472\,\mathrm{J\,K^{-1}\,mol^{-1}}$) is the universal gas constant. A mole of a substance has the volume of that substance containing 6.022×10^{23} molecules; this number of carbon molecules has a mass of 12 g). A gas that obeys such an ideal hyperbola is called ideal.

1.3.2. Vapor Pressure

Consider a beaker full of water at 20 °C with the upper water surface being exposed to the air at atmospheric pressure assumed to be 101.33 kN m^{-2}, the water surface will be exposed to a mixture of air and water because there is a tendency of water molecules to escape from the liquid water surface. The total atmospheric pressure P_A is the sum of the partial water vapor pressure P_V and the pressure due to the air molecules or the air pressure P_G:

$$P_A = P_G + P_V. \tag{1.3.2}$$

Now the vapor pressure of water at 20 °C is 2.34 kN m^{-2} so that, if at any time $P_V < 2.34$ kN m^{-2} then the water will evaporate until $P_V = 2.34$ kN m^{-2} and an equal number of water molecules re-enter the water as leave the surface. The pressure in the water will of course remain equal to the applied atmospheric pressure $P_A = 101.33$ kN m^{-2}.

When the temperature is raised to 100 °C, $P_V = 101.33$ kN m^{-2} and vapor bubbles form within the water and remain in balance with the outside pressure. This condition is called the boiling of the fluid.

1.3.3. The Continuum Concept

It is clear that if we are to analyze the motion of a fluid (gas or liquid) then we should really analyze the motion of each molecule and then sum over all the molecules present accounting for inter-molecular forces and collisions.

In fluid mechanics, as applied to engineering, the gases and liquids we deal with are nearly always at a pressure above atmospheric. There are exceptions such as cavitation in a pump, but it is better to treat these separately. This means that the largest molecular spacing, that of an ideal gas, is about 10^{-7} m and so there are at least 10^{12} molecules in a cubic millimeter of fluid. Thus even with micro-scale instruments, one that can measure fluid properties at a scale of about 1 mm, the sensor will measure that particular property as the average of 10^{12} molecules; the instrument thus measures the average, over all molecules, of say the mean free path, which is the mean velocity, kinetic energy and so on. Such microstructure instruments have been used to measure, for instance, the temperature of water in lakes, rivers and the ocean and even the atmosphere. It is interesting to ask what the smallest scale is at which the temperature varies in such environments. This scale is derived later where we show that even for highly energetic environments this scale is around 1 mm, meaning that between a scale of 10^{-3} m and 10^{-7} m for a gas and 10^{-8} m for a liquid, the mean properties of the molecules do not change. Continuum mechanics thus deals only with the average properties of the molecules and assigns these mean values to the centroid of the averaging volume; these mean properties are well-defined for any averaging volume greater than a few molecular spacing and less than 1 mm. All conservation equations in continuum mechanics deal only with these mean properties and there is no attempt made to derive equations for each molecule and then average; as we shall see the conservation equations are derived, from first principles, for the average properties and applied to volumes larger than 1 mm; such volumes are called fluid particles. It is important to visualize such a fluid particle. It is a collection of molecules over which we average all properties at a particular point in space and at a particular time. However, as time varies, the molecules will not always be the same; molecules continuously escape and enter the fluid particle. In this way, neighboring fluid particles transfer their properties to their neighbors; molecules escaping from a fluid particle with a large momentum into a fluid particle with less mean momentum will cause the slower particle to accelerate; this is the origin of internal friction. Similarly, molecules in a fluid particle at high temperature will have a high level of excitation; when they escape and move into a neighbor with a lower temperature, they will excite the molecules in their new environment and so raise the temperature of the neighbor. This is called diffusion of heat.

1.3.4. Mass of a Fluid

Density ρ: In line with the continuum hypothesis defined above, the density ρ of a fluid is given by:

$$\rho = \lim_{V \to 0} \left(\frac{\textit{Mass of all molecules}}{\textit{Volume, } V} \right) (\mathrm{kg\,m^{-3}}), \tag{1.3.3}$$

where the limit is taken to between 1 mm and 10^{-5} mm. This density may change from point to point or even with time and in general:

$$\rho = \rho(x_1, x_2, x_3, t). \tag{1.3.4}$$

Once the density is defined it is convenient to form specific definitions for related properties that are often encountered in environmental flows.

Specific weight γ ($\mathrm{N\,m^{-3}}$):

$$\gamma = g\rho. \tag{1.3.5}$$

Specific gravity s:

$$s = \frac{\textit{density of a fluid}(\mathrm{kg\,m^{-3}})}{\textit{density of pure water at } 4\,^\circ C\ (=1000)}. \tag{1.3.6}$$

Sigma t σ_t ($\mathrm{kg\,m^{-3}}$):

$$\sigma_t = \rho - 1000. \tag{1.3.7}$$

Specific volume υ ($\mathrm{m^3\,kg^{-1}}$):

$$\upsilon = \frac{1}{\rho}. \tag{1.3.8}$$

1.3.5. Pressure and Compressibility

Consider a fluid sample inside a press where a piston exerts a pressure on the fluid sample. The pressure p exerted by the piston is defined as the force per unit area (F/A). The units of pressure are $\mathrm{N\,m^{-2}}$ or Pascals. The fluid may be compressed while keeping the temperature constant (isothermal compression) by extracting any excess heat, or the container may be insulated and heat prevented from escaping (adiabatic compression); two pressure volume curves result because during adiabatic compression the temperature of the fluid rises and, as it does, so does the pressure.

From common experience, we know that gases are more compressible than liquids; the reason for this lies in the greater molecular spacing in gases. Hence the pressure rises much faster in liquids for the same compression ratio as for gases. The ease with which a fluid may be compressed is quantified by the compressibility of the fluid. The bulk coefficient of compressibility of a fluid, E_υ, is defined as the change in pressure for a unit fractional volume change curves:

$$E_\upsilon = -V\frac{\partial p}{\partial V} = \rho\frac{\partial p}{\partial \rho} = -\upsilon\frac{\partial p}{\partial \upsilon}, \tag{1.3.9}$$

and so E_υ has units of pressure. Obviously E_υ will have a different value depending whether the volume changes are brought about under isothermal or adiabatic conditions; typical values for isothermal compression for water are shown in Table 1.3.1.

1.3.6. Temperature

The temperature of a fluid is a direct measure of the translational and rotational energy of the molecules making up the fluid particle. In practice,

Table 1.3.1 Bulk Modulus of Water (M Pascals) for Isothermal Compression

Pressure M Pascals	Temperature (°C) 0°	20°	49°	93°
0.10133	1965	2179	2264	2057
10.34	1999	2213	2300	2100.3
31.019	2066	2282	2373	2187.8
103.98	2305	2523	2631	2496.3

we do not actually measure the energies directly, but rather establish a scale, called the temperature, which is related to these molecular energies. We introduce this scale, called degrees Kelvin, so that it is zero when the molecular energy in the fluid is zero and 273 K at the "hotness" of ice. The number 273 was chosen so that the temperature of boiling water, at atmospheric pressure, is 100 units larger than that of the temperature of ice. In this way, one unit on the absolute temperature scale corresponds to one unit in the familiar Celsius scale.

Temperature may be measured with many different sensors, all of which rely on some secondary property of the material of the sensor such as an expansion, change of resistance or change of electro potential that can be visualized:

i) Glass thermometer: Uses the expanding mercury in a glass tube.
ii) Bi-metallic strip: Two metals with different coefficients of temperature expansion are glued together; the differential expansion causes the composite material to bend, the degree of displacement can then be measured.
iii) Thermistor: A sintered material that changes its resistance with temperature.
iv) Thermocouple: A junction of two different materials generated a potential dependent on the temperature.
v) Quartz Crystal: A quartz crystal is caused to resonant using an oscillating electrical potential; the oscillation frequency depends on the volume of the crystal, which turns to be a function of the temperature.

1.3.7. Specific Heat

Suppose the fluid is in equilibrium under a certain pressure and at a certain temperature and a small quantity of heat dH per unit mass is added. If the volume is kept constant, the pressure will rise as the input heat raises the temperature and we may define the effectiveness of the heating process by calculating the specific heat:

$$C_v = \left.\frac{dH}{d\theta}\right|_{v=const}, \tag{1.3.10}$$

the units of which are J kg^{-1} °C^{-1}.

Alternatively, we may keep the pressure constant as we add the heat, then the volume will increase slightly; the specific heat at constant pressure is defined as:

$$C_p = \left.\frac{dH}{d\theta}\right|_{p=const}, \tag{1.3.11}$$

which again has units J kg^{-1} °C^{-1}.

When the heat is added at constant pressure, the fluid does external work on the chamber and so only part of the thermal input is stored as thermal energy (temperature) and thus $C_p \geq C_v$. (For liquids $C_p/C_v \approx 1$ and for air $C_p/C_v \approx 1.4$).

1.3.8. Coefficient of Thermal Expansion

Heating a fluid under constant pressure leads to an expansion of the fluid as discussed above. Such an expansion decreases the density of the fluid, which, if in a lake or the ocean, being less dense, will rise to the surface. Such motion is called natural convection and is a very important process in environmental situations; we define a thermal expansion coefficient:

$$\alpha = -\frac{1}{\rho}\frac{d\rho}{d\theta} = \frac{1}{v}\frac{dv}{d\theta}, \tag{1.3.12}$$

the units of which are °C^{-1}.

1.3.9. Viscosity

Consider a rather simple experiment. A fluid is sandwiched between two flat plates and the top plate is set into motion by the application of a force F (see Fig. 1.3.1). It is found experimentally, once steady state is reached, that the stress exerted by the plate causes the fluid to move, such that:

$$\tau = \frac{F}{A} = \mu\frac{U}{h}, \tag{1.3.13}$$

where τ is the shear stress, U is the terminal velocity of the upper plate and h is the plate spacing. The coefficient of proportionality μ (kg m^{-1} s^{-1}) is called the *coefficient of viscosity*; honey has a high and water a relatively low value of the coefficient of viscosity.

We have learned that molecules vibrate randomly about some mean fluid particle motion. In doing this they collide with their neighboring molecules and thus exchange momentum. In a gas an increase in temperature leads to an increased vibration energy and exchange and thus the viscosity increases with increasing temperature. In a liquid increasing the temperature weakens the inter-molecular bonds and this effect predominates, thus the viscosity decreases with increasing temperature. In mechanics, the acceleration of a particles is equal to force per unit mass (1.1.1) and so, if we divide (1.3.13) by the density, we see that quantity μ occurs as the ratio μ/ρ. It is common to absorb the density variations into the coefficient of viscosity and so we define the *kinematic viscosity*:

$$\nu = \frac{\mu}{\rho}, \tag{1.3.14}$$

that has units $m^2\,s^{-1}$.

1.3.10. Surface Tension

At a fluid interface molecules form a bond within the fluid interface that provides the interface with a certain degree of elasticity, known as *surface tension*. The surface tension, defined as the force per unit length of interface, of liquids at an air–liquid interface covers a wide range values depending on the fluid. Typical values of the surface tension of water are presented in Table 1.3.2. When the interface comes into contact with

Table 1.3.2 Typical Values of the Surface Tension of Water

°C	Surface tension, σ Nm^{-1}
0	0.0756
10	0.0742
20	0.0728
30	0.0712
40	0.696
60	0.0662
80	0.0626
100	0.0589

a solid, the interface attaches at an angle, called the *capillary angle*. This angle may cause the fluid to "wet" the solid (e.g. water on glass) or repulse the solid (e.g. mercury on glass). The surface tension stress is this attached to the solid at the capillary angle; if the surface wets the solid, the surface tension will have an upward component and if it repulses the solid it will have a downward component.

Capillary rise in a tube is depicted in Fig. 1.3.4. From free-body considerations, assuming the meniscus is spherical and equating the lifting force created by surface tension to the gravity force leads to the vertical force balance:

$$2\pi r\sigma \cos\theta = \pi r^2 h\gamma, \tag{1.3.15}$$

$$h = \frac{2\sigma \cos\theta}{r\gamma}, \tag{1.3.16}$$

where

σ = surface tension in units of force per unit length
γ = specific weight of liquid
r = radius of tube
h = capillary rise
θ = capillary contact angle

If the tube is clean, $\theta = 0°$ for water and about 140° for mercury. For tube diameters larger than 1/2 in. (12 mm), capillary effects are negligible. If mercury is in contact with water, the surface-tension effect is slightly less

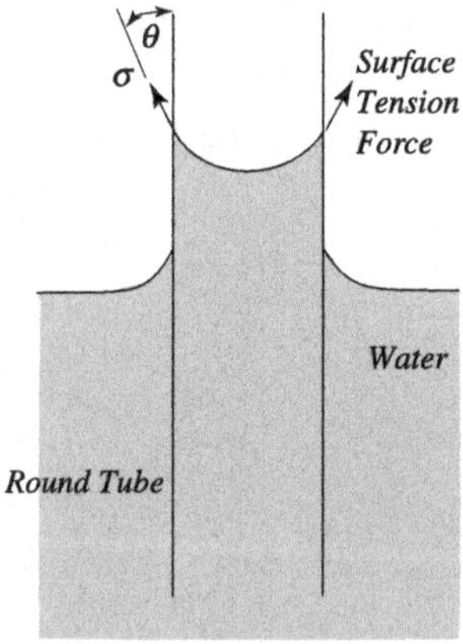

Figure 1.3.4 Capillary rise in a tube.

than when in contact with air. Surface tension decreases slightly with increasing temperature.

Surface-tension effects are generally negligible in most environmental flows; however, they may be important in problems involving capillary rise, the formation of drops and bubbles, the breakup of liquid jets, and in hydraulic model studies where the model is small, all of which are important in water treatment processes.

1.3.11. The Equation of State of Air

At normal environmental temperatures dry air follows the perfect gas law (1.3.1) very closely. When the air is moist as is the case for air in the meteorological boundary layer above a lake, estuary or coastal sea we must take into account the influence of the water vapor. Since the relative proportions of gases that are found in dry air are essentially constant (volume ratios, nitrogen 78.1%, oxygen 21.0% and carbon dioxide 0.9%) the state of the air may be described by giving only the concentration of water vapor. The specific humidity q is defined as the mass of water vapor per unit mass of air including the water vapor.

The equation of state for water vapor is again given by the perfect gas law (1.3.1), which may be written as:

$$p_w = R_w T \rho_w , \tag{1.3.17}$$

where $R_w = 461.50\ \mathrm{m^2\,s^{-2}\,{}^\circ C^{-1}}$ and T is the absolute temperature.

For a mixture of air and water vapor, we have the total thermodynamic pressure is the sum of the partial pressure, so that:

$$p = p_a + p_w, \tag{1.3.18}$$

and from the definition of specific humidity it follows:

$$\rho_w = \rho q, \tag{1.3.19}$$

where conservation of mass implies

$$\rho = \rho_a + \rho_w. \tag{1.3.20}$$

Substituting (1.3.1) and (1.3.17) into the partial pressure equation (1.3.18) and using (1.3.19) and (1.3.20) to eliminate ρ_a and ρ_w leads to:

$$p = \rho T_a\{(1-q)R_a + R_w q\}. \tag{1.3.21}$$

Now recognizing that R_w and R_a are constants in ratio,

$$\frac{R_w}{R_a} = 1.60779, \tag{1.3.22}$$

then equation (1.3.21) may be rewritten as:

$$p = R_a T_v \rho, \tag{1.3.23}$$

where

$$T_v = T_a\{1 + 0.60779q\}, \tag{1.3.24}$$

is called the virtual temperature (i.e. the temperature of dry air with the same density as the wet air).

1.3.12. The Equation of State of Water

The equation of state for water containing solutes depends on the relative concentration of the solutes as well as the concentration of the total dissolved substances. For ocean water the ratio of solutes is remarkably constant (ionic mass ratio chloride 55%, sodium 30%, sulphide 8%, magnesium 4%, potassium 1% and calcium 1%) so the state may be defined by specifying the total mass of dissolves salts. This is called the salinity S and is defined as the total mass of solute divided by the mass solution. In practice, the salinity is calculated from the temperature and conductivity that defines a practical salinity (see UNESCO). Given the salinity S, the temperature θ and the pressure p, the equation of state defines all other variables such as the density in terms of these.

Traditionally this has been done by introducing the secant bulk modules of elasticity of the water:

$$E_v(S,\theta,p) = \rho(S,\theta,0)\frac{p}{(\rho(S,\theta,0) - \rho(S,\theta,p))}, \tag{1.3.25}$$

where p is the pressure above atmospheric and θ is the temperature in degrees Celsius.

Rewriting (1.3.25) we can get an expression for the density ρ in terms of the pressure p and the secant modulus $E\vartheta$:

$$\rho(S,\theta,p) = \frac{\rho(S,\theta,0)}{1 - \dfrac{p}{E_v(S,\theta,p)}}. \tag{1.3.26}$$

The density of seawater at one standard atmosphere ($p = 0$) can be determined from

$$\rho(S,\theta,0) = \rho_w + (b_0 + b_1\theta + b_2\theta^2 + b_3\theta^3 + b_4\theta^4)S + (c_0 + c_1\theta + c_2\theta^2)S^{3/2} + d_0 S^2, \tag{1.3.27}$$

where

$b_0 = 8.24493 \times 10^{-1}$ $\quad c_0 = -5.72466 \times 10^{-3}$

$b_1 = -4.0899 \times 10^{-3}$ $\quad c_1 = 1.0227 \times 10^{-4}$

$b_2 = 7.6438 \times 10^{-5}$ $\quad c_2 = -1.6546 \times 10^{-6}$

$b_3 = -8.2467 \times 10^{-7}$

$b_4 = 5.38753 \times 10^{-9}$ $\quad d_0 = 4.8314 \times 10^{-4}$

The density of the reference pure water is given by:

$$\rho_w = a_0 + a_1\theta + a_2\theta^2 + a_3\theta^3 + a_4\theta^4 + a_5\theta^5, \tag{1.3.28}$$

where

$a_0 = 999.842594$
$a_1 = 6.793952 \times 10^{-2}$
$a_2 = -9.095290 \times 10^{-3}$
$a_3 = 1.001685 \times 10^{-4}$
$a_4 = -1.120083 \times 10^{-6}$
$a_5 = 6.536332 \times 10^{-9}$

The secant bulk modulus (E_v) of seawater is given by:

$$E_v(S,\theta,p) = K(S,\theta,0) + Ap + Bp^2 \tag{1.3.29}$$

where

$$E_v(S,\theta,p) = K_w + (f_0 + f_1\theta + f_2\theta^2 + f_3\theta^3)S + (g_0 + g_1\theta + g_2\theta^2)S^{3/2}, \tag{1.3.30}$$

where

$f_0 = 54.6746$ $\quad g_0 = 7.944 \times 10^{-2}$

$f_1 = -0.603459$ $\quad g_1 = 1.6483 \times 10^{-2}$

$f_2 = 1.09987 \times 10^{-2}$ $\quad g_2 = -5.3009 \times 10^{-4}$

$f_3 = -6.1670 \times 10^{-5}$

$$A = A_w + (i_0 + i_1\theta + i_2\theta^2)S + j_0 S^{3/2}, \qquad (1.3.31)$$

where

$$i_0 = 2.2838 \times 10^{-3} \qquad j_0 = 1.91075 \times 10^{-4}$$
$$i_1 = -1.0981 \times 10^{-5}$$
$$i_2 = -1.6078 \times 10^{-6}$$

$$B = B_w + (m_0 + m_1\theta + m_2\theta^2)S, \qquad (1.3.32)$$

$m_0 = -9.9348 \times 10^{-7}$
$m_1 = 2.0816 \times 10^{-8}$
$m_2 = 9.1697 \times 10^{-10}$

The pure water terms of the secant bulk modulus are given by:

$$K_W = e_0 + e_1\theta + e_2\theta^2 + e_3\theta^3 + e_4\theta^4, \qquad (1.3.33)$$

$e_0 = 19652.21$
$e_1 = 148.4206$
$e_2 = -2.327105$
$e_3 = 1.360477 \times 10^{-2}$
$e_4 = -5.155288 \times 10^{-5}$

$$A_w = h_0 + h_1\theta + h_2\theta^2 + h_3\theta^3, \qquad (1.3.34)$$

$h_0 = 3.239908$
$h_1 = 1.43713 \times 10^{-3}$
$h_2 = 1.16092 \times 10^{-4}$
$h_3 = -5.77905 \times 10^{-7}$

$$B_w = k_0 + k_1\theta + k_2\theta^2, \qquad (1.3.35)$$

$k_0 = 8.50935 \times 10^{-5}$
$k_1 = -6.12293 \times 10^{-6}$
$k_2 = 5.2787 \times 10^{-8}$

The above equation of state is valid for $S = 0$–42; $\theta = -2$–$40\,°\mathrm{C}$; $p = 0$–$10^8\ \mathrm{kg\,m^{-1}\,s^{-2}}$ (approximately to a depth of 10 km or 100 bar) whenever the ionic concentration ratios are the same as those of seawater.

The above relationships are valid for the great majority of lakes around the world, but there are numerous water bodies where these equations do not hold. In such circumstances, fundamental measurements must be carried out to determine the equation of state.

1.4. FLOW DOMAINS, SCALING AND MODELING

In a fluid flow, state variables may be visualized as a function of a particular fluid particle and the observer follows a particular particle with time. This is called a Lagrangian reference system. The Lagrangian reference system is easy to visualize, but hard to match to observations at a point, as it is not known, *a priori*, which particle will be at the exact location where the observation is being carried out. Alternatively, we may imagine a particular state variable being given at each point in space as a function of time; in this instance the value of the state variable at a point and particular time is the value of the fluid particle that is at that point at that time; such an array of values, functions of space and time are called fields in an Eulerian reference system.

For instance, consider the velocity of a fluid. The exact definition of this is the velocity of the fluid particle that is at a particular point at a particular time. If, by some means, it is possible to identify the particle that is at every point in a domain at each time, then it is possible to write the fluid velocity as a function of position and time; a vector field. In general if A is a scalar, $\boldsymbol{B}$ is a vector and $\boldsymbol{C}$ is a tensor, defined on a domain Ω, we write:

$$A = A(x_1, x_2, x_3, t), \tag{1.4.1}$$

$$\boldsymbol{B} = \boldsymbol{B}(x_1, x_2, x_3, t), \tag{1.4.2}$$

$$\boldsymbol{C} = \boldsymbol{C}(x_1, x_2, x_3, t). \tag{1.4.3}$$

For $\boldsymbol{B}$ we must define three scalar functions B_1, B_2, and B_3 and for $\boldsymbol{C}$ we must specify nine scalar functions C_{11}, C_{22}, C_{33}, C_{12}, C_{13}, C_{21}, C_{23}, C_{31}, C_{32}. Such physical quantities are called scalar, vector and tensor fields defined on the domain Ω.

It is often advantageous to have a measure or scale of the magnitude of a field, and scaling of physical variables defining fields is a very important concept in environmental fluid dynamics. It is a technique for determining one single quantity for the magnitude, size or scale of a field, even though the field varies over a domain and time.

This concept is best introduced with an example. Consider a body of a certain shape and size immersed in a fluid domain Ω as shown in Fig. 1.4.1. Suppose the size of the body along the longest axis is L. We may be interested in the flow of the fluid around the body; intuitively the flow field will "feel" the body in a domain roughly the "size" of the body itself, i.e. order L. If (x_1, x_2, x_3) are the reference space coordinates, then if we measure the size of the domain in units of L, we can say the scale of (x_1, x_2, x_3) is L. This is written as:

$$x_1 \sim L, \tag{1.4.4}$$

$$x_2 \sim L, \tag{1.4.5}$$

$$x_3 \sim L, \tag{1.4.6}$$

or another notation is to use the symbol O standing for "order"

$$x_1 = O(L), \tag{1.4.7}$$

$$x_2 = O(L), \tag{1.4.8}$$

$$x_3 = O(L). \tag{1.4.9}$$

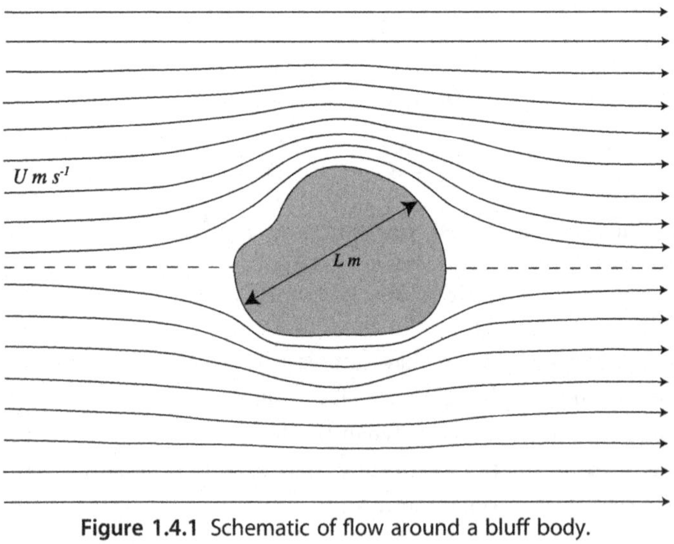

Figure 1.4.1 Schematic of flow around a bluff body.

The exact technical definition of the O symbol is explained in books on asymptotic expansions and is given by:

$$A = O(B), \tag{1.4.10}$$

if

$$\lim_{\alpha \to \alpha_c} \frac{B}{A} = \beta, \tag{1.4.11}$$

where β is a constant at each point in the whole domain and α is a parameter that is made to approach a limit α_c. The exact nature of this limiting procedure will become clearer below. Here we imagine this as a process where a limit is approached and in that limit if the variable B is doubled anywhere, it will double everywhere. To be concrete, if the flow in Fig. 1.4.1 is increased at the extreme right by say a factor of two then the velocity at each point will increase by a factor of two.

Thus, in our minds we can map all geometrically similar bodies onto a single body of unit scale by introducing the new variables:

$$x_1^* = \frac{x_1}{L} = O(1), \tag{1.4.12}$$

$$x_2^* = \frac{x_2}{L} = O(1), \tag{1.4.13}$$

$$x_3^* = \frac{x_3}{L} = O(1). \tag{1.4.14}$$

In other words, if L is changed by a certain factor, the coordinates x_1, x_2, x_3 will scale by the same factor. The scaling (1.4.12)–(1.4.14) can thus also be viewed as a linear transformation of the coordinate system (x_1, x_2, x_3) by the factor L. When two configurations may be mapped onto the same unit problem via such a simple linear transformation we say the two configurations are geometrically similar.

Now consider the velocity field of fluid flowing around the body in the above example. Suppose upstream, far removed from the body, the velocity of the water is U m s^{-1} (Fig. 1.4.1) and the velocity at the point A is (v_1^A, v_2^A, v_3^A) and at B it is (v_1^B, v_2^B, v_3^B). It is intuitively obvious that if we increase U by a certain percentage then, if the flow field around the body remains unchanged, the velocities at A and B (or any other point) will also increase by the same percentage. If this is so we can say for any arbitrary point A:

$$(v_1^A, v_2^A, v_3^A) \sim U. \tag{1.4.15}$$

Once this has been established we may define a dimensionless velocity field:

$$(v_1^{*A}, v_2^{*A}, v_3^{*A}) = \left(\frac{v_1^A}{U}, \frac{v_2^A}{U}, \frac{v_3^A}{U}\right), \tag{1.4.16}$$

that has the property $\boldsymbol{v}^* = O(1)$.

Such scaling of the original physical variables naturally leads to variables that are both non-dimensional and order one. We must, however, put a caveat on this simple definition of scaling. It is only in trivial cases, such as geometric similarity, that the above holds without any proviso. Indeed when (1.4.16) does apply without any further restriction, then in reality the variables (velocities in this case) are simply proportional to the velocity scale U. In environmental fluid dynamics, this is usually only the case under some limiting flow condition as already alluded to in (1.4.11). Many examples come to mind, low or large velocity, small or large stratification, small or large rotation, etc. It is usual to write these limiting conditions in the form of our dimensionless π's, defined above in §1.2. The exact definition of the (1.4.16) must then also involve the statement $\pi \Rightarrow \pi_0$, where π_0 is some limit of the flow configuration. Obviously, there are as many limiting cases as there are non-dimensional numbers. Technically we shall thus say:

$$X \sim Y, \tag{1.4.17}$$

in the limit $\pi \rightarrow \pi_0$ if:

$$\lim_{\pi \rightarrow \pi_0} \frac{X}{Y} = C, \tag{1.4.18}$$

where C is a constant independent of the limiting procedure. Note, however, that the magnitude of C has not been specified and C may be of any magnitude not necessarily close to one. On the other hand good scaling procedures usually yield values of C close to one. Most often it is customary to drop the emphasis on the limiting condition and assume that it is implied. We shall follow this convention and only emphasis the limit when there is some ambiguity or there is a special need for the emphasis.

Once we have found the appropriate scale for all the physical variables (fields) of a particular problem we may use these to form non-dimensional variables that are both non-dimensional and *O(1)*. This is a powerful tool for simplifying problems as it allows direct comparison of the different influences in a problem and so the assessment of which terms are the most important; it is rare in fluid mechanics that we can solve the equations of motion by retaining all the influences, simplifications in the equations are nearly always necessary.

In order to make this concept more concrete consider once again our simple problem of flow over a submerged body, the scale of which is L. The force on this body, due to the fluid drag is obviously a complicated combination of drag due to viscous stress acting on the surface of the body and the pressure forces induced by the flow. We can simplify this problem by looking at limiting case of small and large fluid velocities. A fluid particle, assumed to be of unit volume at the point A at any particular time is thus acted upon by a pressure force $\boldsymbol{F_P}$, an inertia force $\boldsymbol{F_I}$ and a viscous force $\boldsymbol{F_\nu}$ and since, by Newton's law (1.1.1) these forces are in equilibrium we may write:

$$\boldsymbol{F_P} + \boldsymbol{F_\nu} = \boldsymbol{F_I}, \tag{1.4.19}$$

which is a vector equation valid at each general point A in the flow and represents a closed force polygon. In view of the discussion above let the scales of these forces be designated by S_P, S_ν and S_I and introduce the new non-dimensional order one variables:

$$\boldsymbol{F_P^*} = \frac{\boldsymbol{F_P}}{S_P}, \tag{1.4.20}$$

$$\boldsymbol{F_\nu^*} = \frac{\boldsymbol{F_\nu}}{S_\nu}, \tag{1.4.21}$$

$$\boldsymbol{F_I^*} = \frac{\boldsymbol{F_I}}{S_I}. \tag{1.4.22}$$

The starred variables are, by construction, order one non-dimensional variables that can now be substituted into (1.4.19) which, after division by the scale S_I becomes:

$$\left(\frac{S_P}{S_I}\right)\boldsymbol{F_P^*} + \left(\frac{S_\nu}{S_I}\right)\boldsymbol{F_\nu^*} = \boldsymbol{F_I^*}. \tag{1.4.23}$$

The coefficients

$$\frac{S_P}{S_I} = \pi_1 \quad \text{and} \quad \frac{S_\nu}{S_I} = \pi_2, \tag{1.4.24}$$

are non-dimensional groups that form the coefficients of the force polygon. Remembering that the $*$ quantities are $O(1)$ meaning that, the relative magnitudes of each of the terms are given by the magnitude of the π's, and further that the force polygon shape, determined by the relative magnitudes of the π, is fixed globally by the magnitudes of the π's. This last point is the most important feature of this argument as it allows us to say something about the force polygons globally. In particular if we have two situations, called for definiteness model and prototype, with geometrically similar domains, then these two flows will be "dynamically similar" if the force polygons at any general point A are similar. From (1.4.23) it is clear that, in our simple problem of flow over a body, this is achieved when:

$$\left(\frac{S_P}{S_I}\right)_{model} = \left(\frac{S_P}{S_I}\right)_{prototype}, \qquad \left(\frac{S_\nu}{S_I}\right)_{model} = \left(\frac{S_\nu}{S_I}\right)_{prototype}. \tag{1.4.25}$$

We shall take up the concept of dynamic similarity in more detail in the next section. The technique of scaling may also be used to simplify the problem if we remember that the simplest force balance is one where only two forces balance at each point and the force polygon collapses to a line with one force opposing the other at every point in the flow. In our simple problem, we have three forces so we must, *a priori*, decided which two we wish to have in balance and what limit we must consider for the third force to be a minor influence. Consider, as an example, the case where the velocity of the fluid is small so that we would expect viscous forces to be larger than the inertia forces $S_\nu > S_I$, or in the form of limits:

$$\frac{S_I}{S_\nu} \rightarrow 0. \tag{1.4.26}$$

Under such circumstances (1.4.23) reduces to:

$$\boldsymbol{F}_P^* + \left(\frac{S_\nu}{S_P}\right)\boldsymbol{F}_\nu^* = 0, \tag{1.4.27}$$

because the only two forces remaining are the pressure force and the viscous drag force. In order to avoid a trivial case where only one force remains and thus it itself must be zero, it is necessary that:

$$\frac{S_v}{S_P} = 1. \tag{1.4.28}$$

Any constant value other than 1 would be equally acceptable, but 1 was chosen, without loss of generality, as factors other than 1 can be incorporated into the non-dimensional variables. Equation (1.4.27) thus reduces to the simple equation:

$$\boldsymbol{F}_P^* + \boldsymbol{F}_v^* = 0. \tag{1.4.29}$$

The limit (1.4.26) thus defines the scales of the variables (1.4.28) as well as the simplified force balance (1.4.29); all are in the form of force ratios.

In order to be specific, consider the scales of these forces. From the definition of the coefficient of viscosity, §1.3, we see that the viscous force on a fluid particle of unit volume is given by:

$$S_v \sim \frac{\mu U}{L^2}. \tag{1.4.30}$$

The scale for the pressure gradient force is simply the pressure scale P, yet unknown, divided by the length scale L:

$$S_P \sim \frac{P}{L}. \tag{1.4.31}$$

From (1.4.28) it follows:

$$P = \frac{\mu U}{L}. \tag{1.4.32}$$

The inertia force scale is given by:

$$S_I = \frac{\rho_0 U^2}{L}, \tag{1.4.33}$$

so that the limiting condition (1.4.26) becomes

$$\mathrm{Re} = \frac{LU}{\nu} \to 0, \tag{1.4.34}$$

which is the familiar Reynolds number, Re, criterion for slow viscous flow. From the above, it can be seen that scaling is a very powerful tool, that allows great simplifications to be made, and equally importantly allows quantification of the conditions inherent in such simplifications.

Carrying out a specific field or prototype measurements and then generalizing to the overall situation is also called modeling; any activity that leads to a generalized model of the prototype is called modeling. The model may be in the form of a mathematical equation, an empirically derived curve on a graph, a computer algorithm, a physical scaled down replica or even, as is the most common application, a simplified version of the prototype situation. The latter provides the insight for good experimental design. As a general rule, one form of modeling usually has great advantages over the others for any particular problem. It is always wise to exploit this advantage. Often, two or more methods of modeling must be used, one for different parts of the problem.

For instance, in designing a diffuser outfall for discharging effluent from a waster water treatment plant into the ocean, it is normally most efficient to use a mathematical model for the estimating jet mixing near the exit port of the diffuser, a physical laboratory model for the intermediate zone of mixing where the water column stability effects the dispersal of the effluent, by the domain flow in general does not impact on this dispersion and a numerical model for the far field mixing predictions where the characteristics of the overall domain flow are important; as we shall see later these zones are conveniently identified by first doing a scaling analysis.

1.5. DYNAMIC SIMILARITY

In §1.4, we discussed scaling the motion of a fluid particle under the influence of only three forces, viscous, inertia and pressure. In a more general case, the equation of motion (momentum conservation) may be written as

$$\boldsymbol{F}^{(1)} + \boldsymbol{F}^{(2)} + \boldsymbol{F}^{(3)} + \cdots + \boldsymbol{F}^{(n)} = \rho \frac{\mathbf{D}\boldsymbol{v}}{\mathbf{D}t} = \boldsymbol{F}^{(I)}, \tag{1.5.1}$$

where $\boldsymbol{F}^{(1)}$, $\boldsymbol{F}^{(2)}$, $\boldsymbol{F}^{(3)}$,...,$\boldsymbol{F}^{(n)}$ are the forces on a fluid particle of unit volume and $\boldsymbol{F}^{(1)}$ is the inertial force of the accelerating particle. Now at any time in the particle's flow history (any point x_1, x_2, x_3, t) equation (1.5.1) geometrically represents a force polygon. As discussed in §1.4, we say that we have dynamic similarity if, at each point in the flow field, the force polygon in the model is similar to the corresponding force polygon in the prototype.

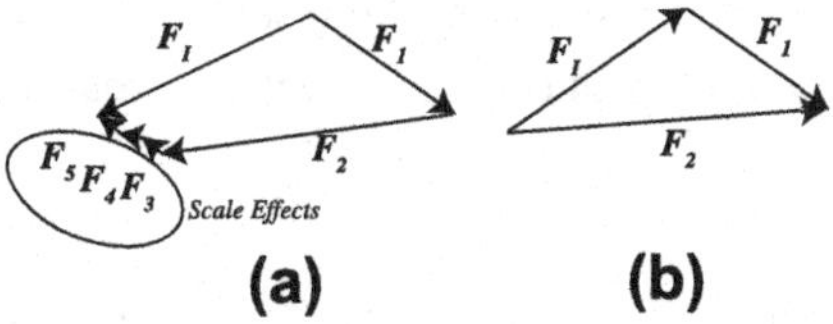

Figure 1.5.1 Force polygon at a typical point in a flow. The forces $\boldsymbol{F_1}$, $\boldsymbol{F_2}$, $\boldsymbol{F_I}$, are clearly much larger than the forces $\boldsymbol{F_3}$, $\boldsymbol{F_4}$, $\boldsymbol{F_5}$ as shown in (a) and may thus be neglected in a general force balance as shown in (b).

Often many of the forces in equation (1.5.1) are much smaller than the dominating two or three forces responsible for the overall force balance. For example, the force polygon shown in Fig. 1.5.1(a), where $\boldsymbol{F}^{(3)}$, $\boldsymbol{F}^{(4)}$, $\boldsymbol{F}^{(5)}$ are the small forces, is approximated very well by that shown in Fig. 1.5.1(b) for the three major forces. Often it is enough to have dynamic similarity in the major forces and treat the error in the minor forces as unimportant scale effects.

The aim in modeling is thus to produce a flow field in a model that is dynamically similar to the one in the prototype with only negligible scale effects. The method to do this is to reduce the prototype flow and the model flow to a unit non-dimensional configuration.

Suppose in the prototype (x_1', x_2', x_3') are the physical coordinates, $\boldsymbol{v}'$ is the velocity and $\boldsymbol{F}^{(1)}$, $\boldsymbol{F}^{(2)}$, ... are the force vectors. The magnitude of (x_1', x_2', x_3') will be given by the overall dimension of the fluid domain L_p, and let V_p be the magnitude of v_p', $M_p^{(j)}$ be the magnitude of $\boldsymbol{F}^{(j)}$ and I_P be the magnitude of the inertia force.

Hence we can introduce non-dimensional variables that also have magnitudes of order unity:

$$(x_1, x_2, x_3) = \left(\frac{x_1'}{L_p}, \frac{x_2'}{L_p}, \frac{x_3'}{L_p}\right), \tag{1.5.2}$$

$$\boldsymbol{u} = \frac{\boldsymbol{v}'}{U_p}, \tag{1.5.3}$$

$$\boldsymbol{f}^{(j)} = \frac{\boldsymbol{F}^{(j)}}{M_p^{(j)}}; \quad \boldsymbol{f}^{(I)} = \frac{\boldsymbol{F}^{(I)}}{I_p}. \tag{1.5.4}$$

This scaling transforms the prototype equation (1.5.1) into the equation:

$$M_p^{(1)}\boldsymbol{f}^{(1)} + M_p^{(2)}\boldsymbol{f}^{(2)} + \cdots + M_p^{(n)}\boldsymbol{f}^{(n)} = I_p\boldsymbol{f}^{(I)}, \tag{1.5.5}$$

valid in a unit domain. Suppose I_p is the largest force per unit volume, then it is useful to divide (1.5.5) by I_p allowing the relative magnitudes of each term to be identified:

$$\frac{M_p^{(1)}}{I_p}\boldsymbol{f}^{(1)} + \frac{M_p^{(2)}}{I_p}\boldsymbol{f}^{(2)} + \cdots + \frac{M_p^{(n)}}{I_p}\boldsymbol{f}^{(n)} = \boldsymbol{f}^{(I)}. \tag{1.5.6}$$

The relative magnitude of the various forces is thus given by the term in brackets preceding each unit force $\boldsymbol{f}^{(j)}$ and the equation is dimensionless.

We can now carry out the same scaling for the model problem (Fig. 1.5.1) and arrive at the force balance equation.

$$\frac{M_m^{(1)}}{I_m}\boldsymbol{f}^{(1)} + \frac{M_m^{(2)}}{I_m}\boldsymbol{f}^{(2)} + \cdots + \frac{M_m^{(n)}}{I_m}\boldsymbol{f}^{(n)} = \boldsymbol{f}^{(I)} \tag{1.5.7}$$

The scaling has in both cases reduced the geometric domain to one of unit size and all the boundary conditions have been scaled to be identical. Thus, the solution to the prototype and model are dynamically similar provided the coefficients to equations (1.5.6) and (1.5.7) are identical:

$$\frac{M_p^{(j)}}{I_p} = \frac{M_m^{(j)}}{I_m}; \quad j = 1, \ldots, n \tag{1.5.8}$$

1.6. HYDROSTATIC PRESSURE

As already seen in §1.3, pressure in a fluid is defined as the force per unit area, perpendicular to a surface. Given that all fluids have a non-zero density and if there is no motion, then the weight of a column of fluid must be in equilibrium with a force at the bottom of the column; this force, divided by the area of the fluid column, is called the gravitational or *hydrostatic pressure.* Since a fluid at rest cannot support a shearing stress, this mechanical pressure is always directed normal to the area.

To understand how pressure acts on bodies submerged in the fluid, we must first investigate how the hydrostatic pressure varies throughout a fluid domain. We do this in two steps, first we show that pressure is a scalar uniquely defined at a point and acting equally in all directions and second we show, as already foreshadowed in §1.2, that pressure increases with depth of fluid; a direct consequence of the weight of a fluid.

Pressure is a Scalar

First, we show that pressure in a fluid at rest, acts equally in all directions. To see this consider a two-dimensional triangular shaped fluid particle as shown in Fig. 1.6.1. Suppose the angle θ is arbitrary and let the pressure on the sloping side be p, the horizontal pressure on the vertical side at the origin O be $p_1^{(0)}$ and the vertical pressure on the horizontal side at the origin O be $p_3^{(0)}$.

If we note that, for fluid at rest, the pressure acts normally to a surface then the pressure distribution along the vertical left face is given by:

$$p_1 = p_1^{(0)} + \left.\frac{\partial p_1}{\partial x_3}\right|^{(0)} x_3 + O(x_3^2), \tag{1.6.1}$$

and the pressure distribution along the bottom horizontal face is, similarly:

$$p_3 = p_3^{(0)} + \left.\frac{\partial p_3}{\partial x_1}\right|^{(0)} x_1 + O(x_1^2). \tag{1.6.2}$$

The pressure on the sloping face can, once again, be obtained through a Taylor expansion about the origin. Let $p^{(0)}$ be the pressure at the bottom left corner of the triangle, then the pressure along the sloping face is given by:

$$p = p^{(0)} + \left.\frac{\partial p}{\partial x_1}\right|^{(0)} x_1 + O(x_1^2) + \left.\frac{\partial p}{\partial x_3}\right|^{(0)} x_3 + O(x_3^2), \tag{1.6.3}$$

where x_1 and x_3 are taken to lie in the sloping face. The equation for the sloping face is given by:

$$x_3 = \left(d_s \sin\theta\left(1 - \frac{x_1}{d_s \cos\theta}\right)\right), \tag{1.6.4}$$

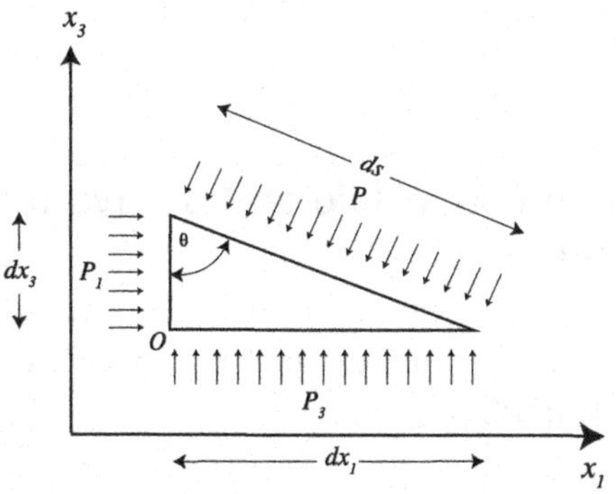

Figure 1.6.1 Pressure on a small fluid particle.

where d_s is the length of the sloping face, assumed to be small. Substituting (1.6.4) into (1.6.3) leads to the desired expression for the pressure along the sloping face:

$$p = p^{(0)} + \left.\frac{\partial p}{\partial x_1}\right|^{(0)} x_1 + O(x_1^2) + \left.\frac{\partial p}{\partial x_3}\right|^{(0)} d_s \sin\theta\left(1 - \frac{x_1}{d_s \cos\theta}\right)$$

$$+O\left(\left(d_s \sin\theta\left(1 - \frac{x_1}{d_s \cos\theta}\right)\right)^2\right). \tag{1.6.5}$$

Since the fluid is at rest, the pressure forces must balance the gravity forces. So equating the pressure forces in the horizontal direction:

$$p_1^{(0)} = p^{(0)} + \frac{1}{2}\left\{\frac{\partial p}{\partial x_3}\sin^2\theta + \frac{\partial p}{\partial x_1}\cos\theta\sin\theta - \frac{\partial p_1}{\partial x_3}\sin^2\theta\right\}d_s, \tag{1.6.6}$$

that in the limit as $d_s \to 0$, becomes:

$$p_1^{(0)} = p^{(0)}. \tag{1.6.7}$$

A similar force balance in the vertical direction, now including the weight of the triangular fluid particles $(O(d_s^2))$ and again taking the limit $d_s \to 0$, leads to the results:

$$p_3^{(0)} = p^{(0)}. \tag{1.6.8}$$

Combining (1.6.7) and (1.6.8) shows that:

$$p^{(0)} = p_1^{(0)} = p_3^{(0)}. \tag{1.6.9}$$

So far we have only considered a two-dimensional triangular fluid particle. A simple generalization to a three-dimensional trihedron shows that:

$$p^{(0)} = p_1^{(0)} = p_2^{(0)} = p_3^{(0)}. \tag{1.6.10}$$

Variation of Pressure in a Fluid at Rest: Hydrostatic Pressure Variation

Consider a fluid at rest as shown in Fig. 1.6.2. In §1.2 we already showed, from dimensional reasoning that the pressure in a fluid at rest must vary according to the relationship:

$$p = p^{(0)} + \pi_0 \rho g h, \tag{1.6.11}$$

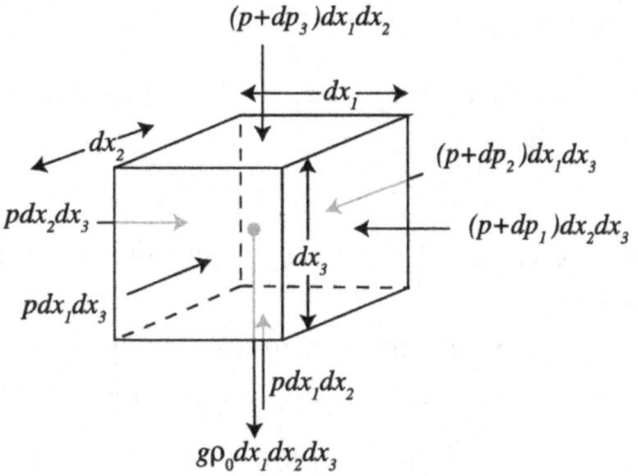

Figure 1.6.2 Infinitesimal pressure variation in a fluid at rest.

where π_0 is an unknown coefficient. We now give a dynamical explanation to show that:

$$\pi_0 = 1. \tag{1.6.12}$$

This is easily achieved by considering the static equilibrium of the small fluid cube of fluid with one corner at the point $(x_1^{(0)}, x_2^{(0)}, x_3^{(0)})$. Summing the forces in the x_1 direction (Fig. 1.6.2) yields:

$$d_{p_1} = \frac{\partial p}{\partial x_1} \mathrm{d}x_1 = 0; \Rightarrow \frac{\partial p}{\partial x_1} = 0. \tag{1.6.13}$$

Summing the forces in the x_2 direction yields:

$$d_{p_2} = \frac{\partial p}{\partial x_2} \mathrm{d}x_2 = 0; \Rightarrow \frac{\partial p}{\partial x_2} = 0. \tag{1.6.14}$$

Summing the forces in the x_3 direction yields:

$$\frac{\partial p}{\partial x_3} = -\rho g. \tag{1.6.15}$$

Comparing this result with equation (1.6.11) leads to the conclusion that $\pi_0 = 1$. It is to be noted that in deriving equation (1.6.15) we have not made any assumptions regarding the density ρ; the fluid may be compressible, incompressible and the density constant or variable; the result expressed by (1.6.15) is valid in general.

Suppose the density $\rho = \rho_0$ and is constant, then we may integrate equation (1.6.15) to obtain

$$p = p_0(x_1) - \rho_0 g x_3. \tag{1.6.16}$$

However, from equation (1.6.13), $\dfrac{\partial p}{\partial x_1} = 0$, so p_0 must be a constant equal to the atmospheric pressure p_a if the origin of the x_3 axis is taken at the surface of the fluid. Thus,

$$p = p_a - \rho_0 g x_3. \tag{1.6.17}$$

The distribution given by (1.6.17) is a linear function of depth, the same as derived before in equation (1.6.11).

The pressure variation in the atmosphere may be derived, on the assumption that the air is at rest, by noting that air obeys the perfect gas law:

$$\rho = p/RT. \tag{1.6.18}$$

However, before we can derive the pressure variation we need to specify the air temperature as a function of height. To illustrate the methodology, we shall assume that the temperature is constant, a good approximation in the first 500 m. Then, combining (1.6.15) and (1.6.18) yields:

$$\frac{\partial p}{\partial x_3} = -\frac{pg}{RT_0}, \tag{1.6.19}$$

where T_0 is the temperature of the air. Simple integration leads to the result:

$$p = p_0 e^{-\frac{g}{RT}\left(x_3 - x_3^{(0)}\right)}, \tag{1.6.20}$$

which shows that the pressure in an isothermal atmosphere decreases exponentially and not linearly as in the constant density case.

As a further example, for a better representation of the actual atmosphere, let us suppose that the air in the atmosphere had settled adiabatically; then it is shown in books on thermodynamics, that the temperature would not be constant, but rather we have a thermodynamic pressure relationship:

$$\rho = \rho_0\left(\frac{p}{p_a}\right)^{\frac{1}{\gamma}}, \tag{1.6.21}$$

where $\gamma = C_P/C_v$.

Substituting equation (1.6.21) into (1.6.15) leads to:

$$\frac{\partial p}{\partial x_3} = -\rho_0 g\left(\frac{p}{p_a}\right)^{\frac{1}{\gamma}}, \tag{1.6.22}$$

Simple integration yields:

$$\frac{p}{p_a} = \left(1 - \left(\frac{\gamma - 1}{\gamma}\right)\frac{\rho_0 g}{p_a}x_3\right)^{\frac{\gamma}{\gamma-1}}. \tag{1.6.23}$$

1.7. PRESSURE FORCES ON A SURFACE

Now that we have shown that pressure in a fluid at rest is a well-defined scalar acting as a perpendicular force to a surface within a fluid, it is possible to calculate the forces on submerged surfaces.

a) *Vertical Wall of Unit Width*

Consider the force on a vertical wall as shown in Fig. 1.7.1. The pressure distribution is hydrostatic and the density ρ of the fluid is assumed to be constant. The pressure at any depth is given by:

$$p = P_A + \rho g(h - x_3), \tag{1.7.1}$$

so that the net force, per unit width of surface, will act horizontally (normal to the vertical surface) and be given by the integral (area of the pressure triangle):

$$F = \int_0^h p \, dx_3. \tag{1.7.2}$$

Substituting equation (1.7.1) into (1.7.2) yields the result:

$$F = \frac{\rho g h^2}{2} + P_A h, \tag{1.7.3}$$

directed horizontally against the vertical wall. Hence we see that the total force is equal to the area of the wall ($h \times 1$) multiplied by the pressure at the centroid of the area $\left(\frac{\rho g h}{2} + P_A\right)$. As we will see later, this is a general result true for any shaped surface.

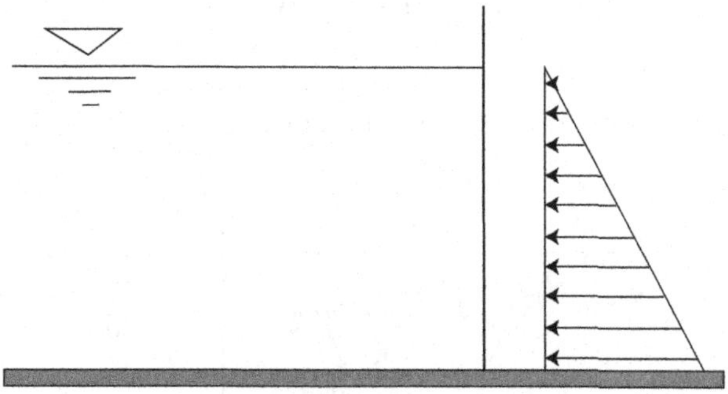

Figure 1.7.1 Pressure force on a vertical wall.

The height x_{3p} of this net force is given by taking moments about the origin:

$$Fx_{3p} = \int_0^h x_3 p \, dx_3. \tag{1.7.4}$$

Again, substituting for the pressure variation from (1.7.1), for the simple case where $p_A = 0$, results in:

$$x_{3p} = \frac{h}{3}. \tag{1.7.5}$$

which represents the distance to the centroid of the pressure distribution.

b) *Sloping Wall of Unit Width*

Instead of the surface, or wall, being vertical with the pressure force acting horizontally, we now consider a wall, of length L that is sloping to the horizontal at an angle θ, as shown in Fig. 1.7.2.

Consider first the total, normal, pressure force on the body:

$$F = \int_0^L p \, ds, \tag{1.7.6}$$

where s is the distance measured from the bottom origin of the surface up along the surface. The integration can be changed to one over the vertical by noting:

$$ds = \frac{dx_3}{\sin\theta}, \tag{1.7.7}$$

so that:

$$F = \int_0^h \frac{p}{\sin\theta} \, dx_3, \tag{1.7.8}$$

where h is the vertical extent of the sloping wall.

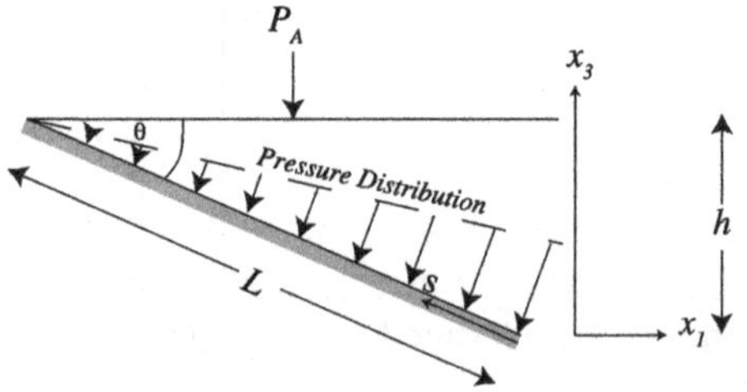

Figure 1.7.2 Pressure force on a sloping wall.

Substituting for pressure from equation (1.7.1) leads to:

$$F = \int_0^h \frac{P_A + \rho g(h - x_3)}{\sin\theta}\,dx_3, \qquad = \frac{P_A h}{\sin\theta} + \frac{\rho g h^2}{2\sin\theta}, \tag{1.7.9}$$

so that:

$$F = \left(P_A + \frac{\rho g h}{2}\right)L, \tag{1.7.10}$$

which is in the same form as in equation (1.7.3).

Now the component of force, F_1, in the x_1 direction is given by:

$$F_1 = F\sin\theta = \left(P_A + \frac{\rho g h}{2}\right)h, \tag{1.7.11}$$

and in the vertical direction we have

$$F_3 = F\cos\theta = \left(P_A + \frac{\rho g h}{2}\right)L. \tag{1.7.12}$$

$$(= P_A L + \text{weight of fluid above the surface})$$

c) *Sloping Wall with Arbitrary Width*

Consider the net total force on the upper surface of a plane surface (Fig. 1.7.3):

$$F = \int_A p\,dA = \int_A (P_A + \rho g x_3)\,dA, \tag{1.7.13}$$

where now x_3 is taken as directed vertically down.

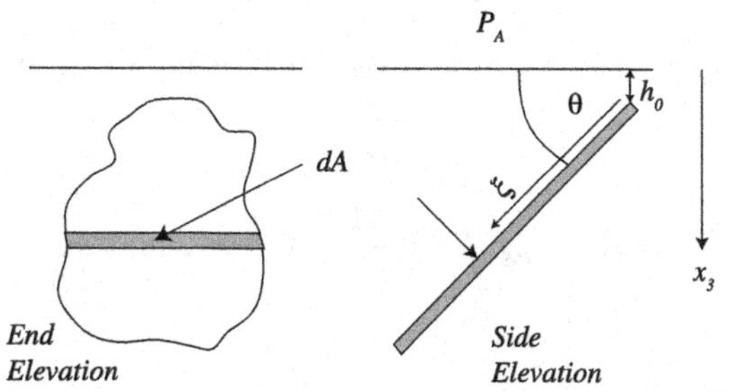

Figure 1.7.3 Pressure force on a plane surface or arbitrary shape.

Introducing a local coordinate ζ, where

$$x_3 = x_3^{(0)} + \zeta \sin \theta, \tag{1.7.14}$$

equation (1.7.13) becomes

$$F = \int_A (P_A + \rho g x_3^{(0)} + \rho g \sin \theta \zeta) \mathrm{d}A. \tag{1.7.15}$$

Thus,

$$F = (P_A + \rho g x_3^{(0)}) A + \rho g \sin \theta \int_A \zeta \, \mathrm{d}A. \tag{1.7.16}$$

Defining:

$$\overline{\zeta} = \int_A \frac{\zeta \mathrm{d}A}{A}, \tag{1.7.17}$$

as the distance to the centroid of the sloping wall, the total force, F, may be written as:

$$F = P_C A, \tag{1.7.18}$$

where

$$P_C = P_A + \rho g x_3^{(0)} + \overline{\zeta} \rho g \sin \theta, \tag{1.7.19}$$

is the pressure at the centroid of the area.

Once again the components of force in the x_1 and x_3 directions are obtained by resolving the force, F, into the x_1 and x_3 directions:

$$F_1 = P_C A \sin \theta = P_C A_1, \tag{1.7.20}$$

$$\begin{aligned} F_3 &= P_C A \cos \theta = P_C A_3 \\ &= \text{Weight of water subtended by } A. \end{aligned} \tag{1.7.21}$$

The center of pressure ζ_p may be obtained by taking moments about the point c (Fig. 1.7.3).

$$F\zeta_p = \int_A \zeta p \, \mathrm{d}A = \rho g \int_A \zeta \left(\frac{p_A}{\rho g} + x_3^{(0)} + \zeta \sin \theta \right) \mathrm{d}A, \tag{1.7.22}$$

which implies that:

$$\zeta_p = \frac{\left(\frac{P_A}{\rho g} + x_3^{(0)}\right)\bar{\zeta} + \sin\theta \frac{I_{cc}}{A}}{\left(\frac{P_A}{\rho g} + x_3^{(0)}\right) + \bar{\zeta}\sin\theta}, \qquad (1.7.23)$$

where $I_{cc} = \int_A \zeta^2 \, dA$ is the moment of inertia of the area A about an axis through the point c.

d) *Curved Surface of Arbitrary Width*

Instead of a plane surface consider now a curved surface as illustrated in Fig. 1.7.4.

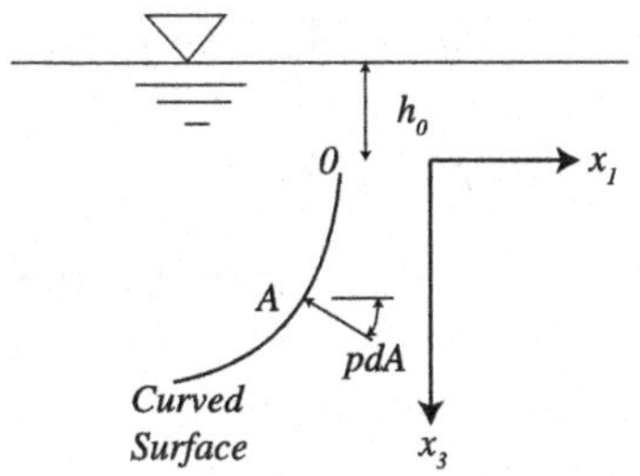

Figure 1.7.4 Pressure force on a curved surface.

The force in the x_1 direction on the elemental area dA is given by:

$$dF_1 = \int_A \rho g (x_3 + x_3^{(0)}) dA_1, \qquad (1.7.24)$$

where $dA_1 = dA \cos\theta$ and θ is the angle between the normal to the area dA and the x_1 axis. Hence, the total force in the x_1 direction on the curved surface is given by:

$$F_1 = \rho g x_3^{(0)} \int_A dA_1 + \rho g \int_A x_3 \, dA_1. \qquad (1.7.25)$$

Defining the area subtended in the x_1 direction:

$$A_1 = \int_A dA_1, \qquad (1.7.26)$$

and the distance to the area centroid:

$$\overline{x}_3 = \frac{\int_A x_3 \, dA_1}{A_1}, \tag{1.7.27}$$

leads to the expression:

$$F_1 = p_c A_1, \tag{1.7.28}$$

where p_c is the pressure at the centroid of the area $(= \rho g x_3^{(0)} + \overline{x}_3 \rho g)$.

The pressure $p_0 = \rho g x_3^{(0)}$ is often referred to as the over-pressure.

The force F_3 in the x_3 direction may be obtained in a similar fashion:

$$\begin{aligned} F_3 &= \int_A \rho g (x_3 + x_3^{(0)}) dA_3 \\ &= \rho g x_3^{(0)} A_3 + \rho g \int_A x_3 dA_3 \\ &= \rho g (V_0 + V + \rho g), \end{aligned} \tag{1.7.29}$$

where $V + V_0$ is the volume subtended by the curved surface (see Fig. 1.7.5).

Hence we see that the total vertical pressure force on the curved surface A is equal to the weight of the volume of water above the curved surface. On the other hand, the net force in the x_1 direction is equal to the net pressure force on the plane vertical surface A_1 and is given by equation (1.7.20). This may be more easily seen from an overall force balance of the volume V and V_1 as shown in Fig. 1.7.6.

Line of action: The position and direction of the resultant force may once again be obtained from a moment calculation. It is not critical about which point the moment is taken, but often one point is more convenient

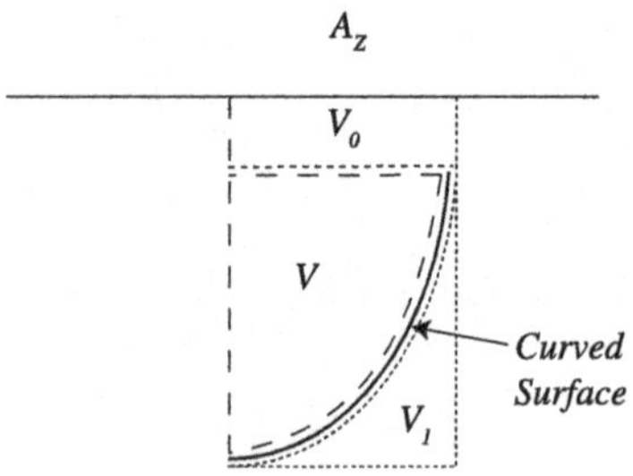

Figure 1.7.5 Definition of pressure volumes.

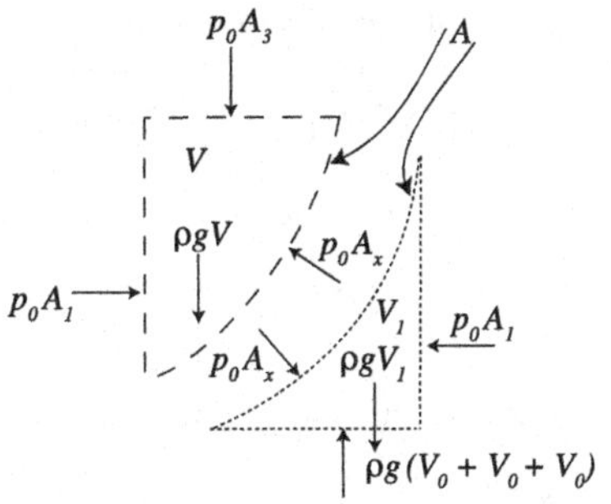

Figure 1.7.6 Force balance on the volume elements subtended by the area.

than another. Taking moments about the point O (Fig. 1.7.5) and consider the equilibrium between the force F_1 and the force distribution $p\mathrm{d}A_1$.

$$\begin{aligned} F_1 x_3^p &= \int_A x_3 p\,\mathrm{d}A_1 \\ &= \int_A x_3 (x_3^{(0)} + x_3)\rho g\,\mathrm{d}A_1 \\ &= \rho g x_3^{(0)} \bar{x}_3 A_1 + \rho g I_{33}(A_1), \end{aligned} \tag{1.7.30}$$

where

$$I_{33}(A_1) = \int_A x_3^2\,\mathrm{d}A_1. \tag{1.7.31}$$

Hence,

$$x_3^p = \frac{\rho g a_0 \bar{x}_3 + \rho g I_{31}/A}{p_c}, \tag{1.7.32}$$

which is identical to the results for the plane surface equation (1.7.23).

Now the line of action in the x_3 direction may be obtained in the same way:

$$\begin{aligned} F_1 x_1^p &= \int_A x_1 p\,\mathrm{d}A_3 \\ &= \int_A x_1 (x_3^{(0)} + x_3)\rho g\,\mathrm{d}A_3 \\ &= \rho g x_3^{(0)} \int_A x_1\,\mathrm{d}A_3 + \rho g \int_A x_1 x_3\,\mathrm{d}A_3 \\ &= \rho g x_3 \bar{x}_1 A_{x_3^{(0)}} + \rho g I_{13}. \end{aligned} \tag{1.7.33}$$

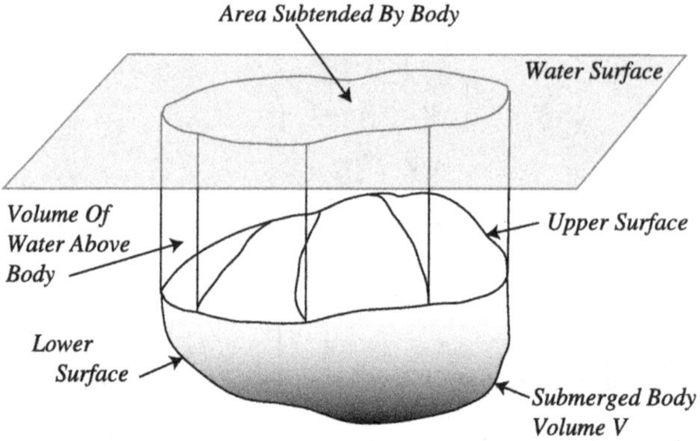

Figure 1.7.7 The partitioning of a submerged body.

Hence,

$$x_1^p = \frac{x_3^{(0)}\overline{x}_1 A_3 + I_{13}}{(V_0 + V)}. \tag{1.7.34}$$

Archimedes Principle: This famous theorem states: "The force on a submerged body is equal to the weight of the volume of fluid displaced".

The proof of this statement rests on a double application of equation (1.7.29). Consider an arbitrary shaped body of volume V as shown in Fig. 1.7.7 and divide the surface area of the body into an upper and a lower part, the separation line being the outer tangent line of a vertical projection.

From (1.7.29), the pressure force on the area Au is equal to $\rho g Vu$ acting down. The pressure force on area A_1 is equal to $\rho g(Vu + V)$ acting up. By subtraction, we see that the net force on the submerged body is $\rho g V$.

1.8. CONTROL VOLUMES

In §1.3, we introduced the concept of continuum mechanics and defined a fluid particle; the important concept here was that the properties ascribed to a fluid particle, such as mass, momentum, energy, temperature, etc., may be obtained by averaging the properties of each molecule in the fluid particle over all the molecules in the fluid particle. In doing so care must to be taken that the fluid particle is large enough to contain many molecules in order to produce a well-defined average, yet small enough that variations in these averages are not affected by the choice of the fluid particle size or shape. The average properties are then ascribed to the centroid of the fluid particle. We saw that a fluid particle could be defined with a size of around

1 mm, large enough to contain millions of molecules, but smaller than the smallest variation in motion properties for environmental flows. When finding the solution to a fluid problem we need to make a decision about the level of detail we wish to resolve. On the one hand, we could endeavor to solve for the path of each fluid particle, on the other hand, we could be content with extracting only bulk properties averaged over the whole fluid domain. The smaller the entity of our focus, the more detail of the flow is revealed, but the more difficult it is to extract the solution. The entity of our focus is called the control volume; a control volume the size of a fluid particle provides maximum detail, a control volume the size of the whole fluid domain provides the minimum detail, but is most easily solved. Then there is also the matter whether we choose a control volume that moves with the fluid and has not mass flux across its boundaries, or we choose a control volume fixed in space, where the flux across the boundaries must be explicitly accounted for.

In general, application of the conservation principles of mass, momentum and energy yields information for the flow at the scales of the control volume or larger. Little information may be obtained for scales of motion smaller than the control volume. There are two types of control volumes:

1) Eulerian, where boundaries are fixed in space with fluid moving in and out at the open boundaries.

2) Lagrangian, where the volume moves with the fluid and always contains the same fluid particles, with no mass flux across the boundaries of the control volume.

In an Eulerian control volume, we speak of fluid particles moving across boundaries, and in Lagrangian control volumes we speak of a volume containing always the same fluid particles. In both these statements there is the inherent assumptions that we are speaking of fluid particles and not molecules; molecules will cross boundaries taking with them mass, energy and momentum. So the concept of continuum flow, in its simplest form, is only valid for a simple fluid, one such as water, with no solutes. This requires that when molecules migrate across boundaries, an equal number of molecules migrate in both directions across boundaries. Properties such as heat (vibrational energy of molecules) and momentum (related to momentum of molecules) are explicitly accounted for at such boundaries. Hence mass of control volumes, be they Eulerian or Lagrangian, is conserved, but heat, momentum and energy need boundary transfers in order to have conservation.

In order to make some of these ideas more specific consider the motion of water in a lake. First, suppose we are interested only in an attribute of the flow, or a budget in the lake as a whole, then it is sufficient to bound the lake water by the bottom, the surface of the lake and sections across each tributary and outflow. Clearly such a control volume, Eulerian as it is fixed in space, allows a budget of any particular attribute; the rate of change of the attribute in the control volume is equated to the flux of that attribute across the bounding surfaces of the control volume. A balance of the attribute, over such a control volume, allows an equation to be written that can be used to calculate a particular flux, given all others are known. For instance, if we know the amount of water flowing into a lake and the rate of change of the volume of the water in the lake, then conservation of water allows us to calculate the rate of outflow. However, little information is gained about the behavior of the motion within the control volume (i.e. motion within the lake) from such a global balance. Therefore, the control volume may be viewed as a filter through which we view the variations; variations smaller than the control volume are filtered out.

On the other hand, applying our conservation principles to every fluid particle in the lake will naturally yield a description of the motion at all scales, from the smallest shearing motion to the circulation within the lake as a whole, but finding a solution to all the particle motions is usually not possible, even with modern computers. The reason for this is simply understood if we accept that a typical lake has a dimension of say, 30 m deep, 20,000 m long and 1000 m wide. Such a lake would therefore contain 6×10^{17} fluid particles, each moving so as to conserve momentum, mass and energy.

Eulerian control volumes have the advantage that they are easily visualized and easily described mathematically by specifying the bounding closed surface:

$$S(x_1, x_2, x_3) = 0, \tag{1.8.1}$$

where S is a function describing the bounding surface and (x_1, x_2, x_3) are Cartesian coordinates. The disadvantage of the Eulerian control volume is that it requires a knowledge of the flow at the bounding surfaces, which may or may not be known, *a priori*; often the flow across a boundary is not uniform in attributes and the control volume analysis will not yield this variation. It will not be possible to resolve this from a control volume balance as it involves scales of motion smaller than the control volume scale itself.

The Lagrangian control volume formulation does not suffer from this complication. However, a Lagrangian control volume deforms with the motion and the shape is thus not known until the problem has been solved; this can often be more difficult than the difficulty of the boundary fluxes encountered in the Eulerian formulation. The Lagrangian control volume is, most commonly, used in the theoretical development of the equations of motion and the Eulerian when solving practical engineering problems. A Lagrangian control volume is defined, at all times, by describing a closed surface at an arbitrary initial instant in time, t_0:

$$S(\xi_1, \xi_2, \xi_3) = 0, \tag{1.8.2}$$

where S is a function defining a closed surface and (ξ_1, ξ_2, ξ_3) are a set of initial coordinates.

Now suppose the fluid particle motion is described by the set of equations:

$$x_i = x_i(\xi_1, \xi_2, \xi_3, t); \quad i = 1, 2, 3, \tag{1.8.3}$$

where $\{x_i\}$ is the position, at time t, of a fluid particle that originated at $\{\xi_i\}$ at time t_0, then (1.8.3) may be inverted to yield the position of a particle at time t_0, given its current position $\{x_i\}$:

$$\xi_i = \xi_i(x_1, x_2, x_3, t). \tag{1.8.4}$$

The coordinates $\{\xi_i\}$ are called material coordinates and it is assumed that this inverse exists at all times. Substituting (1.8.4) into (1.8.2) leads to a description of the control volume surface at all times t. The curves traced out by (1.8.3), as time varies, are called *particle paths*.

A Lagrangian control volume description is, therefore, conceptually very simple; it is a volume containing a constant collection of fluid particles defined by the initial volume. Such a volume may be large if we are interested in bulk variations only or it may be a single fluid particle if the whole flow field is to be resolved. However, in the simplicity of this definition also lies the disadvantage as already stated; the volume description assumes that the particle path description (1.8.3) is known.

A useful strategy is to use the advantages of both the Lagrangian and Eulerian descriptions. This may be done by using a large Lagrangian control volume to formulate the conservation equations and then to transform these statements into an Eulerian control volume the size of a fluid particle. As we shall see, in this way the initial formulation is intuitive and the final equations are referenced to an Eulerian coordinate system making them amenable to manipulations.

The transformation between a conservation statement applied to a large Lagrangian control volume and an infinitesimal Eulerian control volume may be derived quite generally. Suppose $A(x_1, x_2, x_3, t)$ and $B(x_1, x_2, x_3, t)$ are scalar fields and $\{C_i(x_1, x_2, x_3, t)\}$ is a vector field, defined in a domain Ω and $V(t)$ is a Lagrangian control volume moving within the domain. As we shall see later in chapter 2, all the conservation laws for mass, momentum, energy and the second law of thermodynamics may be written in the form:

$$\frac{\mathrm{d}}{\mathrm{d}t}\left\{\int_{V(t)} A\mathrm{d}V\right\} \geq \int_{V(t)} B\mathrm{d}V + \int_{S(t)} C_i n_i \mathrm{d}S, \qquad (1.8.5)$$

where $V(t)$ is an arbitrary Lagrangian control volume, $\{n_i\}$ is the unit normal at the control volume surface $S(t)$ and repeated subscripts implies a sum over that subscript.

Application of the transport theorem (Appendix) allows us to move the time derivative inside the integral sign in the left hand term, leaving the time derivative to operate only on the integrand and no longer on the change induced by the moving control volume. This leads to,

$$\int_{V(t)} \frac{\partial A}{\partial t}\mathrm{d}V + \int_{S(t)} A v_n \mathrm{d}S \geq \int_{V(t)} B\mathrm{d}V + \int_{S(t)} C_i n_i \mathrm{d}S, \qquad (1.8.6)$$

where v_n is the component of velocity of the surface normal to the surface. Since we are dealing with a Lagrangian control volume we may write:

$$v_n = v_i n_i, \qquad (1.8.7)$$

where $\{v_i\}$ is the fluid velocity at the surface of the control volume. Substituting (1.8.7) into (1.8.6) and using Gauss' theorem (Appendix) to change the surface integrals to volume integrals, equation (1.8.6) becomes:

$$\int_{V(t)} \left\{\frac{\partial A}{\partial t} + v_i A_{,i} + A v_{i,i} - B - C_{i,i}\right\}\mathrm{d}V \geq 0. \qquad (1.8.8)$$

Now the control volume $V(t)$ is completely arbitrary so that (1.8.8) implies that the integrand is itself greater or equal to zero:

$$\frac{\partial A}{\partial t} + v_i A_{,i} \geq B + C_{i,i} - Av_{i,i}. \tag{1.8.9}$$

The proof of this last step is easily seen by counter example. Suppose the integrand is not greater than zero on a sub-volume $V_s(t)$. Since A, B and C are continuous, we choose $V(t)$ equal to $V_s(t)$ in (1.8.8) and a contradiction results requiring that the integrand be zero everywhere. It is convenient to introduce a shorthand notation:

$$\frac{\mathrm{D}A}{\mathrm{D}t} = \frac{\partial A}{\partial t} + v_i A_{,i}, \tag{1.8.10}$$

that represents the time derivative of A following the motion and is called the *material time derivative.* The material time derivative is thus the sum of the time rate of change of the variable A at a point in space plus the change due to A changing as it moves or advects in space; the latter will only be non-zero if A varies spatially in the direction of motion.

Substituting (1.8.10) into (1.8.9) leads to:

$$\frac{\mathrm{D}A}{\mathrm{D}t} + Av_{i,i} \geq B + C_{i,i}\,. \tag{1.8.11}$$

The connection between a Lagrangian and an Eulerian control volume is thus clear. Consider the differences between equations (1.8.5) and (1.8.6). The first equation is dependent on whether or not the volume $V(t)$ moves with the fluid, however, none of the terms in equation (1.8.6) depend on the motion of the control volume, as this dependence has been explicitly separated out into the term:

$$\int_{S(t)} Av_n \mathrm{d}S,$$

which represents the flux of the variable A out of the surface.

1.9. INTRODUCTION TO THE KINEMATICS OF FLOW

In §1.8, we discussed the two ways of analyzing fluid motions; the Lagrangian perspective where the focus is on a particular fluid particle and the Eulerian approach where all dynamical descriptions are in terms of variables fixed on a particular point in space. The connection, from a dynamical point of view, is provided by (1.8.3) with the underlying difference being that Lagrangian particles move through an Eulerian point in space as the motion proceeds. As pointed out in §1.8, the advantage of the Lagrangian frame of reference is that Newton's laws apply directly to a fluid particle, but the forces on the particle are not known *a priori*. The advantage of the Eulerian frame of reference is that the boundaries of a control volume are known at all times and so the forces on the control volume can be calculated. The disadvantage of the Eulerian reference frame is that boundaries are usually far removed from the individual fluid particles and so the motion, on scales smaller than the control volume domain cannot be resolved. The underlying gap in both approaches is the *a priori* knowledge of the kinematics of the motion. Here we present a brief introduction to the key features of the kinematics of motion.

Consider two fluid particles that are separated, at time $t = t_0$, by a distance $\{d\xi_i\}$ as shown in Fig. 1.9.1, both moving through the control volume according to the path description given by (1.8.3) with the inverse given by (1.8.4). The separation at time t_1 will thus be given, to first order, by:

$$dx_i = \frac{\partial x_i}{\partial \xi_j} \partial \xi_j. \tag{1.9.1}$$

The quantity $\frac{\partial x_i}{\partial \xi_j}$ is called the *displacement gradient tensor* and follows directly once (1.8.3) is known. Similarly, the velocity of a particle P relative to a particle Q becomes at t_1, to first approximation:

$$dv_i = \frac{\partial v_i}{\partial x_j} dx_j, \tag{1.9.2}$$

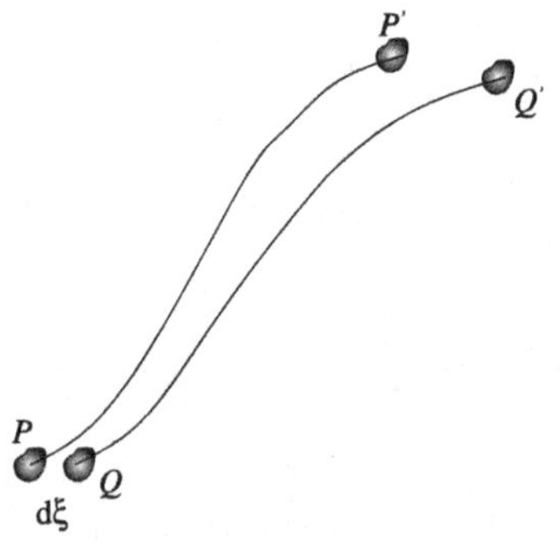

Figure 1.9.1 Fluid particles traveling along fluid paths.

where $\frac{\partial v_i}{\partial x_j}$ is called the *velocity gradient tensor* (i.e. the change of the ith component of velocity in the jth direction).

For simplicity of writing, we again introduce the summation notation (detailed in Appendix) together with the abbreviated form of the spatial derivative; a comma followed by the index of the spatial coordinate. A simple partitioning allows us to write:

$$\frac{\partial v_i}{\partial x_j} = v_{i,j} = \frac{1}{2}\left(v_{i,j} + v_{j,i}\right) + \frac{1}{2}\left(v_{i,j} - v_{j,i}\right) = d_{ij} + \Omega_{ij}, \qquad (1.9.3)$$

where d_{ij} is called the rate of strain tensor and Ω_{ij} is called the rate of rotation tensor.

Before continuing let us illustrate the physical meaning of d_{ij} and Ω_{ij} and see why the above names have been attached to these quantities. Consider first Ω_{ij}. Suppose we introduce the quantity $\{\zeta_i\}$, called the vorticity of the velocity field, defined by:

$$\zeta_i = e_{ijk}v_{k,j}, \qquad (1.9.4)$$

then it follows directly from the permutation properties of e_{ijk} that,

$$\zeta_i = \frac{1}{2}e_{ijk}\left(v_{k,j} - v_{j,k}\right) = e_{ijk}\Omega_{kj} = -e_{ijk}\Omega_{jk}. \qquad (1.9.5)$$

Now if we multiply (1.9.5) by $e_{\alpha\beta i}$ and sum over i it follows directly that:

$$\Omega_{ij} = -\frac{1}{2}e_{ijk}\zeta_k, \qquad (1.9.6)$$

where we have used the identity:

$$e_{ijk}e_{klm} = \delta_{il}\delta_{jm} - \delta_{im}\delta_{jl}. \qquad (1.9.7)$$

Substituting (1.9.6) and (1.9.4) into (1.9.3) leads to the result:

$$\mathrm{d}v_i = d_{ij}\mathrm{d}x_j - \frac{1}{2}e_{ijk}\zeta_k\mathrm{d}x_j. \qquad (1.9.8)$$

The second term is the cross-product of the vorticity vector with the particle separation vector; the second part of (1.9.8) is thus a velocity component induced by a relative rotation with an angular velocity $\left\{\frac{1}{2}\zeta_k\right\}$; the vorticity is proportional to the angular velocity of the two particles relative to each other.

Now in order to find an interpretation for the rate of strain tensor d_{ij}, consider the rate of change of the length increment $\{dx_i dx_i\}^{\frac{1}{2}}$ separating any two particles. By (1.9.1) we may write:

$$dx_i dx_i = \frac{\partial x_i}{\partial \xi_j}\frac{\partial x_i}{\partial \xi_k}\partial\xi_j\partial\xi_k, \tag{1.9.9a}$$

and differentiating (1.9.2) with respect to time, following the motion, leads to:

$$dv_i = \frac{\partial v_i}{\partial \xi_j}\partial\xi_j = \frac{\partial v_i}{\partial \xi_j}\frac{\partial \xi_j}{\partial x_\beta}dx_\beta = \frac{\partial v_j}{\partial x_\beta}dx_\beta. \tag{1.9.9b}$$

Similarly, differentiating (1.9.9a) with respect to time, following the motion:

$$\frac{1}{2}\frac{D(dx_i dx_i)}{Dt} = v_{i,\beta}dx_\beta dx_i = \left(\frac{v_{i,\beta}+v_{\beta,i}}{2}\right)dx_\beta dx_i = d_{i\beta}dx_\beta dx_i. \tag{1.9.10}$$

This equation implies that the rate of elongation of the square of the element length is given by half $d_{i\beta}$ times the magnitude $dx_\beta dx_i$. If we divide both sides by $\{dx_i dx_i\}^{\frac{1}{2}}$ then (1.9.10) becomes:

$$\frac{1}{ds}\frac{D(ds)}{Dt} = d_{ij}\frac{dx_i}{ds}\frac{dx_j}{ds}, \tag{1.9.11}$$

where we have changed the index β to j as it is a dummy index and

$$ds = \{dx_i dx_i\}^{\frac{1}{2}}.$$

To see this more clearly, take two fluid particles, separated by a small distance parallel to the x_1 axis, then only dx_1 is non-zero and

$$\frac{1}{ds}\frac{D(ds)}{Dt} = d_{11}, \tag{1.9.12}$$

but since then $ds = dx_1$, this becomes:

$$\frac{1}{dx_1}\frac{D(dx_1)}{Dt} = d_{11}, \tag{1.9.13}$$

so that d_{11} is the longitudinal rate of strain of an element parallel to the x_1 axis. Interpretation of the other components d_{ij} may be found similarly and it may be shown that d_{ij} is the rate at which elements along ith and jth directions are sheared with respect to each other.

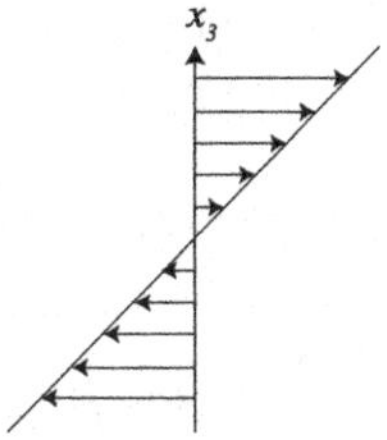

Figure 1.9.2 Simple plane shear flow.

As an example of this discussion, consider the simple plane shear flow shown in Fig. 1.9.2. The velocity field given by:

$$v_i = \{\alpha x_3, 0, 0\}, \tag{1.9.14}$$

and the vorticity field follows from the definition (1.9.4):

$$\zeta_i = \{0, \alpha, 0\}, \tag{1.9.15}$$

leading to the rate of deformation tensor:

$$d_{ij} = \begin{pmatrix} d_{11} & d_{12} & d_{13} \\ d_{21} & d_{22} & d_{23} \\ d_{31} & d_{32} & d_{33} \end{pmatrix} = \begin{pmatrix} 0 & 0 & \alpha/2 \\ 0 & 0 & 0 \\ \alpha/2 & 0 & 0 \end{pmatrix}. \tag{1.9.16}$$

It is easy to show that (1.9.11) becomes:

$$\frac{1}{\mathrm{d}s}\frac{\mathrm{D}(\mathrm{d}s)}{\mathrm{D}t} = \alpha\frac{\mathrm{d}x_1}{\mathrm{d}s}\frac{\mathrm{d}x_3}{\mathrm{d}s}. \tag{1.9.17}$$

The implications of (1.9.17) are illustrated in Fig. 1.9.2, where it is seen that the element $\mathrm{d}x_3$ is only rotated by the motion and not extended, $\mathrm{d}s$ is extended and rotated at a rate $\alpha/2$ and $\mathrm{d}x_1$ remains unchanged by the motion.

1.10. BULK CONSERVATION OF MASS

As indicated above, in §1.8, we shall first deal with simple fluids for which the mass of a Lagrangian *V(t)* volume, containing the same fluid particles, is conserved. This may be written mathematically simply as:

$$M = \int\limits_{V(t)} \rho(x_1, x_2, x_3, t)\mathrm{d}V = Const, \tag{1.10.1}$$

where $\rho(x_1, x_2, x_3, t)$ is the density of the fluid at the point $\mathbf{x}$ and time t, $V(t)$ is the control volume defined by the moving closed surface S and $\mathrm{d}V$ is the incremental volume element.

Using (1.8.6) this may be changed to an Eulerian control volume:

$$\frac{\mathrm{d}M}{\mathrm{d}t} = \int\limits_{V(t)} \frac{\partial \rho(x_1, x_2, x_3, t)}{\partial t} \mathrm{d}V + \int\limits_{S(t)} \rho(x_1, x_2, x_3, t) v_i n_i \, \mathrm{d}S = 0. \tag{1.10.2}$$

This is now valid for any $V(t)$, moving with the fluid or fixed in space.

We shall now illustrate with a series of examples how the bulk conservation law (1.10.2) may be applied to some simple practical problems.

Water Flowing Through a Pipe Contraction: Suppose water, assumed to be incompressible, is flowing through a pipe contraction as shown in Fig. 1.10.1.

The diameter of the pipe at section 1 is D_1 and at section 2 is D_2. Let $v_1^{(1)}$ be the x_1 component of velocity at section 1 and $v_1^{(2)}$ corresponding velocity at section 2. The control volume is shown and is effectively made up of two open sections 1 and 2 and the closed pipe wall. If it is assumed that the density of the water is constant, equal to ρ_0, then conservation states:

Rate of mass flowing in at section 1* = *Rate of mass leaving at section 2 (1.10.3)

The rate of mass flowing with the fluid is generally called the mass flux. From Fig. 1.10.1, we may write (1.10.3) mathematically in the form:

$$\int\limits_{S_2} \rho_0 v_1^{(2)} \mathrm{d}S - \int\limits_{S_1} \rho_0 v_1^{(1)} \mathrm{d}S = 0. \tag{1.10.4}$$

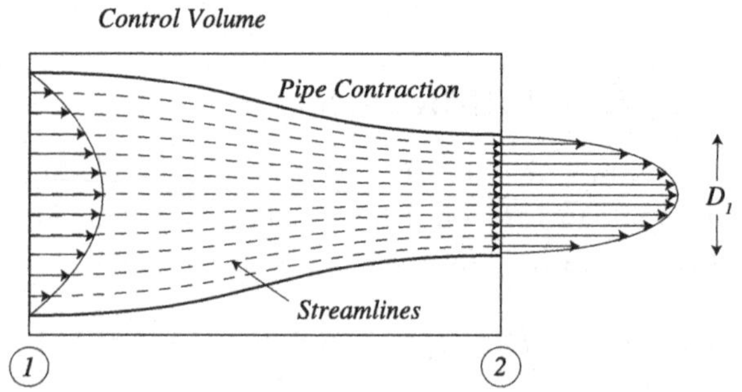

Figure 1.10.1 Flow of water through a pipe contraction.

The sign of the inflowing mass flux, by (1.8.7), is negative and the outflow at section 2 is positive.

The difficulty with (1.10.4) is thus seen immediately; unless we know the velocity distribution across the areas S_1 and S_2 we cannot evaluate the integrals in (1.10.4). Unfortunately, even by invoking conservation of momentum, these velocity distributions can only be found by decreasing the size of the control volume down to the fluid particle size; using the large control volume shown in Fig. 1.10.1 does not allow us to proceed further. Using the fluid particle description, we show in §5.6 that for a round pipe and laminar flow the velocity distribution is parabolic so we may write:

$$v_1^{(1)} = U^{(1)}\left(1 - \frac{r^2}{R_1^2}\right), \tag{1.10.5}$$

where $U^{(1)}$ is the maximum velocity at the centerline of the pipe, r is the radial distance from the centerline, R_1 is the radius of the pipe at section 1 and where we have assumed the flow has axial symmetry. A similar expression holds for $v_1^{(2)}$. The integrals in (1.10.4) are over the cross-sectional areas and, because the velocities $v_1^{(1)}$ and $v_1^{(2)}$ are only dependent on the radial distance r, the integral is most easily derived, by changing to a radial coordinate system.

The mass flow, through the annulus between r and dr becomes:

$$\mathrm{d}M_R = \rho_0 v_1^{(2)} 2\pi r \mathrm{d}r. \tag{1.10.6}$$

The total mass flux through the pipe at section 2 will be given by the integral:

$$M_R^{(2)} = \int_0^{R_2} 2\pi r \rho_0 v_1^{(2)} \mathrm{d}r. \tag{1.10.7}$$

Substituting from (1.10.5), but applied to section 2 yields:

$$M_R^{(2)} = 2\pi\rho_0 U^{(2)} \int_0^{R_2} \left(r - \frac{r^3}{R_2^2}\right)\mathrm{d}r. \tag{1.10.8}$$

Carrying out the integration yields the expression for $M_R^{(2)}$:

$$M_R^{(2)} = \frac{\pi\rho_0 U^{(2)} R_2^2}{2}. \tag{1.10.9}$$

Conservation of mass (1.10.4) then states

$$\frac{\pi \rho_0 U^{(2)} R_2^2}{2} = \frac{\pi \rho_0 U^{(1)} R_1^2}{2}. \tag{1.10.10}$$

It is customary to introduce the discharge velocity U_d defined by:

$$U_d^{(i)} = \frac{Q_q^{(i)}}{\pi R_i^2}; \quad i = 1, 2, \tag{1.10.11}$$

where πR_i^2 is the area of the pipe at section i and

$$Q^{(i)} = \int_0^R 2\pi r v_1^{(i)} \, dr = \frac{\pi U^{(i)} R_i^2}{2}, \tag{1.10.12}$$

is the volume flux. Substituting (1.10.12) into (1.10.11) we see for laminar flow:

$$U_d^{(i)} = \frac{U^{(i)}}{2}. \tag{1.10.13}$$

In summary, when the fluid is incompressible and of constant density, conservation of mass reduces to conservation of volume and, when the flux is laminar so that the velocity is known, the mass flux can be simply evaluated as a function of the maximum centerline velocity or the discharge velocity.

Flow from a Tank: Consider a circular large tank that is being emptied through a small circular pipe as shown in Fig. 1.10.2. Suppose R_T is the tank radius and a is the radius of the pipe through which the water is emptying. Again for definiteness, we shall assume that the flow in the pipe is laminar so that the velocity profile at section 1 is given by (1.10.5).

The assumed Eulerian control volume is shown in Fig. 1.10.3, the only open boundary is at section 1, but now the mass inside the control volume is obviously changing with time as the tank empties.

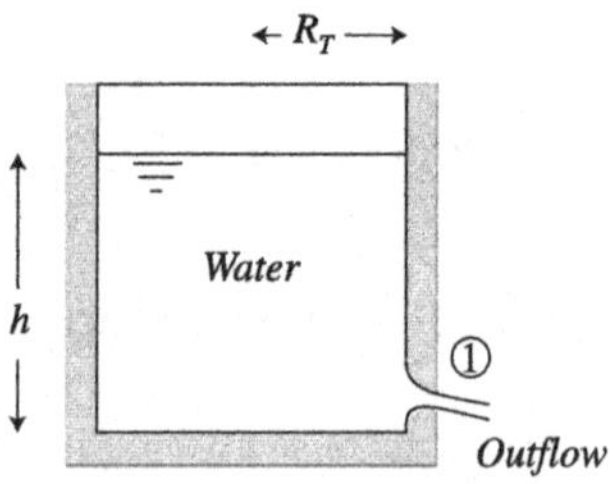

Figure 1.10.2 Schematic of water flowing out of a tank.

Conservation of mass now states:

Rate of increase of mass inside the control volume

$$= \textbf{\textit{Rate of mass flux into the control volume}} \tag{1.10.14}$$

Applying (1.10.14) to a control volume encompassing the water volume:

$$\frac{\mathrm{d}}{\mathrm{d}t}(\rho_0 \pi R_T^2 h) = -\int_0^a \rho_0 v_1^{(1)} 2\pi r \mathrm{d}r. \tag{1.10.15}$$

where section (1) is taken at the outflow.

Using the result (1.10.12) leads an expression for the depth h of water in the tank:

$$\frac{\mathrm{d}h}{\mathrm{d}t} = \frac{-a^2 U_d^{(1)}}{R_T^2}. \tag{1.10.16}$$

It is clear from (1.10.16) that we cannot proceed any further and get an expression for the rate of depth change until we have a relationship between the discharge velocity $U_d^{(1)}$ and the depth of water h, in the tank. We shall derive such an expression in §1.11 and §1.12 when we deal with conservation of momentum and energy.

We shall foreshadow these sections by presenting an intuitive energy argument and so derive the desired exit velocity. If we neglect any frictional energy losses, then what is happening when the water flows out of the tank is that the rate of change of potential energy of the water in the tank is being lost in the form of the flux of kinetic energy flux of the outflow.

Now the potential energy of the water in the tank relative to the height of the outflow pipe is given by:

$$PE = g\rho_0 \pi R_T^2 h \frac{h}{2}, \tag{1.10.17}$$

since the center of gravity of the water is located at a height of $h/2$. On the other hand, the kinetic energy flux, $F_{KE}^{(1)}$, through section 1 is given by the rate, $2\pi r v_1 \, \mathrm{d}r$, at which kinetic energy, $\left(\frac{1}{2}\right)\rho_0 v_1^2$, is lost through the open boundary:

$$F_{KE}^{(1)} = \int_0^a \frac{1}{2}\rho_0 (v_1^{(1)})^3 2\pi r \mathrm{d}r. \tag{1.10.18}$$

Substituting (1.10.5) yields the expression:

$$F_{KE}^{(1)} = \rho_0 \pi \left(U_d^{(1)}\right)^3 a^2. \tag{1.10.19}$$

Equating the time derivative of (1.10.17)–(1.10.19) leads to:

$$g\rho_0 \pi R_T^2 h \frac{dh}{dt} = -\rho_0 \pi a^2 \left(U_d^{(1)}\right)^3. \tag{1.10.20}$$

Substituting from (1.10.16) allows an expression for $U_d^{(1)}$ to be written:

$$U_d^{(1)} = (gh)^{1/2}, \tag{1.10.21}$$

an expression that which we will encounter many more times later in this book. Substituting (1.10.21) into (1.10.16) and integrating with respect to time leads to an expression for the depth of water in the tank:

$$h = h_0 \left(1 - \frac{a^2 g^{\frac{1}{2}} t}{h_0^{\frac{1}{2}} R_T^2}\right)^2, \tag{1.10.22}$$

where h_0 is the depth of water at time $t = 0$ and from which we see that the tank will be empty in a time

$$t = \frac{R_T^2 h_0^{1/2}}{a^2 g^{1/2}}. \tag{1.10.23}$$

1.11. BULK CONSERVATION OF MOMENTUM

Similarly to the discussion on conservation of mass in §1.10, conceptually conservation of momentum is again most easily visualized when applied to a Lagrangian control volume as shown in Fig. 1.11.1.

Suppose the control volume $V(t)$ contains a fluid that has a density $\rho(\underset{\sim}{x}, t)$ so that the body force per unit volume is ρg acting downwards as shown in Fig. 1.11.1. If $\boldsymbol{\tau}$ is the net stress vector acting on the surface of the control volume then the net force due to this stress per area dS is equal to $\boldsymbol{\tau}$ dS. Conservation of momentum then may be simply written in component form:

$$\frac{d}{dt}\left\{\int_{V(t)} \rho v_i dV\right\} = -\int_{V(t)} \rho g_i \delta_{i3} dV + \int_{S(t)} \tau_i dS, \tag{1.11.1}$$

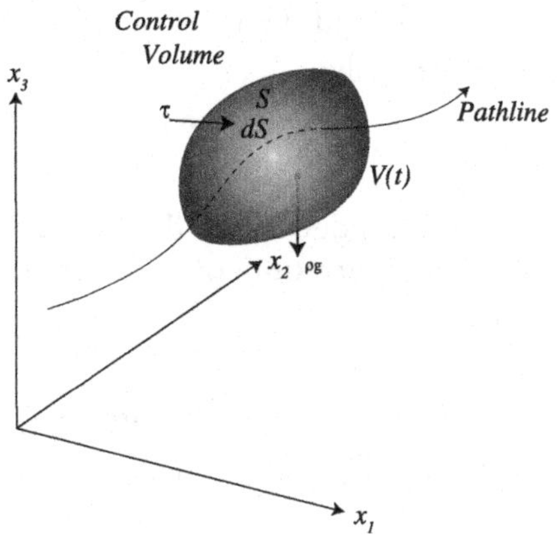

Figure 1.11.1 A Lagrangian control volume *V*(t) moving with the fluid.

where δ_{i3} is called the Kronecker delta:

$$\delta_{ij} = \begin{cases} 1 & i = j \\ 0 & otherwise \end{cases}, \tag{1.11.2}$$

so that the first term of the right hand side of (1.11.1), is non-zero only when $i = 3$ as gravity acts only in the negative x_3 direction. Equation (1.11.1) states that the rate of change of momentum of all the fluid in our control volume is equal to the force due to the weight or gravity plus the net force due to the stresses acting on the surface of the control volume; this is simply Newton's 2nd law (1.1.1).

The surface stress $\boldsymbol{\tau}$ is the stress the fluid immediately outside the bounding surface, *S(t)*, exerts on the fluid inside the control volume. From hydrostatics, discussed in §1.6, we know that when there is no motion:

$$\boldsymbol{\tau} = -p(\boldsymbol{x}, t)\hat{\boldsymbol{n}}, \tag{1.11.3}$$

where $\hat{\boldsymbol{n}}$ is the outward normal to the surface S($\boldsymbol{x}$,t) at any point on the surface and $p(\underset{\sim}{x}, t)$ is the hydrostatic pressure. On the other hand, when the fluid is moving, there is also a viscous stress due to the viscosity of the fluid (§1.3). As seen in §1.3, this stress is proportional to the shear of the motion and the coefficient of proportionality is the coefficient of viscosity μ, a property of the fluid dependent on the fluid temperature. In §2 we develop the full functional dependence, called the constitutive equation, between the fluid shear or rate of strain and the internal fluid stress τ, but here it is sufficient to write, quite generally:

$$\boldsymbol{\tau} = -p(\boldsymbol{x}, t)\hat{\boldsymbol{n}} + \overline{\boldsymbol{\tau}}, \tag{1.11.4}$$

where $\overline{\boldsymbol{\tau}}$ is the stress in the fluid due to the fluid motion.

By analogy to §1.10, conservation of momentum for the Eulerian control volume V reads:

$$\int_V \frac{\partial(\rho v_i)}{\partial t} \mathrm{d}V = -\int_{S_1+S_2} \rho v_i(v_j \hat{n}_j)\mathrm{d}S - \int_V \rho g \delta_{i3} \mathrm{d}V + \int_{S_1+S_2+S_3} \overline{\tau}_i \mathrm{d}S - \int_{S_1+S_2+S_3} p\hat{n}_i \mathrm{d}S \tag{1.11.5}$$

Rate of change of momentum in the control volume V (now fixed in space)
= the rate of inflow of momentum + force due to gravity
+ force due to surface stresses
+ pressure force on the surface of the control volume. (1.11.6)

Once again to make these ideas more concrete we apply (1.11.5) to a series of simple examples.

Force on a pipe angle: Consider water flowing through a pipe bend as shown in Fig. 1.11.2 from a tank at section 1 and exiting horizontally into the atmosphere at section 2.

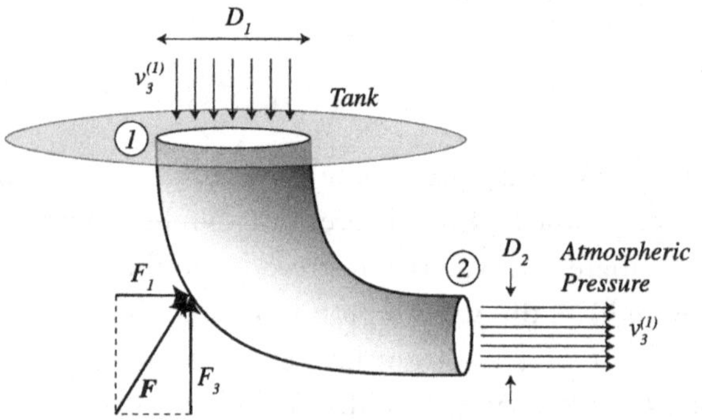

Figure 1.11.2 Flow through a pipe bend with a contraction.

Water is assumed to flow through the pipe bend with a discharge $Q\,\mathrm{m}^3\,\mathrm{s}^{-1}$ and the density of the water is assumed to be constant. The bracket holding the bend in place must exert, via the pipe walls, a force on the fluid that will arrest the incoming vertical momentum towards section 2.

We place an Eulerian control volume over the pipe bend, with an open boundary at 1 and 2 and pipe bend forming the remainder of the control volume. The force $\boldsymbol{F}$ exerted by the pipe on the fluid may now be obtained by solving the equation of conservation of mass (1.10.2) and the horizontal and vertical conservation of momentum (1.11.5). The sign of the mass and momentum fluxes are automatically accounted for if we note that the normal $\hat{\boldsymbol{n}}$ always points out of the control volume; then if the fluid is exiting at the boundary, the flux contribution is positive and when the flux is entering a boundary it is negative. Noting further, that the flow is assumed to be steady, then (1.10.2) and (1.11.5) become:

Conservation of mass:

$$\rho_0 \frac{v_3^{(1)} \pi D_1^2}{4} = \rho_0 \frac{v_1^{(2)} \pi D_2^2}{4}. \tag{1.11.7}$$

Conservation of vertical momentum:

$$0 = F_3 - \rho_0 v_3^{(1)} \frac{\pi D_1^2}{4} v_3^{(1)} - p^{(1)} \frac{\pi D_1^2}{4}. \tag{1.11.8}$$

Conservation of horizontal momentum:

$$0 = F_1 - \rho_0 v_1^{(2)} \frac{\pi D_2^2}{4} v_1^{(2)}, \tag{1.11.9}$$

where it assumes that the contribution from atmospheric pressure cancels and where:

$$\boldsymbol{F} = F_1 \hat{\boldsymbol{i}}_1 + F_3 \hat{\boldsymbol{i}}_3, \tag{1.11.10}$$

$$\boldsymbol{v}^{(1)} = -v_3^{(1)} \hat{\boldsymbol{i}}_3, \tag{1.11.11}$$

$$\boldsymbol{v}^{(2)} = -v_1^{(2)} \hat{\boldsymbol{i}}_1, \tag{1.11.12}$$

and $v_3^{(1)}$ and $v_1^{(2)}$ are assumed constant across the pipe cross-section.

The gravitational body force, due to the weight of the fluid, merely adds a vertical force equivalent to the weight of the fluid in the bend; this is not included here.

From (1.11.7), (1.11.8) and (1.11.9) it follows:

$$F_3 = \rho_0 \frac{\pi D_1^2}{4} (v_3^{(1)})^2 + p^{(1)} \frac{\pi D_1^2}{4}, \tag{1.11.13}$$

and

$$F_1 = \rho_0 \frac{\pi D_1^2}{4 D_2^4} (v_3^{(1)})^2, \tag{1.11.14}$$

indicating that the force components are both proportional to the square of the velocity. However, to fully determine the vertical force F_3, we still need to find an expression of the entry pressure.

If we now assume that there are no energy losses in the pipe bend then we can use (1.10.21) to write:

$$v_1^{(2)} = (gh)^{1/2}, \tag{1.11.15}$$

where h is the depth of water in the tank. Using (1.11.7) and (1.11.15) we may write the force components (1.11.13) and (1.11.14) in terms of the head h of water in the tank:

$$F_3 = \rho_0 \frac{g \pi D_1^2 h}{4} \left(1 + \frac{D_2^4}{D_1^4} \right), \tag{1.11.16}$$

$$F_1 = \rho_0 \frac{g \pi D_1^2 h}{4}. \tag{1.11.17}$$

Hydraulic jump: Another example of the power and simplicity of the bulk Eulerian control volume is illustrated by its application to the hydraulic jump; a phenomena where fast flowing flow in a channel or river abruptly changes depth and proceeds in a more tranquil fashion. We will discuss such flows more fully in §5.15, but here we use the bulk conservation of mass and momentum to present the underlying theory of such hydraulic jumps. When water flows in an open channel or culvert at high speed with a depth h_1 the flow depth may change abruptly or jump to a new, larger, depth h_2 and continue at a lower speed. Such jumps are common occurrences in nature and are accompanied by large energy losses, but momentum is conserved. To see this consider the hydraulic jump shown in Fig. 1.11.3.

The Eulerian control volume is chosen, as shown in Fig. 1.11.3, to be of unit width and to encompass the jump (change in water level) and with an

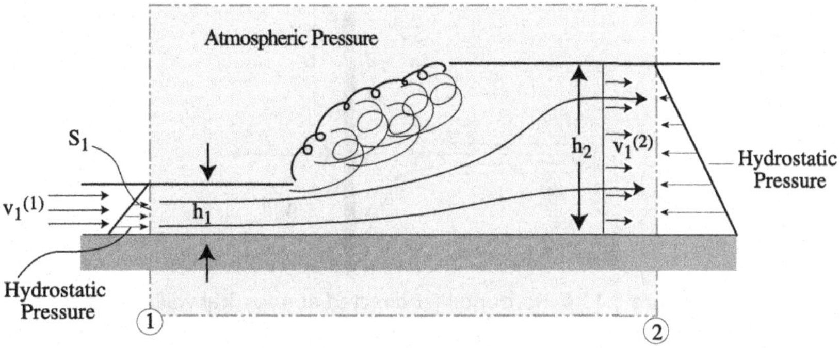

Figure 1.11.3 Schematic of a hydraulic jump.

open boundary at S_1 and S_2. If the flow is assumed to be steady, the conservation of mass equation (1.10.2) becomes:

$$0 = \rho_0 v_1^{(1)} h_1 - \rho_0 v_1^{(2)} h_2 , \tag{1.11.18}$$

and the x_1 momentum conservation equation (1.11.5) yields:

$$0 = \frac{1}{2}\rho_0 g h_1^2 - \frac{1}{2}\rho_0 g h_2^2 + \rho_0 (v_1^{(1)})^2 - \rho_0 (v_1^{(2)})^2, \tag{1.11.19}$$

where it was assumed that the pressure distribution at the open boundaries S_1 and S_2 is hydrostatic (flow is parallel), the velocity is uniform, the frictional resistance on the bottom is negligible, the flow is two-dimensional and the atmospheric pressure distribution balances over the control volume.

Using (1.11.18) we may eliminate $v_1^{(2)}$ from (1.11.19) and we get:

$$0 = \frac{1}{2} g h_1^2 - \frac{1}{2} g h_2^2 + (v_1^{(1)})^2 - \frac{h_1^2}{h_2^2}(v_1^{(1)})^2. \tag{1.11.20}$$

Defining:

$$\beta = \frac{h_2}{h_1}, \tag{1.11.21}$$

and

$$F_1^2 = \frac{(v_1^{(1)})^2}{g h_1}, \tag{1.11.22}$$

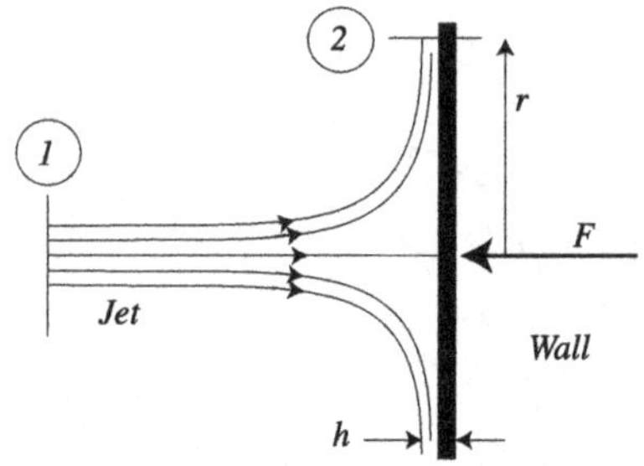

Figure 1.11.4 Horizontal jet directed at a vertical wall.

it is not difficult to show, by solving (1.11.20), for the ratio β that:

$$\beta = \frac{1}{2}\left(\left(1 + 8F_1^2\right)^{1/2} - 1\right). \tag{1.11.23}$$

The solution (1.11.23) shows that there exists a depth h_2 conjugate to initial depth h_1, given by (1.11.23) for which the flow satisfies both conservation of mass and momentum, provided $F_1^2 \geq 1$. As we shall show in §5, there is an associated energy loss whenever a hydraulic jump occurs.

Force due to a horizontal jet: Consider the water-jet from a fire hose being directed horizontally at a vertical wall as shown in Fig. 1.11.4.

Suppose the nozzle has a diameter at the exit of 0.05 m and the jet formed is cylindrical. If the jet is aimed exactly perpendicularly at the wall, the water will spread radially over the wall. With these assumptions, conservation of mass and momentum allows us to calculate the total force, F, exerted on the wall by the jet.

A suitable control volume is shown in Fig. 1.11.4. From symmetry we may assume that the water spreads radially equally in all directions so the x_1 momentum equation (1.11.5) states:

$$-\rho_0 v_1^{(1)} v_1^{(1)} \frac{\pi D_1^2}{4} = -F. \tag{1.11.24}$$

Substituting the numerical values for the velocity $v_1^{(1)}$, the diameter D_1, and assuming the density ρ_0 of water is 1000 kg m^{-3} yields:

$$F = 49\,\text{N}. \tag{1.11.25}$$

1.12. BULK CONSERVATION OF ENERGY

In this book, we confine our attention mostly to two fluids, water and air. Water may be assumed to be incompressible when encountered in most environmental applications such as catchments, rivers, ground water, lakes, estuaries and the coastal ocean. In deep lakes or the ocean, depths are deep enough to require us to account for the compressibility of water, but we can do this in a simplified way not requiring the tracking of thermal energy due to expansion or compression as distinct from mechanical energy (potential and kinetic energy). Air on the other hand is quite compressible, but we shall focus only on the meteorological boundary layer in the first few 1000 m and so once again we may assume that the air is incompressible; the implication of this assumption is discussed in some depth in §2.

With this simplification and the assumption that the energy gained and lost by internal friction due to the viscosity is negligibly small, the conservation of mechanical energy for a Lagrangian control volume *V(t)* as shown in Fig. 1.11.1 becomes (see §2.5 for the generalization to total energy conservation):

$$\frac{\mathrm{d}}{\mathrm{d}t}\int\limits_{V(t)}\frac{1}{2}\rho v_i v_i \mathrm{d}V = -\int\limits_{V(t)}\rho g\delta_{i3}v_i \mathrm{d}V + \int\limits_{S(t)}\tau_i v_i\,\mathrm{d}S, \qquad (1.12.1)$$

where $\int\limits_{V(t)}\frac{1}{2}\rho v_i v_i \mathrm{d}V$ is the total kinetic energy of the fluid particles inside *V(t)*, $-\int\limits_{V(t)}\rho g\delta_{i3}v_i \mathrm{d}V$ is the rate of working against gravity as fluid particles within the control volume move against gravity and $\int\limits_{S(t)}\tau_i v_i \mathrm{d}S$ is the rate at which the surfaces stresses do work on the fluid inside the control volume *V(t)*. In words (1.12.1) states

Rate of increase of kinetic energy of all the fluid particles inside $V(t)$ = minus the rate of doing work lifting fluid against gravity plus the rate at which surface stresses do work on the control volume. (1.12.2)

If we note:

$$\delta_{i3}v_i = v_3, \qquad (1.12.3)$$

and from (1.11.4) we have:

$$\tau_i v_i = -p\hat{n}_i v_i + \overline{\tau}_i v_i, \qquad (1.12.4)$$

then we may write (1.12.1) in the form:

$$\frac{\mathrm{d}}{\mathrm{d}t}\int_{V(t)}\frac{1}{2}\rho v_i v_i \mathrm{d}V = -\int_{V(t)}\rho g v_3 \mathrm{d}V - \int_{S(t)} p v_i n_i \mathrm{d}S + \int_{S(t)} \overline{\tau}_i v_i \mathrm{d}S, \tag{1.12.5}$$

which is a scalar equation. Each term has a simple interpretation:

$$\frac{\mathrm{d}}{\mathrm{d}t}\int_{V(t)}\frac{1}{2}\rho v_i v_i \,\mathrm{d}V = \text{Rate of increase of kinetic energy of all the particles contained by the control volume.} \tag{1.12.6}$$

$$-\int_{V(t)}\rho g v_3 \mathrm{d}V = \text{Rate at which the gravity works on the fluid. Alternatively this can be viewed as the rate of decrease of potential energy.} \tag{1.12.7}$$

$$-\int_{S(t)} p v_i n_i \,\mathrm{d}S = \text{Rate at which the pressure outside the control volume works on the control volume.} \tag{1.12.8}$$

$$\int_{S(t)} \overline{\tau}_i v_i \,\mathrm{d}S = \text{Rate at which the outside shear forces work on the control volume.} \tag{1.12.9}$$

Once again the Lagrangian control volume allows a conceptual simple formulation of conservation of energy; however, practical problems are more easily solved by choosing an Eulerian control volume V, fixed in space and allowing for fluxes of energy across open boundaries. For control volume fixed in space, (1.12.5) becomes:

$$\begin{aligned}\int_V \frac{1}{2}\frac{\partial(\rho v_i v_i)}{\partial t}\mathrm{d}V + \int_{S_1+S_2}\frac{1}{2}\rho v_i v_i (v_j n_j)\mathrm{d}S = -\int_V \rho g v_3 \,\mathrm{d}V \\ - \int_{S_1+S_2} p v_i n_i \,\mathrm{d}S + \int_{S_1+S_2}\overline{\tau}_i v_i \mathrm{d}S.\end{aligned} \tag{1.12.10}$$

There is no contribution to the surface integrals over S because the fluid does not have a velocity normal to the surface S; the surface S is impervious.

Also, as stated at the outset of this section internal energy gains and losses have been neglected in this introductory section, these are accounted for later in the book.

Bernoulli's equation: A very important application of the energy conservation law is the derivation of what is known as Bernoulli's law for steady flow of a fluid with no internal friction; $\mu = 0$. This states that the kinetic energy plus the potential per unit volume plus the pressure remains constant as the fluid particle moves along a streamline:

$$\frac{(v_i^{(1)})^2}{2g} + \frac{p^{(1)}}{\rho_0 g} + x_3^{(1)} = \frac{(v_i^{(2)})^2}{2g} + \frac{p^{(2)}}{\rho_0 g} + x_3^{(2)}, \tag{1.12.11}$$

where the bracketed superscript refers to the section of stream-tube (Fig. 1.12.1) and all other variable have their usual meaning. The above interpretation is retrieved if one multiplies (1.12.11) by $\rho_0 g$.

We choose a control volume that is made up of a stream-tube (a set of streamlines passing through a closed curve as shown in Fig. 1.12.1). Applying (1.12.10) to this control volume and assuming that the closed curve encompasses a small area of fluid so that we may assume the velocities at the two open boundaries are constant over the enclosed areas S_1 and S_2, yields:

$$-\frac{1}{2}\rho_0 (v_i^{(1)})^2 q^{(1)} + \frac{1}{2}\rho_0 (v_i^{(2)})^2 q^{(2)} = -\rho_0 g \int_V v_3 \mathrm{d}V + p^{(1)} q^{(1)} - p^{(2)} q^{(2)}, \tag{1.12.12}$$

where $q^{(1)}$ and $q^{(2)}$ are the volume fluxes at section S_1 and S_2 and the pressure is constant over S_1 and S_2, respectively.

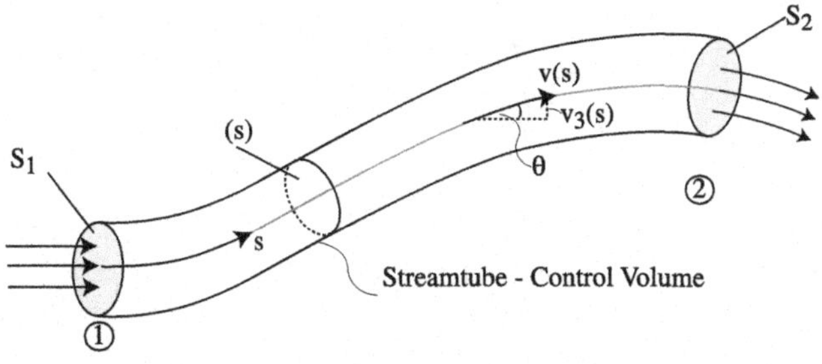

Figure 1.12.1 Schematic of a stream-tube.

Given that the control volume is a stream-tube (the surface S is composed of streamlines and so no flow crosses S), the flow is steady and the density of the fluid is assumed constant, conservation of mass reduces to:

$$q^{(1)} = q^{(2)} = q. \tag{1.12.13}$$

Dividing (1.12.12) by $\rho_0 g q$ and separating the variables from section S_1 and S_2 leads to:

$$\frac{(v_i^{(1)})^2}{2g} + \frac{p^{(1)}}{\rho_0 g} = \frac{(v_i^{(2)})^2}{2g} + \frac{p^{(2)}}{\rho_0 g} + \int_V \frac{v_3}{q} \mathrm{d}V. \tag{1.12.14}$$

Let A(s) be the area of the stream-tube and s is the distance from S_1 along the stream-tube, then:

$$\mathrm{d}V = A(s)\mathrm{d}s, \tag{1.12.15}$$

so that we may also write:

$$q = \frac{A(s)v_3(s)}{\sin\theta}, \tag{1.12.16}$$

where $\sin\theta$ is the local angle the streamlines in the stream-tube make with the horizontal:

$$\sin\theta = \frac{v_3(s)}{v(s)}, \tag{1.12.17}$$

and where $v(s)$ is the speed of the fluid at s.

The integral in (1.12.14) may thus be written:

$$\begin{aligned}
\int_V \frac{v_3}{q}\mathrm{d}V &= \int_{x_3^{(1)}}^{x_3^{(2)}} \frac{V_3(s)\sin\theta}{A(s)v_3(s)} A(s)\,\mathrm{d}S \\
&= \int_{x_3^{(1)}}^{x_3^{(2)}} \sin\theta \mathrm{d}S \\
&= \int_{x_3^{(1)}}^{x_3^{(2)}} \mathrm{d}x_3 \\
&= x_3^{(2)} - x_3^{(1)},
\end{aligned} \tag{1.12.18}$$

so that (1.12.14) becomes:

$$\frac{(v_i^{(1)})^2}{2g} + \frac{p^{(1)}}{\rho_0 g} + x_3^{(1)} = \frac{(v_i^{(2)})^2}{2g} + \frac{p^{(2)}}{\rho_0 g} + x_3^{(2)}, \tag{1.12.19}$$

which is the desired result. This equation is called Bernoulli's law and applies to flow along a streamline in steady flow in a fluid with no internal stresses; such fluids are called inviscid.

It is common to call:

$\frac{(v_i^{(1)})^2}{2g}$: the velocity head at section S_1

$\frac{p^{(1)}}{\rho_0 g}$: the pressure head at section S_1

$x_3^{(1)}$: the elevation head at section S_1

$H^{(1)} = \frac{(v_i^{(1)})^2}{2g} + \frac{p^{(1)}}{\rho_0 g} + x_3^{(1)}$: the total head at section S_1.

Venturi Meter: Consider flow in a tube that has a contraction as shown in Fig. 1.12.2; such contractions are commonly used to measure the discharge in a pipe. To see this we apply the conservation laws between sections S_1 and S_2.

We shall assume that the velocity at section S_1 and S_2 are uniform, conservation of mass implies:

$$\rho_0 v_1^{(1)} A_1 = \rho_0 v_1^{(2)} A_2, \tag{1.12.20}$$

where ρ_0 is the density of the fluid, A_1 and A_2 are the areas of the tube at sections S_1 and S_2, respectively.

Now we may apply Bernoulli's law along a streamline flowing past A and B:

$$\frac{(v_1^{(1)})^2}{2g} + \frac{p_A}{\rho_0 g} + h = \frac{(v_1^{(2)})^2}{2g} + \frac{p_B}{\rho_0 g}, \tag{1.12.21}$$

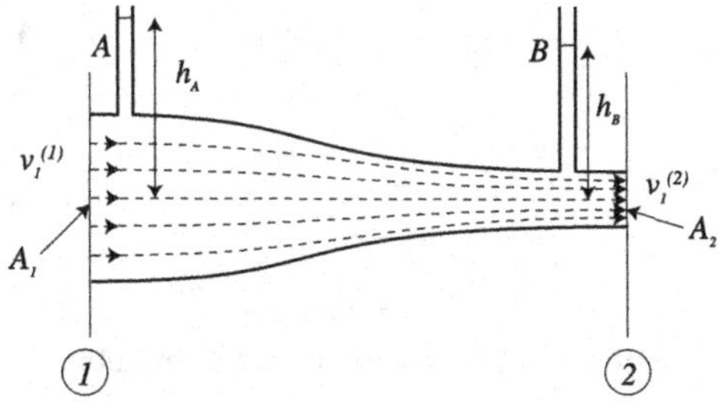

Figure 1.12.2 Schematic of a Venturi Meter.

where p_A and p_B are the pressure at A and B, respectively. Given that the streamlines at section S_1 and section S_2 are parallel we may write:

$$p_A = \rho_0 g h_A, \tag{1.12.22}$$

$$p_B = \rho_0 g h_B, \tag{1.12.23}$$

Substituting (1.12.22) and (1.12.23) into (1.12.21) and using (1.12.20) yields the desired result:

$$\frac{(v_1^{(1)})^2}{2g}\left(1 - \frac{A_1}{A_2}\right) = h_B - h_A - h. \tag{1.12.24}$$

Equation (1.12.24) may be rearranged to read:

$$v_1^{(1)} = \left\{ 2g\frac{h_A + h - h_B}{\left(\frac{A_1}{A_2} - 1\right)} \right\}^{1/2}, \tag{1.12.25}$$

and we see that a contraction in a pipe offers a means of measuring the velocity in the pipe.

Pitot Tube: The velocity of a flow may be measured, using the same principle, with a device as shown in Fig. 1.12.3, called a Pitot tube.

The device consists of a small hole at S in the control tube forming a stagnation point (and thus stagnation pressure) in the flow and a series of small holes at O in the outside tube measuring the pressure in the fluid flow.

Bernoulli's equation may be applied from A to S:

$$\frac{(v_1^{(A)})^2}{2g} + \frac{p_A}{\rho_0 g} = \frac{p_S}{\rho_0 g} = \frac{\rho_0 g h_S}{\rho_0 g} = h_S. \tag{1.12.26}$$

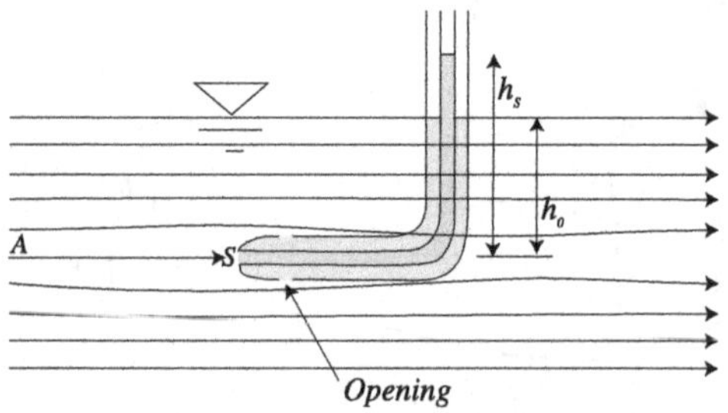

Figure 1.12.3 Schematic of a Pitot Tube.

Now the pressure at A:

$$p_A = \rho_0 g h_0, \qquad (1.12.27)$$

so that:

$$v_1^{(A)} = \{2g(h_S - h_0)\}^{1/2}. \qquad (1.12.28)$$

The device is called a Pitot static tube.

1.13. SOLVING PROBLEMS USING THE CONSERVATION LAWS

Above we have derived the three conservation laws of mass (§1.10), momentum (§1.11) and energy (§1.12) for rather large or bulk, control volumes and given some simple applications. Before leaving this introductory material, we provide a few solutions to some fundamental fluid configurations, to further illustrate the power of the bulk conservation equations.

Cavity outflow. Consider water running out of a rectangular duct as shown in Fig. 1.13.1.

Once again we shall assume that the water velocities at section S_1 and S_2 are constant over the cross-section that the flow is steady and that the viscosity of the water is small so that its effect may be neglected.

Conservation of mass (1.10.2) applied between sections S_1 and S_2 implies that:

$$\rho_0 h_1 v_1^{(1)} = \rho_0 h_2 v_1^{(2)} = \rho_0 q. \qquad (1.13.1)$$

Conservation of x_1 momentum (1.11.5) yields

$$-\rho_0 v_1^{(1)} q + \rho_0 v_1^{(2)} q = \frac{1}{2}\rho_0 g h_1^2 - \frac{1}{2}\rho_0 g h_2^2 + p^{(1)} h_1, \qquad (1.13.2)$$

where it was assumed that atmospheric pressure acts uniformly on all sides and so does not need to be included and $p^{(1)}$ is an unknown upstream

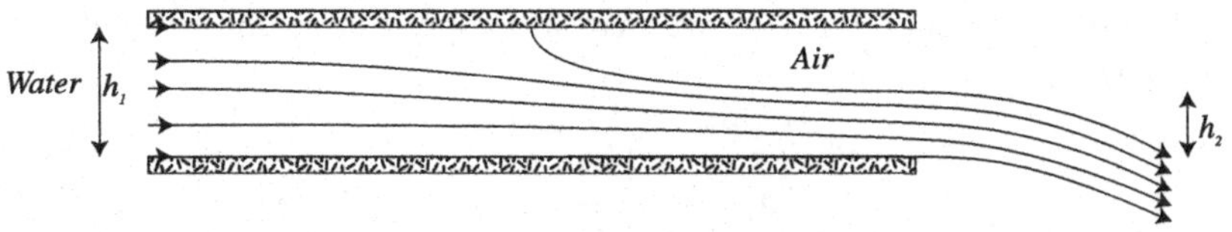

Figure 1.13.1 Schematic of cavity flow.

pressure (Fig. 1.13.1). As seen from (1.13.1) and (1.13.2) there are three unknowns ($v_1^{(2)}$, h_2 and $p^{(1)}$), but only two equations. The required additional equation is the Bernoulli's equation (1.12.17) applied from A to C (Fig. 1.13.1).

$$\frac{(v_1^{(1)})^2}{2g} + \frac{p^{(1)}}{\rho_0 g} + h_1 = \frac{(v_1^{(2)})^2}{2g} + h_2. \tag{1.13.3}$$

From (1.13.3) and (1.13.1) we may show

$$\frac{p^{(1)}}{\rho_0 g} = (h_2 - h_1) + \left(\frac{h_1^2}{h_2^2} - 1\right)\frac{(v_1^{(1)})^2}{2g}. \tag{1.13.4}$$

This together with (1.13.1) may be substituted into (1.13.2) resulting in an equation for the ratio $h_2/h_1 = \delta$:

$$\delta^4 - 2\delta^3 - (F_1^2 - 1)\delta^2 + 2F_1^2\delta - F_1^2 = 0, \tag{1.13.5}$$

where $F_1 = \dfrac{v_1^{(1)}}{(gh_1)^{1/2}}$.

Obviously $\delta = 1$ is a solution so that by long division (1.13.5) becomes

$$(\delta - 1)(\delta^3 - \delta^2 - F_1^2\delta - F_1^2) = 0. \tag{1.13.6}$$

It is seen by substitution, $\delta = 1$ is again a solution of the cubic part of (1.13.6) so that (1.13.6) may be written as:

$$(\delta - 1)^2(\delta^2 - F_1^2) = 0. \tag{1.13.7}$$

Expanding the second term yields:

$$(\delta - 1)(\delta - 1)(\delta - F_1)(\delta + F_1) = 0. \tag{1.13.8}$$

The only viable solution is

$$\delta = F_1. \tag{1.13.9}$$

Assuming no energy is lost, we may also apply Bernoulli's law between points B and C:

$$h_1 = \frac{(v_1^{(2)})^2}{2g} + h_2. \tag{1.13.10}$$

This may be rewritten in the form:

$$\delta^3 - \delta^2 + \frac{F_1^2}{2} = 0. \tag{1.13.11}$$

Substituting for F_1 from (1.13.9) leads to the equation:

$$\delta^3 - \delta^2 + \frac{1}{2}\delta^2 = 0, \tag{1.13.12}$$

which has the solution:

$$\delta = \frac{1}{2}. \tag{1.13.13}$$

We see from (1.13.9) and (1.13.13) that in general there is only one flow condition:

$$F_1 = \frac{1}{2}, \tag{1.13.14}$$

for which a steady air pocket can exist; for $F_1 > \frac{1}{2}$ the tube will flow full and for $F_1 < \frac{1}{2}$ the air pocket will progressively move upstream.

Spreading of an oil slick, assuming no energy losses. Consider an oil slick of density ρ_0 floating on seawater with density ρ_S spreading against a current $v_1^{(1)}$ as shown in Fig. 1.13.2.

Under what conditions will the current $v_1^{(1)}$ arrest the spreading oil slick and what is the corresponding underflow depth h_2?

The solution to this problem is similar to the cavity flow problem discussed above and the similarity illustrates some important principles. Once again we shall make the following assumptions:

(a) The fluids are inviscid.

(b) The fluids are incompressible.

(c) The velocities $v_1^{(1)}$ and $v_1^{(2)}$ are uniform over the depth.

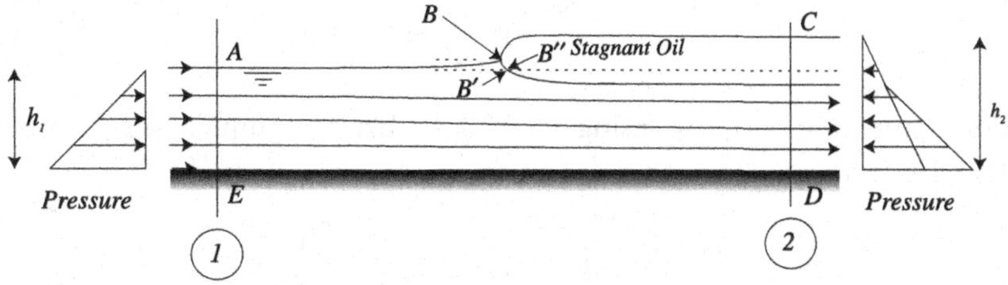

Figure 1.13.2 Gravitational overflow of a fluid with no mixing.

(d) The oil is stagnant.
(e) There is no energy loss.
When the oil slick is arrested, the seawater will also be stationary at the stagnation point B and the pressure build up causes the seawater to rise a height δ_1 as shown in Fig. 1.13.2.

Conservation of mass (1.10.2) applied to the control volume reduces to:

$$\rho_S h_1 v_1^{(1)} = \rho_S h_2 v_1^{(2)}, \tag{1.13.15}$$

where the superscripts designate the entrance and exit sections of the control volume.

Similarly conservation of x_1 momentum (1.11.5) applied to the same control volume yields:

$$-\rho_S (v_1^{(1)})^2 h_1 + \rho_S (v_1^{(2)})^2 h_2 = \frac{1}{2}\rho_S g h_1^2 - \frac{1}{2}\rho_0 g (h_1 + \delta_2)^2 - \frac{1}{2}\Delta\rho g h_2^2, \tag{1.13.16}$$

where δ_2 is the height the oil surface floats above the seawater (oil is lighter), the density difference $\Delta\rho = \rho_S - \rho_0$ and the last two terms are the pressure on section S_2 as shown in Fig. 1.13.2; the adjustments of the free surface height δ_1 and δ_2 take the place of the over-pressure $p^{(1)}$ in cavity flow example.

If we assume no energy is lost, a third equation may be obtained by applying Bernoulli's equation between A and C:

$$\frac{(v_1^{(1)})^2}{2g} + h_1 = h_2 + \frac{\rho_0(h_1 - h_2 + \delta_2)}{\rho_S} + \frac{(v_1^{(2)})^2}{2g}. \tag{1.13.17}$$

Equations (1.13.16) and (1.13.17) may be simplified by introducing the internal Froude number

$$F_i^{(1)} = \frac{v_1^{(1)}}{\left(g h_1 \dfrac{\Delta\rho}{\rho_S}\right)^{\frac{1}{2}}}, \tag{1.13.18}$$

and eliminating $v_1^{(2)}$ by using (1.13.15). Brief manipulation changes (1.13.16) into:

$$\frac{\delta_2 \rho_0}{\Delta\rho h_1} = \frac{(1 - \delta_r)}{2}\left\{\left(1 - \delta_r\right) - \frac{2}{3}(F_i^{(1)})^2\right\}, \tag{1.13.19}$$

and (1.13.17) into:

$$\frac{\delta_2 \rho_0}{\Delta \rho h_1} = (1 - \delta_r) + \frac{1}{2} \frac{(F_i^{(1)})^2}{\delta_r^2} (\delta_r^2 - 1), \qquad (1.13.20)$$

where $\delta_r = h_2/h_1$ and where we have ignored terms $O(\delta_1^2; \delta_2^2)$. Equating the two independent equations (1.13.19) and (1.13.20), leads to the result analogous to (1.13.9):

$$\delta_r = F_i^{(1)}. \qquad (1.13.21)$$

In this example, applying Bernoulli from B to C (Fig. 1.13.2) does not lead to an independent equations as in example (1.13.1) because it introduces the further unknown δ_1. To proceed we assume that the flow in the seawater near the point B is stagnant so that the pressure at the points B' and B'' are equal and are given by the hydrostatic pressure. This implies:

$$\rho_S g \delta_1 = \rho_0 g \delta_2. \qquad (1.13.22)$$

The height δ_1 may be obtained in terms of $v_1^{(1)}$ by applying Bernoulli's law between A and B:

$$\frac{(v_1^{(1)})^2}{2g} = \delta_1, \qquad (1.13.23)$$

which may be rewritten in the form,

$$\frac{\delta_1}{h_1} = \frac{1}{2} \frac{\Delta \rho}{\rho_S} (F_i^{(1)})^2. \qquad (1.13.24)$$

Substituting (1.13.21) into (1.13.20) yields:

$$\frac{\delta_2}{h_1} = \frac{1}{2} \frac{\Delta \rho}{\rho_S} (\delta_r - 1)^2. \qquad (1.13.25)$$

Equating (1.13.24) and (1.13.25) and using (1.13.21) leads to the result:

$$\delta_r = \frac{1}{2}, \qquad (1.13.26)$$

which is identical to (1.13.13) for the case where we have an air cavity instead of an oil slick and where we define an internal Froude number instead of a surface Froude number.

The above analysis shows that there exists a solution that satisfied conservation of mass, momentum and energy for an arrested oil slick where the oil occupies half the depth. The height δ_2 follows immediately from (1.13.25) and (1.13.26)

$$\frac{\delta_2}{h_1} = \frac{1}{8}\frac{\Delta\rho}{\rho_S}. \tag{1.13.27}$$

The result obtained here is somewhat counter intuitive, as oil slicks are known to spread out over the water in thin layers; the explanation for this apparent contradiction lies in our assumption that no energy is lost, inherent in the application of Bernoulli's law.

Spreading of buoyant plumes with energy losses. We now take a closer look at the spreading of a buoyant fluid over another, more dense ambient fluid in order to see the effect of energy losses near the propagating front. This example is the same configuration to that presented in the above oil slick example, except now we shall allow for energy losses. The flow configuration is shown in Fig. 1.13.3.

Now that we are allowing for energy losses we can no longer apply Bernoulli from A to C and further, since energy losses usually involve some sort of turbulent mixing, it would seem reasonable to also allow for mixing of the two fluids. However, we shall postpone this additional complication to later. For now, we shall seek a solution under the same assumption as in the oil slick example; the provision for energy losses being the only generalization.

The conservation of mass and momentum remains unchanged to (1.13.15) and (1.13.16), but we repeat these here for convenience:

Conservation of mass (1.10.2) implies:

$$\rho_S h_1 v_1^{(1)} = \rho_S h_2 v_1^{(2)}. \tag{1.13.28}$$

Conservation of x_1 momentum:

$$-\rho_S (v_1^{(1)})^2 h_1 + \rho_S (v_1^{(2)})^2 h_2 = \frac{1}{2}\rho_S g h_1^2 - \frac{1}{2}\rho_0 g (h_1 + \delta_2)^2 - \frac{1}{2}\Delta\rho g h_2^2. \tag{1.13.29}$$

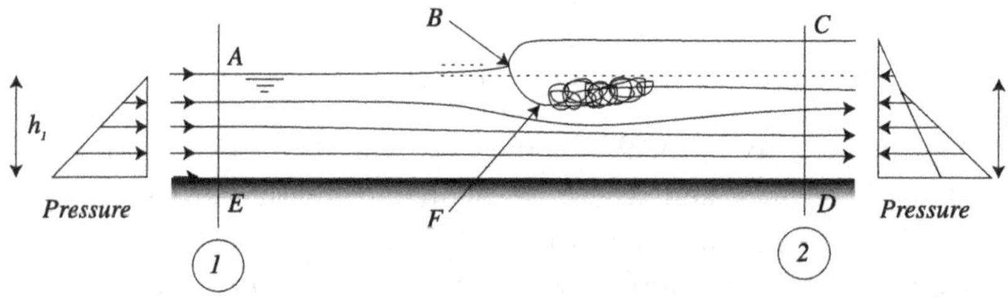

Figure 1.13.3 Gravitational overflow with mixing.

Once again we may assume pressure equality at B so that:

$$\rho_S g \delta_1 = \rho_0 g \delta_2, \tag{1.13.30}$$

and Bernoulli's law applies from $A \to B$ as there are no losses between these points:

$$\frac{(v_1^{(1)})^2}{2g} + h_1 = h_1 + \delta_1. \tag{1.13.31}$$

Now we introduce the internal Froude number:

$$F_i^2 = \frac{(v_1^{(1)})^2}{gh_1 \dfrac{\Delta\rho}{\rho_S}(1-\delta_r)} = \frac{(F_i^{(1)})^2}{1-\delta_r}, \tag{1.13.32}$$

and the ratio

$$\delta_r = \frac{h_2}{h_1}. \tag{1.13.33}$$

From (1.13.29), (1.13.30) and (1.13.31), it follows:

$$\delta_2 = \frac{\Delta\rho h_1(1-\varepsilon)}{2\rho_0}(F_i^{(1)})^2. \tag{1.13.34}$$

Substituting (1.13.33) and (1.13.30) into (1.13.29) and dividing by $\Delta\rho g h_1^2(1-\delta_r)$ leads to the solution:

$$F_i^2 = \frac{\delta_r(1-\delta_r)}{(2-\delta_r)} + \mathrm{O}\left(\frac{\delta_2}{h_1}\right)^2. \tag{1.13.35}$$

This is the same relationship as established for the oil slick example and for the hydraulic jump.

We now apply the conservation of energy equation (1.12.10) to the control volume and allow energy losses:

$$-\frac{1}{2}\rho_S(v_1^{(1)})^3 h_1 + \frac{1}{2}\rho_S(v_1^{(2)})^3 h_2 = \frac{1}{2}\rho_S g h_1^2 v_1^{(1)} - \left\{\rho_0 g(h_1 - h_2 + \delta_2)\, h_2 + \frac{1}{2}\rho_S g h_2^2\right\} v_1^{(2)} - \rho_S g \int_V v_3 \,\mathrm{d}V + \sum, \tag{1.13.36}$$

where the integral is taken over the whole control volume and Σ is the energy dissipation rate in the turbulent region.

In order to proceed we need to evaluate the integral in (1.13.36). In order to do this, we note the conservation of mass in two dimensions applied to an infinitesimal control volume as shown in Fig. 1.13.4 for an incompressible fluid:

$$\frac{\partial v_1}{\partial x_1} + \frac{\partial v_3}{\partial x_3} = 0. \quad (1.13.37)$$

A full derivation of this result is given in the next section, but (1.13.37) follows directly from Fig. 1.13.4.

Consider now

$$-\rho_S g \int_V v_3 \, \mathrm{d}V = -\rho_S g \int_{x_1^{(1)}}^{x_1^{(2)}} \int_0^h v_3 \, \mathrm{d}x_3 \, \mathrm{d}x_1, \quad (1.13.38)$$

but from (1.13.37) it follows that:

$$v_3 = -\int_0^{x_3} \frac{\partial v_1}{\partial x_1} \mathrm{d}x_3. \quad (1.13.39)$$

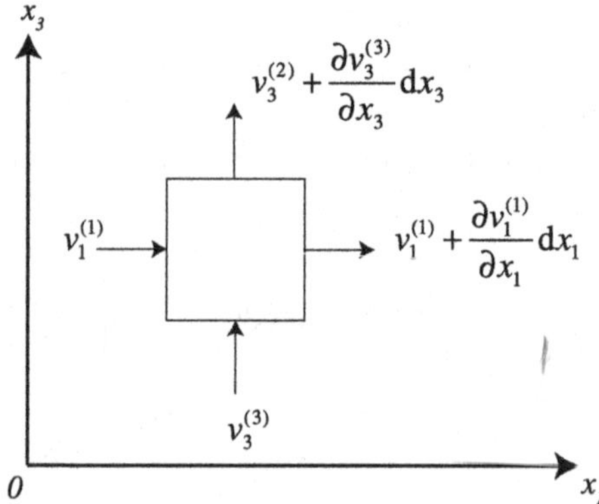

Figure 1.13.4 Infinitesimal area element.

If it is now assumed that $\frac{\partial v_1}{\partial x_1}$ is not a function of x_3 and substituting from (1.13.38) and carrying out the integration yields:

$$-\rho_S g \int_V v_3 \, \mathrm{d}V = \frac{1}{2}\rho_S g v_1^{(1)} h_2^2 - \frac{1}{2}\rho_S g v_1^{(2)} h_2, \tag{1.13.40}$$

where we have made use of the identity:

$$\frac{\partial}{\partial x_1}(h^2 v_1) = h^2 \frac{\partial v_1}{\partial x_1} + 2 v_1 h \frac{\partial h}{\partial x_1}. \tag{1.13.41}$$

The interpretation of (1.13.36) is clear if we note that $\frac{1}{2}\rho_S g h_1^2$ is the potential energy of the fluid entering the control volume and multiplying this by $v_1^{(1)}$ yields the rate of input of potential energy at S_1.

Substituting for δ_2 from (1.13.34) and for $v_1^{(2)}$ from (1.13.28) into (1.13.36) and dividing the result by $v_1^{(1)} h_1^2 \Delta\rho g(1-\delta_r)$ leads to the equation:

$$F_i^2 = 2\delta_r^2 + \frac{2\delta_r^2 \sum}{\Delta\rho g h_1 q(1-\delta_r)}. \tag{1.13.42}$$

For the case where $\Sigma = 0$, combining (1.13.42) and the momentum equation (1.13.35) again leads to:

$$\delta_r = \frac{1}{2} \quad \text{or} \quad \delta_r = 1, \tag{1.13.43}$$

which is the result obtained previously.

To see the impact of energy loss, suppose the buoyant surface overflow is vigorous and we lose the total incoming kinetic energy such that:

$$\sum = \frac{1}{2}\rho_S (v_1^{(1)})^3 h_1. \tag{1.13.44}$$

This may be substituted into (1.13.41) and then combined with (1.13.35) to yield:

$$\delta_r = \frac{F_i}{(2+F_i)^{1/2}}, \tag{1.13.45}$$

which approaches unity as F_i increases; this is the case where the flow becomes more and more energetic.

Another limiting case may be derived by noting that when h_1 becomes large and the flow is vigorous (F_i large) $(h_1 - h_2)/h_1$ becomes small, so that $v_1^{(1)} \approx v_1^{(2)}$. For this case we may apply Bernoulli's law from E to D, so that:

$$\frac{\rho_S g h_1}{\rho_S g} = \frac{\rho_0 g(\delta_2 + h_1 - h_2)}{\rho_S g} + \frac{\rho_S g h_2}{\rho_S g}. \tag{1.13.46}$$

Substituting for δ_2 from (1.13.34) reveals:

$$F_i^2 = 2, \tag{1.13.47}$$

so that (1.13.45) becomes

$$\delta_r = \frac{1}{\sqrt{2}}, \tag{1.13.48}$$

which means that $(h_2 - h_1)/h_1 = d/h_1 \approx 0.71$, and we see that d becomes progressively smaller with increasing energy loss.

In general if we let,

$$L = \frac{2\sum}{\Delta\rho g h_1 q(1 - \delta_r)}, \tag{1.13.49}$$

which is the rate of energy loss, dimensionalized with a rate of potential energy change, we get the equation:

$$(2 + L)\delta_r^2 - (3 + 2L)\delta_r + 1 = 0, \tag{1.13.50}$$

by equating (1.13.41) and (1.13.35). The solution to this is given by:

$$\delta_r = \frac{3 + 2L}{2(2 + L)}\left\{1 - \left(1 - \frac{4(2 + L)}{(3 + 2L)^2}\right)^{1/2}\right\}, \tag{1.13.51}$$

which again shows that as L increases, $\delta_r \to 0$.

Withdrawal into a submerged sink. Consider a buoyant fluid overlying a slightly heavier fluid, such as would be the case in a lake or ocean due to the slightly warmer temperatures of the surface layers.

As shown schematically in Fig. 1.13.5, a discharge q located in a semi-infinite fluid of density ρ_2 over which lies a layer of less dense (density equal to ρ_1) fluid of thickness h_1.

As the discharge q is turned on, suppose a cusp forms at the point B, a distance d above the sink. To a very good approximation, if the flow is tangential to the vertical at B, the velocity at B:

$$v_B = \frac{q}{2\pi d}, \tag{1.13.52}$$

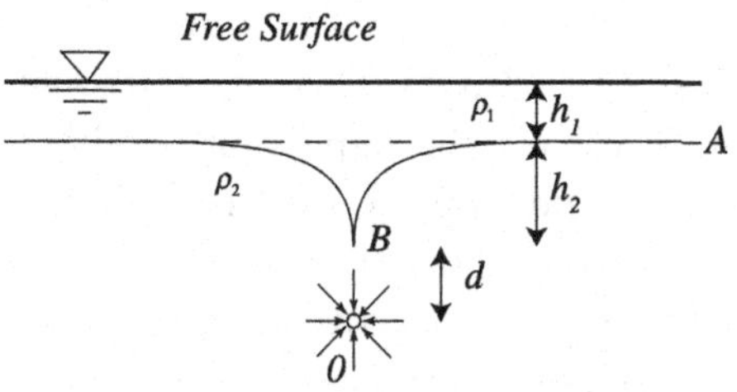

Figure 1.13.5 Schematic of sink flow under a density interface.

so applying Bernoulli's law from $A \to B$ yields:

$$h_2 + \frac{\rho_1 g h_1}{\rho_2 g} = \frac{q^2}{8\pi^2 d^2 g} + d + \frac{\rho_1 g(h_1 + h_2 - d)}{\rho_2 g}, \tag{1.13.53}$$

where it was assumed that the upper fluid is stagnant and the pressure is hydrostatic.

Rearranging (1.13.52) and defining the internal outflow Froude number:

$$F_i^2 = \frac{q^2}{\frac{\Delta\rho}{\rho_2} g h_2^3}, \tag{1.13.54}$$

leads to:

$$F_i^2 = 8\pi^2 \delta_r (1 - \delta_r)^2, \tag{1.13.55}$$

where $\delta_r = (h_2 - d)/h_2$, the ratio of the drawdown depth to the height of the fluid h_2 above the sink.

The relationship (1.13.54) is sketched in Fig. 1.13.6 where it is seen the δ_r increases linearly as F_i^2 increases from 0 to $\delta_r = 1/3(F_i^2 = 32\pi^2/27 = 3.42^2)$. The Bernoulli equation (conservation of energy) does not allow a flow

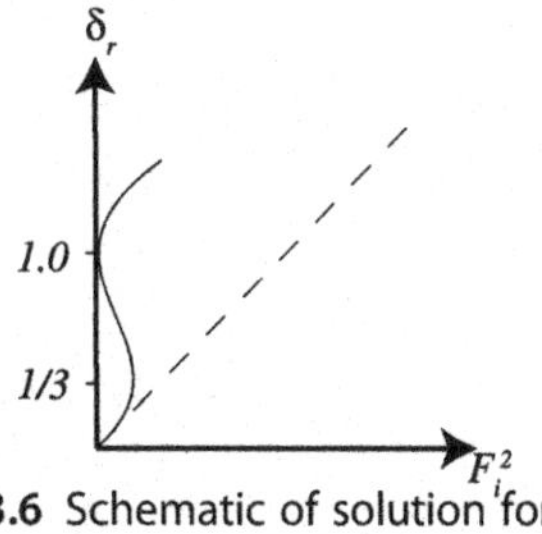

Figure 1.13.6 Schematic of solution for sink flow.

with $F_i > 3.42$; Fig. 1.13.6 shows that once the maximum F_i is reached, $\delta_r = 1$ can be reached with lower discharges. This suggest (though difficult to prove) that, when $F_i = 3.42$ is reached, the cusp (point B) will jump to the sink and fluid from both the upper and lower layer will be withdrawn though the sink. In practice the critical F_i is actually a little smaller than 3.42, but that is a matter for a more advanced treatment; the above provides a first estimate and good insight as to the power of Bernoulli's law.

CHAPTER 2

Equations of Motion: Axiomatic Approach

Contents

In this chapter the reader will find a rigorous development of the fundamental equations of motion. The main focus is on the flow of incompressible fluids, but the equations are developed for any Newtonian fluid. The material in this chapter is intended as a sort of back up, when rigor is required, readers may wish to skip this chapter on their first reading, especially if they are from a non-engineering background. The material in this chapter was developed for a first year graduate course that I have taught at various institutions, in Australia, the US, Italy and Germany.

2.1. CONSERVATION OF MASS

As we saw in §1.10, conservation of mass simply states that the mass of a Lagrangian control volume is conserved. This is correct for a simple fluid such as air or water, where the ratio of the concentration of components is uniform throughout the flow field. When the fluid under consideration is a mixture, such as fresh and salt water, molecular diffusion can transport the

Environmental Fluid Dynamics
ISBN 978-0-12-088571-8, DOI: 10.1016/B978-0-12-088571-8.00002-4

salt component across a material boundary and we need to account separately for this mass exchange; we shall postpone a discussion of this to §2.10. For a simple one component fluid, conservation of mass for a Lagrangian control volume $V(t)$ may be written as (already stated (1.10.1)):

$$\frac{\mathrm{d}}{\mathrm{d}t}\int_{V(t)} \rho(x_1, x_2, x_3, t)\mathrm{d}V = 0, \tag{2.1.1}$$

where $\rho(x_1,x_2,x_3,t)$ is the density of the fluid at the point $\{x_i\}$ and time t and $V(t)$ is the control volume shown in Fig. 2.1.1. Equation (2.1.1) may be compared to (1.8.5) with

$$A = \rho; B = C_i = 0 \tag{2.1.2}$$

and the inequality sets to an equality.

The local differential form of this conservation principle thus becomes, from (1.8.9):

$$\frac{\partial\rho}{\partial t} + (v_i\rho)_{,i} = \frac{\mathrm{D}\rho}{\mathrm{D}t} + \rho v_{i,i} = 0 \tag{2.1.3}$$

The term $\frac{\partial\rho}{\partial t}$ represents the time rate of change of density at a particular point. When coupled, as in (2.1.3), with the advective rate of change $v_i\rho_{,i}$ we get the rate of change of the density $\frac{\mathrm{D}\rho}{\mathrm{D}t}$ following a fluid particle. The remaining term $\rho v_{i,i}$ is the change of density brought about by the divergence of the fluid motion $v_{i,i}$.

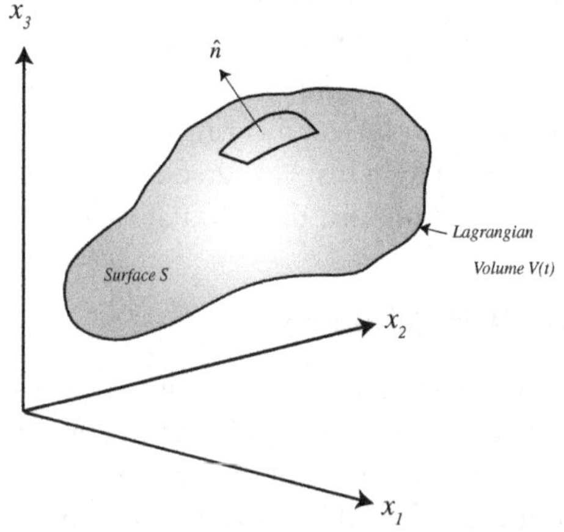

Figure 2.1.1 Lagrangian bulk control volume.

The meaning of this term is best seen by considering an initial fluid particle of volume:

$$dV_0 = d\xi_1, d\xi_2, d\xi_3 \tag{2.1.4}$$

As the particles move according to (1.8.3), the volume will change to dV, given by (Aris, 1962)

$$dV = J\,dV_0, \tag{2.1.5}$$

where J is the Jacobian of the transformation (1.8.3) given by the matrix:

$$J = \begin{vmatrix} \frac{\partial x_1}{\partial \xi_1} & \frac{\partial x_1}{\partial \xi_2} & \frac{\partial x_1}{\partial \xi_3} \\ \frac{\partial x_2}{\partial \xi_1} & \frac{\partial x_2}{\partial \xi_2} & \frac{\partial x_2}{\partial \xi_3} \\ \frac{\partial x_3}{\partial \xi_1} & \frac{\partial x_3}{\partial \xi_2} & \frac{\partial x_3}{\partial \xi_3} \end{vmatrix}. \tag{2.1.6}$$

Thus J is the ratio of the volume of a small material fluid particle to its initial volume and the quantity $\frac{DJ}{Dt}$ represents the rate of change of this normalized elementary volume as the volume follows the motion. In other words $\frac{DJ}{Dt}$ is the fractional change in volume of a Lagrangian fluid particle. An exercise in calculus reveals:

$$\frac{1}{J}\frac{DJ}{Dt} = v_{i,i} \tag{2.1.7}$$

so that the divergence $v_{i,i}$ of the velocity vector is the fractional increment of the volume change of a Lagrangian fluid particle as it moves about in the domain. The term $\rho v_{i,i}$ in (2.1.3), thus represents the change of density due to a change of volume of the fluid brought about, for instance, by a change in pressure.

2.2. THE STRESS TENSOR

Before we can discuss conservation of momentum it is necessary to introduce the concept of a stress tensor. Consider first a simple stress vector $\{\tau_i\}$ that describes the stress at a point on a surface S as indicated in Fig. 2.2.1. The stress at a point is defined as the limit of the total force on the surface element divided by the surface area dS, as the surface area dS approaches

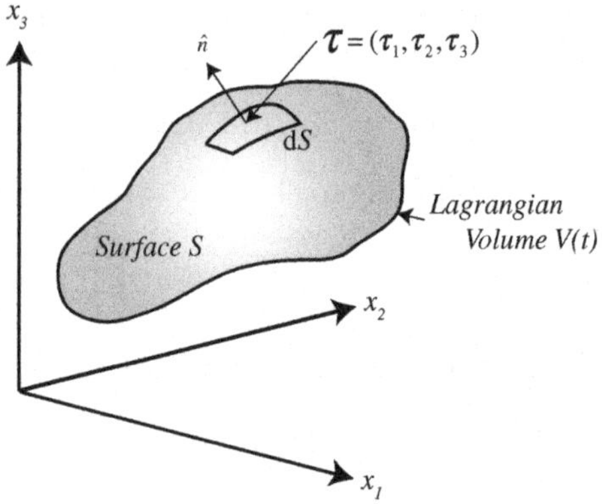

Figure 2.2.1 Schematic of a Lagrangian control volume.

zero. In terms of fluid particles, we can say that this limit takes the area to a size comparable to that of a fluid particle. The definition of a stress thus requires the specification of 9 quantities, the stress vector $\{\tau_i\}$ and the orientation $\{\hat{n}_i\}$ of the surface upon which the stress acts (Fig. 2.2.1).

In the simplest situation, the stress $\{\tau_i\}$ depends on the position $\{x_i\}$, the time t and the orientation of the surface so that:

$$\tau_i = \tau_i(x_1, x_2, x_3, t, \hat{n}_1, \hat{n}_2, \hat{n}_3). \tag{2.2.1}$$

In general the stress could also be a function of other variables such as, for instance, the curvature of the surface as is the case when dealing with surface tension stresses. The simple expression given by (2.2.1), however, suffices here and is generally referred to as the Cauchy stress assumption.

It follows from Newton's third law (for every action there is an equal and opposite reaction) that:

$$\tau_i(x_1, x_2, x_3, t, \hat{n}_1, \hat{n}_2, \hat{n}_3) = -\tau_i(x_1, x_2, x_3, t, -\hat{n}_1, -\hat{n}_2, -\hat{n}_3). \tag{2.2.2}$$

The stress tensor is most easily introduced by considering the equilibrium of the small tetrahedron of fluid as shown in Fig. 2.2.2. Let $\tau(i,j)$ be the stress in the j direction acting on a plane oriented in the ith direction, then the force balance in the ith direction becomes

$$\tau(1,i)\frac{dx_2\, dx_3}{2} + \tau(2,i)\frac{dx_1\, dx_3}{2} + \tau(3,i)\frac{dx_1\, dx_2}{2} = \tau_i\, dS + O(dx^3), \tag{2.2.3}$$

where dx is the magnitude of dx_1, dx_2 and dx_3 and the term of $O(dx^3)$ represents the change of momentum of the fluid inside the tetrahedron.

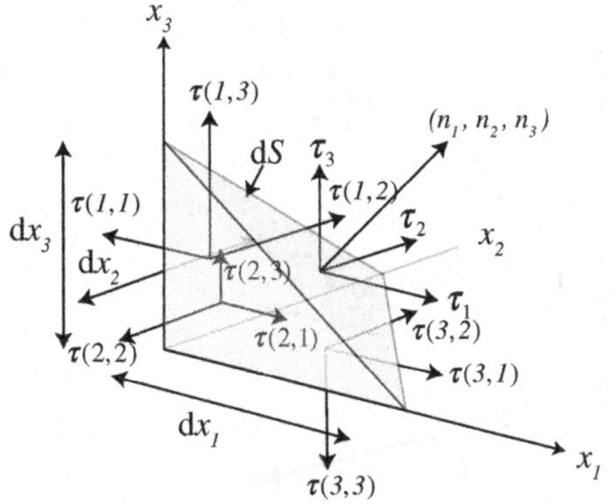

Figure 2.2.2 Infitesimal force tetrahedron.

From the geometry of the tetrahedron the expressions for the unit vector components follow, such that:

$$\hat{n}_1 = \frac{dx_2 dx_3}{2dS}, \tag{2.2.4}$$

$$\hat{n}_2 = \frac{dx_1 dx_3}{2dS}, \tag{2.2.5}$$

$$\hat{n}_3 = \frac{dx_1 dx_2}{2dS}. \tag{2.2.6}$$

Dividing (2.2.3) by dS, taking the limit d$x \to 0$ and substituting from (2.2.4)–(2.2.6) yields the relationship:

$$\tau_i = \tau(j, i)\hat{n}_j. \tag{2.2.7}$$

Now, since $\{\hat{n}_i\}$ and $\{\tau_i\}$ are general vectors it follows from the tensor quotient rule (Appendix) that $\tau(j,i)$ is a second order tensor τ_{ij}.

2.3. CONSERVATION OF LINEAR MOMENTUM

Newton's second law of motion may be applied to a Lagrangian control volume $V(t)$, such as shown in Fig. 2.3.1:

$$F_i = ma_i = \frac{d}{dt}(mv_i), \tag{2.3.1}$$

where $\{F_i\}$ is the net external force acting on the body, $\{v_i\}$ is the velocity of the center of gravity of the body and $\{a_i\}$ is the acceleration of the body.

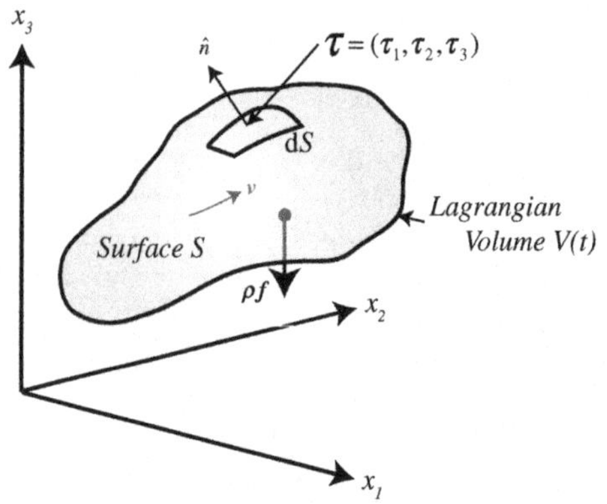

Figure 2.3.1 Forces acting on a Lagrangian control volume $V(t)$.

If $\{f_i\}$ is the body force per unit mass acting at each point of the fluid in the control volume, ρ is the density and $\{\tau_i\}$ is the stress acting on the surface $S(t)$ of the control volume (Fig. 2.3.1), then (2.3.1) becomes:

$$\frac{\mathrm{d}}{\mathrm{d}t}\left\{\int_{V(t)} \rho v_i \,\mathrm{d}V\right\} = \int_{V(t)} \rho f_i \,\mathrm{d}V + \int_{S(t)} \tau_i \,\mathrm{d}S. \tag{2.3.2}$$

Suppose we fix our attention on a particular value of the index i and substitute for the stress vector $\{\tau_i\}$ from (2.2.7) into (2.3.2) then:

$$\frac{\mathrm{d}}{\mathrm{d}t}\left\{\int_{V(t)} \rho v_{(i)} \,\mathrm{d}V\right\} = \int_{V(t)} \rho f_{(i)} \,\mathrm{d}V + \int_{S(t)} \tau_{j(i)} n_j \,\mathrm{d}S, \tag{2.3.3}$$

where the brackets around the index indicate that it is fixed. Direct comparison with (1.8.5) suggests letting,

$$A = \rho v_{(i)}; \quad B = \rho f_{(i)}; \quad C_j = \tau_{j(i)}, \tag{2.3.4}$$

so from (1.8.10) it follows that:

$$\frac{\mathrm{D}}{\mathrm{D}t}(\rho v_i) + \rho v_i v_{j,j} = \rho f_i + \tau_{ji,j}. \tag{2.3.5}$$

This may be simplified by noting that:

$$\frac{\mathrm{D}}{\mathrm{D}t}(\rho v_i) = \rho\frac{\mathrm{D}v_i}{\mathrm{D}t} + v_i\frac{\mathrm{D}\rho}{\mathrm{D}t}, \tag{2.3.6}$$

so that using (2.1.3) and (2.3.5) becomes:

$$\rho \frac{Dv_i}{Dt} = \rho f_i + \tau_{ji,j}. \tag{2.3.7}$$

These are the momentum equations applicable locally at each point in the fluid.

Often in Environmental Fluid Dynamics the body of water, be it a lake, estuary or coastal sea, is large enough for the fluid or internal wave to take an appreciable time to traverse the domain. In such circumstances it may be necessary to modify (2.3.7) to account for the earth's angular rotation $\{\Omega_i\}$ of the coordinate system and derive an equivalent form relative to the rotating reference frame fixed on the surface of the earth. This is most easily done by noting that, at a point at a fixed position $\{x_i^r\}$ in the rotating coordinate system, a stationary fluid particle will actually be moving with a velocity $\{e_{ijk}\Omega_j x_k^r\}$ relative to a coordinate system fixed in space.

Thus if the point is moving with a velocity $\{v_i^r\}$, relative to the rotating frame of reference then, relative to the fixed frame, it will, by addition, have a velocity given by:

$$\frac{dx_i}{dt} = v_i^r + e_{ijk}\Omega_j x_k^r. \tag{2.3.8}$$

The derivative operator thus translates, in the rotating frame of reference, to a derivative plus a cross-product with the rotation vector. Repeating the operation leads to an expression for the acceleration of the particle:

$$\begin{aligned} \frac{d^2(x_i^r)}{dt^2} &= \frac{d^2 x_i^r}{dt^2} + e_{ijk}\Omega_j x_k^r + e_{i\alpha\beta}\Omega_j(v_\beta^r + e_{\beta lm}\Omega_l) \\ &= \frac{d^2 x_i^r}{dt^2} + 2e_{ijk}\Omega_j x_k^r + e_{i\alpha\beta}e_{\beta lm}\Omega_\alpha\Omega_l x_m^r. \end{aligned} \tag{2.3.9}$$

This expression may be simplified by noting that:

$$\begin{aligned} e_{i\alpha\beta}e_{\beta lm}\Omega_\alpha\Omega_l x_m^r &= (\delta_{il}\delta_{\alpha m} - \delta_{im}\delta_{\alpha l})\Omega_\alpha\Omega_l x_m^r \\ &= \Omega_m\Omega_i x_m^r - \Omega_l\Omega_l x_i^r \\ &= \left(\frac{\Omega_\alpha\Omega_\alpha x_m^r x_m^r}{2}\right)_{,i} \end{aligned} \tag{2.3.10}$$

Also the body force term in (2.3.7), may be written in the form:

$$\rho f_i = -g\rho\delta_{i3} = -g\rho x_{3,i}, \qquad (2.3.11)$$

where g is the acceleration due to gravity as, in environmental flows, gravity is the only common body force.

Substituting (2.3.11) and (2.3.10) into (2.3.7) we get:

$$\rho\frac{\mathrm{D}v_i}{\mathrm{D}t} + 2\rho e_{ijk}\Omega_j v_k = -\rho\left(gx_3 - \frac{\Omega_\alpha\Omega_\alpha x_m x_m}{2}\right)_{,i} + \tau_{ji,j} \qquad (2.3.12)$$

where, for convenience, the superscript r has been left off even though (2.3.12) applies in a rotating frame of reference.

The term

$$\rho\frac{\mathrm{D}v_i}{\mathrm{D}t} = \frac{\partial v_i}{\partial t} + v_j v_{i,j}, \qquad (2.3.13)$$

again represents the rate of change of the momentum of a fluid particle as it moves through the domain. The first term $\frac{\partial v_i}{\partial t}$ on the *RHS* of (2.3.13) is the component of acceleration due to a time rate of change of the velocity field relative to the Eulerian coordinate axes. By contrast the term $v_j v_{i,j}$ is the component of acceleration due to spatial variations in the velocity field. If it is remembered that the momentum equation (2.3.1) holds for a Lagrangian fluid particle moving through a particular point in our Eulerian coordinate system then it is clear that particles in a steady flow field normally still possess accelerations as they move from one velocity at one point to a different velocity at the next point. The velocity tensor $v_{i,j}$ accounts for changes of the ith component of the velocity as the particle moves with a velocity v_j in the jth direction.

The term

$$\rho\left(gx_3 - \frac{\Omega_\alpha\Omega_\alpha x_m x_m}{2}\right),$$

is simply the gravitation force plus the centrifugal force acting on a fluid particle of unit volume. The centrifugal force is due to the earth's rotation and we can estimate its magnitude directly by noting that the earth rotates about its axis once a day so that for a north pointing coordinate system

$$\{\Omega_i\} = (0, 0, 7.3 \times 10^{-5} rs^{-1}).$$

Now at the equator, the centrifugal force is largest and acts in exactly the opposite direction to that of the gravity. The magnitude of this force compared to gravity is given by the ratio $\frac{\Omega_3^2 x_1}{g}$. Given that the radius of the earth is approximately 6,378,000 m, this ratio is about 3.5×10^{-3} meaning that gravity is reduced at the equator by about 0.4%. The magnitude at other latitudes is even less. Given that the earth is approximately spherical the centrifugal force will be a little larger at the end of the lake closest to the equator, and the surface of a lake will slope up ever so slightly toward the poles; however, this effect is too small to be of any significance in hydrodynamics of natural systems such as lakes, estuaries or coastal seas. For this reason we shall, from now on, ignore the action of the centrifugal force and assume that the acceleration due to gravity alone acts as a body force.

By contrast the term,

$$2e_{ijk}\Omega_j v_k,$$

is the *i*th component of a force that is again a mass (ρ) times an acceleration $2e_{ijk}\Omega_j v_k$. This is called the Coriolis force. The physical interpretation of this force is best understood by imagining a fluid particle moving in a straight line at a constant velocity relative to an inertial coordinate system. As seen in Fig. 2.3.2, when viewed by an observer relative to the rotating frame of reference, the fluid particle moving due North in the absolute frame of reference appears to veer off to the left in the southern hemisphere.

The acceleration, normal to the path of travel, is given by $2\rho e_{ijk}\Omega_j v_k$, relative to the rotating frame of reference. As Newton's laws are true only in

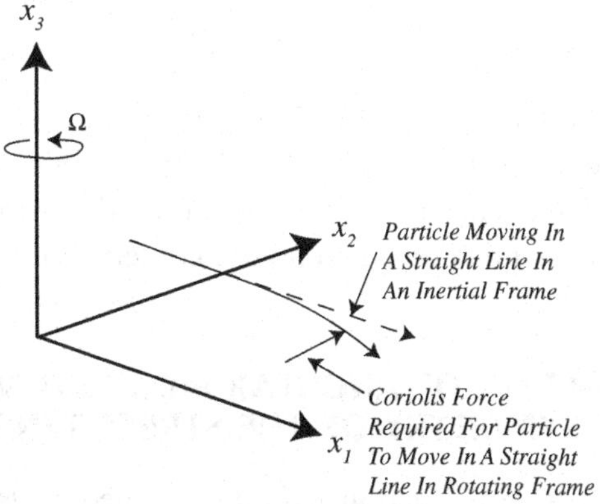

Figure. 2.3.2 Definition of the Coriolis force.

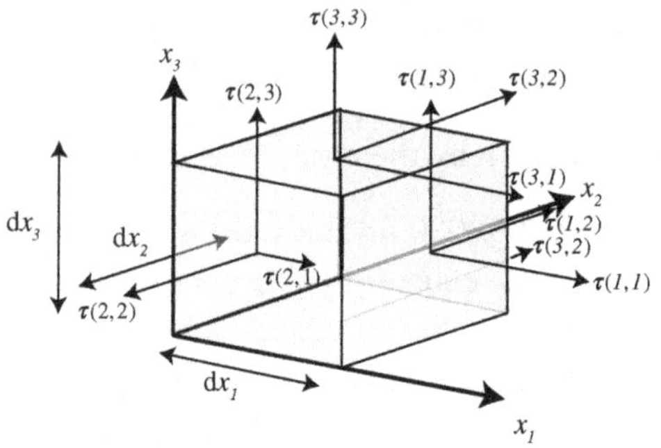

Figure 2.3.3 Stress tensor components on the face of a cube (only front faces shown).

a fixed or absolute reference frame, then for the particle to travel due North relative to the rotating frame of reference a force $2pe_{ijk}\Omega_j v_k$ must be applied to the right to prevent the particle from veering to the left.

The remaining term is the divergence of the surface stress tensor $\tau_{ji,j}$. The reason why it is the divergence of the stress tensor is best understood by considering a small cube of fluid as shown in Fig. 2.3.3.

For simplicity we shall consider only the contribution of the surface forces in the $\{1,0,0\}$ direction (x_1 direction) acting on the three surfaces $\{1,0,0\}$ $\{0,1,0\}$ and $\{0,0,1\}$. As the cube is small it is possible to expand the stresses, shown in Fig. 2.3.3, in a Taylor series. The net force, as contributed by all surfaces stresses, acting on this fluid particle in the $\{1,0,0\}$ direction is thus given by:

$$(\tau_{1i,1} + \tau_{2i,2} + \tau_{3i,3})\mathrm{d}x_1\mathrm{d}x_2\mathrm{d}x_3 + O(\mathrm{d}x^4),$$

so that the net force per unit volume in the limit as $\mathrm{d}x \to 0$ is given by,

$$\tau_{ji,j}.$$

The divergence term thus arises in the momentum equation (2.3.12) because it is the difference in the stress at the two opposing surfaces of the cube that leads to a change in the attribute. Divergence terms arise whenever a net flux of an attribute enters a conservation equation.

2.4. CONSERVATION OF ANGULAR MOMENTUM AND THE SYMMETRY OF THE STRESS TENSOR

The angular momentum equation may be obtained by taking the cross-product of the vector $\{x_i\}$ with the momentum equation (2.3.2). For the purpose of showing that τ_{ij} is symmetric it is sufficient to remain within the fixed frame of reference, so we may write:

$$\frac{\mathrm{d}}{\mathrm{d}t}\int_{V(t)} e_{ijk}\rho x_j v_k \,\mathrm{d}V = \int_{V(t)} e_{ijk}\rho x_j f_k \,\mathrm{d}V + \int_{S(t)} e_{ijk}\rho x_j \tau_k \,\mathrm{d}S. \quad (2.4.1)$$

Equation (2.4.1) is the general form of the angular momentum equation for the Lagrangian control volume $V(t)$.

Once again we can make use of the general Transport Theorem (1.8.5) with

$$A = e_{(i)jk}\rho x_j v_k; \quad B = e_{(i)jk}\rho x_j f_k; \quad C_i = e_{(i)jk}\rho x_j \tau_{lk}\,, \quad (2.4.2)$$

to obtain the local angular momentum equation:

$$\frac{\mathrm{D}}{\mathrm{D}t}\left(e_{(i)jk}\rho x_j v_k\right) + e_{(i)jk}\rho x_j v_k v_{l,l} = e_{(i)jk}\rho x_j f_k + \left(e_{(i)jk}\rho x_j \tau_{lk}\right)_{,l} \quad (2.4.3)$$

The density ρ may be extracted by expanding the first term and then using (2.1.3):

$$\rho e_{ijk} x_j \frac{\mathrm{D}v_k}{\mathrm{D}t} = e_{ijk}\rho x_j f_k + e_{ijk}\rho x_j \tau_{lk,l} \quad (2.4.4)$$

The first term of (2.4.4) represents the rate of change of angular velocity $e_{(i)jk}x_j v_k$ times the mass ρ per unit volume. The second term $e_{(i)jk}\rho x_j f_k$ is the couple, about the origin, applied by the body force and the last term $(e_{(i)jk}\rho x_j \tau_{lk})_{,l}$ represents the net couple applied by the surface stresses.

We may now use (2.4.4) to show that the stress tensor is symmetric so that,

$$\tau_{ij} = \tau_{ji} \quad (2.4.5)$$

This is most easily done by starting with the identity:

$$\int_{S(t)} e_{ijk} x_j \tau_{lk} n_l \,\mathrm{d}S = \int_{V(t)} e_{ijk}\left(x_j \tau_{lk}\right)_l \,\mathrm{d}V = \int_{V(t)} e_{ijk}\left(x_j \tau_{lk,l} + \tau_{jk}\right) \mathrm{d}V. \quad (2.4.6)$$

Substituting for $\tau_{lk,l}$ from (2.3.7) leads to the result:

$$\int_{S(t)} e_{ijk} x_j \tau_{lk} n_l \,\mathrm{d}S + \int_{V(t)} e_{ijk}\rho x_j f_k \,\mathrm{d}V - \int_{V(t)} e_{ijk}\rho x_j \frac{\mathrm{D}v_k}{\mathrm{D}t}\,\mathrm{d}V$$

$$= \int_{V(t)} \frac{e_{ijk}}{2}\left(\tau_{jk} - \tau_{kj}\right) \mathrm{d}V, \quad (2.4.7)$$

where the last term was derived by noting:

$$e_{ijk}\tau_{jk} = e_{ikj}\tau_{kj}. \tag{2.4.8}$$

Eliminating the surface integral in (2.4.7) and using (2.4.4) leads to:

$$\begin{aligned}&\int\limits_{V(t)} \left(e_{ijk}(n_j\tau_{lk})_{,l} - e_{ijk}\rho x_j f_k - e_{ijk}\tau_{jk} - e_{ijk}x_j\tau_{lk,l} + e_{ijk}\rho x_j f_k\right)\mathrm{d}V \\ &= \int\limits_{V(t)} \frac{e_{ijk}}{2}(\tau_{jk} - \tau_{kj})\mathrm{d}V\end{aligned} \tag{2.4.9}$$

Given that the left hand side is zero requires:

$$\int\limits_{V(t)} e_{ijk}(\tau_{jk} - \tau_{kj})\mathrm{d}V = 0 \tag{2.4.10}$$

Now since $V(t)$ is arbitrary it follows that τ_{jk} is symmetric, proving the starting premise. This symmetry property is of great importance as it is intimately connected to the diffusion of vorticity, a concept developed in §2.12.

2.5. CONSERVATION OF ENERGY

We shall adopt the same axiomatic approach to the treatment of thermodynamics as we have done for the momentum equation in §2.3. An energy conservation law, similar in spirit to the momentum law (2.3.2), called the first law of thermodynamics will be assumed for a Lagrangian control volume $V(t)$. However, before doing so we must define the basic forms of energy within such a Lagrangian control volume.

First consider the store of kinetic energy resulting from the motion of the fluid particles within the control volume. The kinetic energy per unit volume is given by $\frac{1}{2}\rho v_i v_i$ so that the total kinetic energy K of the fluid within the control volume is the sum or integral over the control volume $V(t)$:

$$K = \int\limits_{V(t)} \frac{1}{2}\rho v_i v_i \,\mathrm{d}V. \tag{2.5.1}$$

Second, as work is done on the fluid that results in either a change of temperature θ and or specific volume ϑ ((density)-1), the internal energy, reflecting the energy content of the fluid molecules, changes. In general, we may define the specific internal energy e as the internal energy per unit mass:

$$e = e(\theta, \vartheta), \tag{2.5.2}$$

which is a measure of the energy stored in the molecular structure of the fluid. This energy can be recovered as heat by lowering of the temperature or as mechanical energy by allowing the fluid to do work by expansion.

The quantity $e(\theta,\vartheta)$ is an additive property so that the total internal energy E of the fluid, within the Lagrangian control volume $V(t)$, is given by:

$$E = \int\limits_{V(t)} \rho e(\theta, \vartheta) \mathrm{d}V. \tag{2.5.3}$$

Third, each time a fluid particle, within or on the boundary of the control volume, moves and there is a body force and/or surface force, the particle does work or the system does work on the particle; a particle moving upwards against gravity needs to do work, on the other hand, a particle falling under gravity will gain kinetic energy as gravity works on the particle. Hence the rate at which the body force $\{f_i\}$ and the surface stresses $\{\tau_i\}$ do work is given by:

$$W = \int\limits_{V(t)} \rho v_i f_i \mathrm{d}V + \int\limits_{S(t)} v_i \tau_i \,\mathrm{d}S. \tag{2.5.4}$$

Lastly, we have the thermal energy exchanges. There are two sources of thermal energy input to the control volume $V(t)$. First, radiative energy such as solar radiation can penetrate the fluid and be absorbed by the fluid particles as they move through the domain. We shall let q be the energy absorption per unit mass due to such radiation. Second, the control volume may exchange heat with the surrounding fluid by conduction through the surface S. We shall call the transfer per unit area equal to h. Given these two mechanisms, the total rate of change of heat Q is given by,

$$Q = \int\limits_{V(t)} \rho q \mathrm{d}V - \int\limits_{S(t)} h \mathrm{d}S. \tag{2.5.5}$$

The first law of thermodynamics then simply states that the rate of change of the energy store (kinetic and internal energy) is equal to the rate of energy input. That is:

$$\frac{\mathrm{d}}{\mathrm{d}t}(K + E) = W + Q. \tag{2.5.6}$$

Once again we shall assume that the conductive heat flux h is a quantity that is only a function of position $\{x_i\}$, time t and the orientation of the surface $\{n_i\}$ so that:

$$h = h(x_1, x_2, x_3, t, n_1, n_2, n_3). \tag{2.5.7}$$

Since an infinitely thin surface cannot accumulate heat, an equation equivalent to (2.2.2) must apply:

$$h = h(x_1, x_2, x_3, t, n_1, n_2, n_3) = h(x_1, x_2, x_3, t, -n_1, -n_2, -n_3), \tag{2.5.8}$$

and by the same logic as was used to derive (2.2.7), we may show that there exists a vector $\{h_i\}$ such that:

$$h = h_i n_i. \tag{2.5.9}$$

The proof of this assertion is straight forward if we interpret h_i as the heat flux across the ith surface of the tetrahedron similar to that shown in Fig. 2.2.2.

Once again we may apply the general Transport Theorem (1.8.5) by introducing (2.5.1), (2.5.3), (2.5.4) and (2.5.5) into (2.5.6) and at the same time using (2.5.9) and (2.2.7) to define the heat flux and the surface stress in their component form:

$$\frac{\mathrm{d}}{\mathrm{d}t}\left\{\int_{V(t)} \left(\frac{1}{2}\rho v_i v_i + \rho e\right)\mathrm{d}V\right\} = \int_{V(t)} (\rho q + \rho f_i v_i)\mathrm{d}V + \int_{S(t)} (\tau_{ij} v_j - h_i) n_i \,\mathrm{d}S. \tag{2.5.10}$$

Now defining:

$$A = \left(\frac{1}{2}\rho v_i v_i + \rho e\right); \quad B = (\rho q + \rho f_i v_i); \quad C_i = (\tau_{ij} v_j - h_i) \tag{2.5.11}$$

it follows immediately from (1.8.11) that:

$$\frac{\mathrm{D}}{\mathrm{D}t}\left[\rho\left(\frac{1}{2}v_i v_i + e\right)\right] + \left[\rho\left(\frac{1}{2}v_i v_i + e\right)\right] v_{i,i} = (\rho q + \rho f_i v_i) + (\tau_{ij} v_j - h_i)_{,i}. \tag{2.5.12}$$

As before we may remove the second term on the left hand side by moving ρ outside the time derivative by noting the conservation of mass relationship (2.1.3) to yield,

$$\rho \frac{\mathrm{D}\left(\frac{1}{2}v_i v_i + e\right)}{\mathrm{D}t} = \rho q + \rho f_i v_i + \left(\tau_{ij} v_j\right)_{,i} - h_{i,i}, \tag{2.5.13}$$

which is the total energy equation or the differential form of the first law of thermodynamics.

The partitioning between the mechanical and thermal components of energy in (2.5.13) may be achieved by noting that if we multiply (2.3.13) by v_i and neglect the rotation terms, then we obtain an equation that has the kinetic energy as one of its terms:

$$\rho \frac{\mathrm{D}}{\mathrm{D}t}\left(\frac{v_i v_i}{2}\right) = \rho f_i v_i + v_i \tau_{ji,j}. \tag{2.5.14}$$

This is called the mechanical energy equation. Subtracting (2.5.14) from (2.5.13) leads to the thermal energy equation:

$$\rho \frac{\mathrm{D}e}{\mathrm{D}t} = \rho q - d_{ij}\tau_{ji} - h_{i,i}, \tag{2.5.15}$$

showing that the heating by the stresses within the fluid is caused by the deformation of the fluid and not by rotation or translation.

2.6. THE SECOND LAW OF THERMODYNAMICS

The second law of thermodynamics states that there exists a property of the fluid and its state, determined by the temperature and the specific volume, called the specific entropy

$$s = s(\theta, \vartheta), \tag{2.6.1}$$

such that, as the fluid moves, the heating per unit temperature is bounded by the rate of change of s. This is the second law of thermodynamics and may be written as:

$$\frac{\mathrm{d}}{\mathrm{d}t}\int\limits_{V(t)} \rho s \,\mathrm{d}V \geq \int\limits_{V(t)} \frac{\rho q}{\theta}\mathrm{d}V - \int\limits_{S(t)} \frac{h}{\theta}\mathrm{d}S. \tag{2.6.2}$$

We shall make no attempt to justify this, but rather as in the case of conservation of mass, momentum and energy, assume that it is one of the axioms of our work. By analogy with (1.8.5), let

$$A = \rho s; \ B = \frac{\rho q}{\theta}; \ C_i = \frac{-h_i}{\theta} \tag{2.6.3}$$

so that from (1.8.11) we get the result,

$$\frac{\mathrm{D}}{\mathrm{D}t}(\rho s) + \rho s v_{i,i} \geq \frac{\rho q}{\theta} - \left(\frac{h}{\theta}\right)_{,i} \tag{2.6.4}$$

Again expanding the time derivative in (2.6.4) and using conservation of mass (2.1.3) leads to:

$$\rho \frac{\mathrm{D}s}{\mathrm{D}t} \geq \frac{\rho q}{\theta} - \left(\frac{h}{\theta}\right)_{,i} \tag{2.6.5}$$

The physical interpretation of (2.6.5) rests with the definition of the specific entropy; it is that quantity in which the change is greater or equal to the heating per unit temperature. A process in which the equality holds is called reversible since, for such a process, we can take the fluid from a state $A(\theta_A, \vartheta_A)$ to a state $B(\theta_B, \vartheta_B)$ and back to the original A without a net input of energy. A process is called:

$$\text{isothermal if} \quad \frac{\mathrm{D}\theta}{\mathrm{D}t} = 0, \tag{2.6.6}$$

$$\text{adiabatic} \quad \text{if} \quad q = h = 0, \tag{2.6.7}$$

$$\text{isoentropic} \quad \text{if} \quad \frac{\mathrm{D}s}{\mathrm{D}t} = 0, \tag{2.6.8}$$

$$\text{and isoenergetic} \quad \text{if} \quad \frac{\mathrm{D}e}{\mathrm{D}t} = 0. \tag{2.6.9}$$

2.7. THE NAVIER–STOKES CONSTITUTIVE EQUATIONS

So far we have derived the various conservation equations (2.1.3), (2.3.12) and (2.5.13), five in total, together with one inequality (2.6.5), three equations of symmetry for the stress tensor (2.4.5) and one equation of state (2.5.2) or (2.6.1). There are a total of 10 equations for the unknowns; $\rho, v_i, \tau_{ij}, e, h_i, s, \theta$. We are thus short by nine equations for a determined system.

The remaining equations may be derived from a description of the material of the fluid. So far we have not said anything about the relationship between the stress tensor τ_{ij} and the rate of strain tensor $v_{i,j}$ or the relationship between the heat flux vector h_i and the temperature field θ. Naturally, these equations will depend on the properties of the fluid in the domain. Without being too specific, it is possible to derive relationships in a general way from a few axiomatic requirements and, as we will see, the given relationships depend only on a set of well-defined parameters. These general requirements are as follows:

1. The fluid is simple and homogeneous such that:

$$\tau_{ij} = \tau_{ij}(\theta, \vartheta, d_{ij}, \theta_{,i}), \tag{2.7.1}$$

$$h_i = h_i(\theta, \vartheta, d_{ij}, \theta_{,i}). \tag{2.7.2}$$

   In other words, the stress tensor and the heat flux vector are homogeneous, depending only on the position vector $\{x_i\}$ and time t, through the variables $(\theta, \vartheta, d_{ij}, \theta_{,i})$.
2. The fluid is isotropic so that the relationship between the stress tensor and the rate of strain tensor does not depend on the orientation of the coordinate system $\{x_i\}$. In other words, the relationship is invariant under a rotation transformation of the coordinate axes. A similar condition holds for the heat flux vector.
3. The fluid is linear and the stress tensor and the heat flux vector depend only linearly on the rate of deformation tensor and the temperature gradient vector.
4. A fluid at rest experiences only stresses arising from the pressure in the fluid:

$$\tau_{ij} = -p(\theta, \vartheta)\delta_{ij} \tag{2.7.3}$$

5. From our previous work, we have also shown that the stress tensor τ_{ij} must be symmetric as given by (2.4.5).
6. The flux of heat is zero in a fluid of uniform temperature.

Conditions (1), (3) and (5) together imply that

$$\tau_{ij} = -p(\theta,\vartheta)\delta_{ij} + B_{ijkl}(\theta,\vartheta)d_{kl}, \tag{2.7.4}$$

where $B_{ijkl}(\theta,\vartheta)$ are a set of constants, independent of the motion, depending only on θ and ϑ.

Similarly the heat flux vector may be written in the form:

$$h_i = -\kappa(\theta,\vartheta)\theta_{,i}, \tag{2.7.5}$$

where $\kappa(\theta,\vartheta)$ is again a property of the fluid and where we have introduced a negative sign to emphasize that heat flows from hot to cold.

The fourth order tensor $B_{ijkl}(\theta,\vartheta)$ must be isotropic and symmetric in i and j. In Appendix 1 we show that the most general such tensor is given by:

$$B_{ijkl}(\theta,\vartheta) = \lambda(\theta,\vartheta)\delta_{ij}\delta_{kl} + \mu(\theta,\vartheta)\{\delta_{ik}\delta_{jl} + \delta_{il}\delta_{jk}\}, \tag{2.7.6}$$

where again $\mu(\theta,\vartheta)$ and $\lambda(\theta,\vartheta)$ are properties of the fluid.

Substituting (2.7.6) into (2.7.4) leads an expression for the stress tensor:

$$\tau_{ij} = -p(\theta,\vartheta)\delta_{ij} + \lambda(\theta,\vartheta)v_{\alpha,\alpha}\delta_{ij} + 2\mu(\theta,\vartheta)d_{ij}. \tag{2.7.7}$$

Together equations (2.7.5) and (2.7.7) provide the missing equations and we now have a complete set of equations to solve for the unknown flow variables.

2.8. SOME THERMODYNAMIC CONSIDERATIONS

By manipulating the energy equation and the second law of thermodynamics we can shed considerable light on the role of viscous dissipation within the fluid and also derive an equation for the temperature field.

The derivation of these results is simplified by introducing the Helmholz free energy function ψ:

$$\psi(\theta,\vartheta) = e - \theta s. \tag{2.8.1}$$

Differentiating (2.8.1) with respect to time following the motion, substituting for $\mathrm{D}e/\mathrm{D}t$ from the thermal energy equation (2.5.15) and introducing the result into the expression for the second law of thermodynamics (2.6.5) yields:

$$-\frac{\rho}{\theta}\frac{\mathrm{D}\psi}{\mathrm{D}t} + \frac{\tau_{ij}v_{i,j}}{\theta} - \frac{\rho s}{\theta}\frac{\mathrm{D}\theta}{\mathrm{D}t} - \frac{h_i\theta_{,i}}{\theta^2} \geq 0. \tag{2.8.2}$$

However, it follows directly from (2.8.1) that,

$$\frac{D\psi}{Dt} = \frac{\partial\psi}{\partial\theta}\frac{D\theta}{Dt} + \frac{\partial\psi}{\partial\vartheta}\frac{D\vartheta}{Dt}. \tag{2.8.3}$$

Substituting (2.8.3) into (2.8.2) yields:

$$-\frac{\rho}{\theta}\frac{\partial\psi}{\partial\theta}\frac{D\theta}{Dt} + \frac{\tau_{ij}v_{i,i}}{\theta} - \frac{\rho}{\theta}\frac{\partial\psi}{\partial\vartheta}\frac{D\vartheta}{Dt} - \frac{\rho s}{\theta}\frac{D\theta}{Dt} - \frac{h_i\theta_{,i}}{\theta^2} \geq 0. \tag{2.8.4}$$

Now ϑ is the inverse of the density and noting that (2.1.3) allows us to write:

$$\frac{D\vartheta}{Dt} = \frac{1}{\rho^2}\frac{D\rho}{Dt} = \frac{1}{\rho}v_{i,i}. \tag{2.8.5}$$

In order to separate the influence of the pressure we set:

$$\tau_{ij} = \overline{\tau}_{ij} + p\delta_{ij}, \tag{2.8.6}$$

so that (2.8.4) becomes,

$$-\frac{\rho}{\theta}\left(s + \frac{\partial\psi}{\partial\theta}\right)\frac{D\theta}{Dt} - \frac{1}{\theta}\left(p + \frac{\partial\psi}{\partial\vartheta}\right)v_{i,i} + \frac{\overline{\tau}_{ij}v_{j,i}}{\theta} - \frac{h_i\theta_{,i}}{\theta^2} \geq 0. \tag{2.8.7}$$

So far we have made no assumptions concerning the thermodynamic process and $\frac{D\theta}{Dt}$, as well as $v_{i,i}$, are arbitrary. Hence it is necessary that:

$$s = \frac{\partial\psi(\theta,\vartheta)}{\partial\theta}, \tag{2.8.8}$$

and

$$p = -\frac{\partial\psi(\theta,\vartheta)}{\partial\vartheta}, \tag{2.8.9}$$

so that (2.8.7) may be simplified to:

$$\varphi - \frac{h_i\theta_{,i}}{\theta} \geq 0, \tag{2.8.10}$$

where $\phi = \overline{\tau}_{ij}v_{j,i}$ is the dissipation of energy by internal shear stresses.

Equation (2.8.10) may be interpreted in two ways. First, substituting (2.8.8) and (2.8.9) into (2.8.3) leads to:

$$\frac{D\psi}{Dt} = -s\frac{D\theta}{Dt} - p\frac{D\vartheta}{Dt}. \tag{2.8.11}$$

But also from (2.8.1):

$$\frac{D\psi}{Dt} = \frac{De}{Dt} - \theta\frac{Ds}{Dt} - s\frac{D\theta}{Dt}, \tag{2.8.12}$$

and by choosing s and ϑ as the independent variables we may write:

$$\frac{De}{Dt} = -\frac{\partial e}{\partial s}\frac{Ds}{Dt} + \frac{\partial e}{\partial \vartheta}\frac{D\vartheta}{Dt}. \tag{2.8.13}$$

Substituting (2.8.13) into (2.8.12) yields:

$$\frac{D\psi}{Dt} = \left(\frac{\partial e}{\partial s} - \theta\right)\frac{Ds}{Dt} + \frac{\partial e}{\partial \vartheta}\frac{D\vartheta}{Dt} - s\frac{D\theta}{Dt}. \tag{2.8.14}$$

Now equating (2.8.14) and (2.8.11) leads to the result,

$$\left(\frac{\partial e}{\partial s} - \theta\right)\frac{Ds}{Dt} = \left(\frac{\partial e}{\partial \vartheta} + p\right)\frac{D\vartheta}{Dt}. \tag{2.8.15}$$

Again $\frac{D\vartheta}{Dt}$ and $\frac{Ds}{Dt}$ are arbitrary independent variables so that,

$$p = -\frac{\partial e(s, \vartheta)}{\partial \vartheta}, \tag{2.8.16}$$

and

$$\theta = -\frac{\partial e(s, \vartheta)}{\partial s}. \tag{2.8.17}$$

It follows immediately from cross-differentiation of (2.8.16) and (2.8.17) that:

$$\frac{\partial \theta}{\partial \vartheta} = -\frac{\partial p}{\partial s}, \tag{2.8.18}$$

and from (2.8.8) and (2.8.9) that,

$$\frac{\partial s}{\partial \vartheta} = \frac{\partial p}{\partial \theta}. \tag{2.8.19}$$

Now using (2.8.16) and (2.8.17), (2.8.13) may be rewritten as,

$$\frac{De}{Dt} = \theta\frac{Ds}{Dt} - p\frac{D\vartheta}{Dt}. \tag{2.8.20}$$

Combining this relationship with the thermal energy equation (2.5.15) yields an expression for the rate of change of entropy:

$$\rho \frac{\mathrm{D}s}{\mathrm{D}t} = \frac{\rho q}{\theta} - \left(\frac{h_i}{\theta}\right)_{,i} + \frac{1}{\theta}\left(\frac{-h_i \theta_{,i}}{\theta} + \varphi\right). \tag{2.8.21}$$

When we compare (2.8.21) with the second law of thermodynamics (2.6.5) and remember the positive nature of the last term, we see that:

$$\frac{1}{\theta}\left(\frac{-h_i \theta_{,i}}{\theta} + \varphi\right),$$

is the irreversible production of entropy within the fluid.

Second, by introducing the Navier–Stokes relationships for h_i and $\bar{\tau}_{ij}$ from (2.7.5) and (2.7.7) we can show,

$$\varphi = \frac{\mu}{2}(v_{i,j} + v_{j,i})^2 + \lambda (v_{i,i})^2, \tag{2.8.22}$$

and

$$\frac{h_i \theta_{,i}}{\theta} = -\frac{\kappa (\theta_{,i})^2}{\theta}, \tag{2.8.23}$$

indicating that the dissipation due to the deformation and dilation is always positive; viscous action always leads to a heating of the fluid. However, this does not mean that entropy always increases since:

$$\frac{h_i \theta_i}{\theta^2} + \left(\frac{h_i}{\theta}\right)_{,i} = -\frac{\kappa \theta_{,ii}}{\theta^2}, \tag{2.8.24}$$

so that from (2.8.21):

$$\rho \frac{\mathrm{D}s}{\mathrm{D}t} = \frac{\rho q}{\theta} + \frac{\kappa \theta_{,ii}}{\theta} + \frac{\phi}{\theta}, \tag{2.8.25}$$

that clearly shows that if heat is gained by diffusion $\theta_{ii} > 0$ then the entropy increases, whereas if heat is lost then the entropy decreases.

In the next section we use some of the above relations to derive an equation for the temperature field.

2.9. THE TEMPERATURE EQUATION

All natural aquatic systems are thermally stratified to some degree and this stratification leads to a non-zero buoyancy term in the momentum equation

(2.3.12); the density of the water is determined mainly by temperature in most environmental flows except in estuaries, where salinity has a major influence. It is therefore important to derive an equation that describes the temperature field. This may be done by introducing, into the thermal energy equation (2.5.15), some of the thermodynamics relationships from the last section. To see this consider (2.6.1) from which it follows:

$$\frac{\mathrm{D}s}{\mathrm{D}t} = \frac{\partial s}{\partial \theta}\frac{\mathrm{D}\theta}{\mathrm{D}t} + \frac{\partial s}{\partial \vartheta}\frac{\mathrm{D}\vartheta}{\mathrm{D}t} \tag{2.9.1}$$

We define the specific heat, at constant volume by:

$$C_\vartheta = \theta \left.\frac{\partial s}{\partial \theta}\right|_\vartheta, \tag{2.9.2}$$

and the latent heat of expansion by:

$$l_\vartheta = \theta \left.\frac{\partial s}{\partial \vartheta}\right|_\theta, \tag{2.9.3}$$

so that, from (2.9.1) it follows that:

$$\frac{\mathrm{D}e}{\mathrm{D}t} = C_\vartheta \frac{\mathrm{D}\theta}{\mathrm{D}t} + l_\vartheta \frac{\mathrm{D}\vartheta}{\mathrm{D}t} - p\frac{\mathrm{D}\vartheta}{\mathrm{D}t}. \tag{2.9.4}$$

Substituting from (2.9.1) and (2.5.15) yields an equation for the variation of the temperature:

$$\frac{\mathrm{D}\theta}{\mathrm{D}t} = \frac{q}{C_\vartheta} - \left(\kappa_\theta \theta_{,i}\right)_{,i} + \frac{\overline{\tau}_{ij} v_{i,j}}{\rho C_\vartheta} - \frac{l_\vartheta v_{k,k}}{\rho C_\vartheta}, \tag{2.9.5}$$

where we have also used (2.8.5) to eliminate $\frac{\mathrm{D}\vartheta}{\mathrm{D}t}$ and substituted for h_i from (2.7.5) and where:

$$\kappa_\theta = \frac{\kappa}{\rho C_\vartheta}, \tag{2.9.6}$$

is called the thermal diffusion coefficient of the fluid.

In §2.8 we saw that the term $\overline{\tau}_{ij} v_{i,j}$ is the internal dissipation, or heat generation, due to the viscosity of the fluid. Equation (2.9.5) confirms the role of this term and shows that the mechanical work of the motion, which is represented by this term, raises the temperature of the fluid.

The last term $l_\vartheta v_{k,k}$ represents the change of temperature due to expansion or contraction of the fluid in response to a pressure change, conforming to our common experience that adiabatic expansion or compression leads to cooling and heating of the fluid. In general, as will be shown in §2.11, the last two terms of (2.9.5) are considerably smaller than the term $h_{i,i}$ and, for $q = 0$ and constant κ, (2.9.5) reduces to the simple, well known, heat diffusion equation:

$$\frac{D\theta}{Dt} = \kappa_\theta \theta_{,ii}. \tag{2.9.7}$$

2.10. THE SOLUTE EQUATION

The water in most natural systems contains a certain amount of dissolved salts. This may contribute to the buoyancy forces in the momentum equation (2.3.12) and so we now must derive an equation for the behavior of the solute within the water column. A mixture of salt in water behaves in reality as a two-component fluid and in the present rational mechanics approach, should be treated as two intermingling fluids. However, we choose to take, once again, a simple axiomatic approach and assume that if $c(x_1,x_2,x_3,t)$ is the concentration of solute in the water, then by analogy with (2.7.5), the flux $\{F_i\}$ of solute is given by:

$$F_i = -\kappa_c(\theta, \vartheta, c)c_{,i}, \tag{2.10.1}$$

where κ_c is called the diffusivity of salt. Given this transport across a material surface, we can write the conservation of solute equation for a Lagrangian control volume $V(t)$:

$$\frac{d}{dt}\left\{\int_{V(t)} c\, dV\right\} = \int_{S(t)} F_i n_i\, dS. \tag{2.10.2}$$

By analogy with (1.8.5) and letting:

$$A = c \text{ and } C_i = F_i, \tag{2.10.3}$$

we obtain the required equation for the solute concentration c:

$$\frac{Dc}{Dt} + cv_{i,i} = \left(\kappa_c c_{,i}\right)_{,i} \tag{2.10.4}$$

As we shall see later in §2.11, for environmental flows, the divergence term in (2.10.4) is very small and the diffusion coefficient κ_c is only a very weak function of the state variables (θ,v) so that (2.10.4) may be approximated by the simple diffusion equation:

$$\frac{\mathrm{D}c}{\mathrm{D}t} = \kappa_c c_{,ii}. \tag{2.10.5}$$

At first glance (2.10.5) seems to contradict conservation of mass equation (2.1.3); this is discussed in the next section.

2.11. SHALLOW LAYER APPROXIMATIONS

The flow of water and air in most environmental contexts is very slow, compared to the speed of sound and so it would seem reasonable to expect that water may be treated as an incompressible fluid, so that:

$$v_{i,i} = 0. \tag{2.11.1}$$

This introduces considerably simplification to the equations of motion, as we shall show below. It is thus important to verify that such an approximation may be made when dealing with both the airflow in the meteorological boundary layer and the motion of water in natural environmental systems.

To be specific, consider the motion of water in a lake. From the discussion in §1.3, the equation of state for water may be written as:

$$\rho = \rho(\theta, S, p), \tag{2.11.2}$$

where ρ is the density (kg m^{-3}), S is the salinity (psu) and p is the pressure (bar). Equation (2.11.2) states that the density changes (decreases) with increasing temperature, decreasing salinity and decreasing pressure. In order to assess the applicability of the incompressibility assumption we use the conservation of mass equation (2.1.3):

$$v_{i,i} = -\frac{1}{\rho}\frac{\mathrm{D}\rho}{\mathrm{D}t}, \tag{2.11.3}$$

and now estimate the magnitude of $\frac{\mathrm{D}\rho}{\mathrm{D}t}$ from the equation of state (2.11.2). This may be done, by differentiating (2.11.2) such that:

$$\frac{1}{\rho}\frac{\mathrm{D}\rho}{\mathrm{D}t} = \frac{1}{\rho}\frac{\partial\rho}{\partial\theta}\frac{\mathrm{D}\theta}{\mathrm{D}t} + \frac{1}{\rho}\frac{\partial\rho}{\partial S}\frac{\mathrm{D}S}{\mathrm{D}t} + \frac{1}{\rho}\frac{\partial\rho}{\partial p}\frac{\mathrm{D}p}{\mathrm{D}t}. \tag{2.11.4}$$

The coefficients in front of the derivative terms in (2.11.4) represent standard properties of the water:

$$\frac{1}{\rho}\frac{\partial\rho}{\partial\theta} = \alpha, \text{ where } \alpha = \text{coefficient of thermal expansion,} \quad (2.11.5)$$

$$\frac{1}{\rho}\frac{\partial\rho}{\partial S} = \beta, \text{ where } \beta = \text{coefficient of salinity contraction,} \quad (2.11.6)$$

$$\frac{1}{\rho}\frac{\partial\rho}{\partial p} = \frac{1}{E_\vartheta}, \text{ where } E_\vartheta \text{ is the bulk modules of elasticity.} \quad (2.11.7)$$

On the other hand the time derivatives in (2.11.4) are a reflection of the particular motion the fluid (water) is undergoing. In order to estimate the magnitude of the *RHS* of (2.11.4) we shall take a situation that represents some extreme behavior in a lake. Suppose the lake is strongly stratified with a bottom temperature of 15 °C and a surface temperature of 30 °C; the transition occurring over say 10 m. It has been observed that a strong wind can excite large (50 m) internal waves with a period as small as 30 min; these are very extreme estimates. The various terms in (2.11.4) may thus be estimated as follows:

$\alpha \mathrm{D}\theta/\mathrm{D}t$ The coefficient of thermal expansion is 2.5×10^{-4} °C. The value of $\mathrm{D}\theta/\mathrm{D}t$ may be estimated from (2.9.7) as $\kappa_o\theta_{,ii}$, where κ_o is the thermal diffusion coefficient which has a value of $10^{-7}\,\mathrm{m}^2\,\mathrm{s}^{-1}$ at the conditions of our example. The second derivative may be estimated as $15/10^2$ equal to $0.15\,°\mathrm{C\,m}^{-2}$, so that the magnitude of the whole term is approximately equal to 3.7×10^{-12}, very small indeed.

$\beta \mathrm{D}S/\mathrm{D}t$ This term may be estimated in the same way as the term above, but given that the diffusion coefficient κS is about a factor of 100 smaller than κ_o this term is even smaller than that due to $\mathrm{D}\theta/\mathrm{D}t$.

$(1/E\vartheta)\mathrm{D}p/\mathrm{D}t$ The value of the bulk modulus for water at the given temperature and a mean depth of 50 m is 2.2×10^9 Pa. Now if we assume a maximum excursion of the water column of 50 m in 30 min then $\mathrm{D}p/\mathrm{D}t$ is approximately $2.7 \times 10^{-2}\,\mathrm{Nm}^{-2}\,\mathrm{s}^{-1}$ so that the magnitude of the whole term becomes 1.2×10^{-7}, again extremely small.

It is thus clear that for water flowing in a lake, typical of an environmental flow, where the velocities and temperature gradients are relatively small, (2.11.1) is a good approximation to the conservation of mass equation (2.1.3). As we have already seen, (2.11.1) implies that the volume as well as the mass of the water is, to first order, conserved.

The temperature (2.9.8) and salinity (2.10.5) equations are thus important only in their own right and not relevant to the first order mass balance of the fluid; the same approximation may be made when describing the flow of air in the meteorological boundary layer.

The above does, however, not mean that the compressibility of water can be totally neglected. In deep lakes, especially when studying convection processes, the variation of the density with depth must be included, but this enters the equations separately to (2.11.1) as we shall explain later.

We can now summarize the implication of the above approximation (2.11.1). The Navier–Stokes constitutive equations (2.7.7) become:

$$\tau_{ij} = -p\delta_{ij} + \mu(\theta, \vartheta)(v_{i,j} + v_{j,i}), \tag{2.11.8}$$

so that the difference between the mean of the direct stresses τ_{ii} and the pressure p becomes

$$\left(\frac{1}{3}\tau_{ii} + p\right) = 2\mu(\theta, \vartheta)v_{i,i} = 0, \tag{2.11.9}$$

indicating that the average of the direct stresses, called the mechanical pressure, is equal to the thermodynamic pressure at all times whether there is motion or not. Substituting (2.11.8) into the momentum equation (2.3.12) and assuming that the variations in the inertial forces due to changes in the fluid density are small, called the Boussinesq approximation yields:

$$\rho_0\frac{\mathrm{D}v_i}{\mathrm{D}t} + 2\rho e_{ijk}\Omega_j v_k = -\rho g\delta_{3i} - p_i + \mu(\theta, \vartheta)v_{i,jj}. \tag{2.11.10}$$

The thermal energy equation, given that internal dissipation is very small, reduces to (2.9.8), reproduced here for convenience

$$\frac{\mathrm{D}\theta}{\mathrm{D}t} = \kappa_0\theta_{,ii}, \tag{2.11.11}$$

and the salinity equation becomes:

$$\frac{\mathrm{D}S}{\mathrm{D}t} = \kappa_S S_{,ii}. \tag{2.11.12}$$

The mechanical energy equation (2.5.14) may also be simplified and becomes:

$$\rho_0\frac{\mathrm{D}}{\mathrm{D}t}\left(\frac{v_iv_i}{2}\right) = -\rho_0 g v_3 - (pv_i)_{,i} + \mu\left(\frac{v_iv_i}{2}\right)_{,jj}, \tag{2.11.13}$$

$$\rho \frac{\mathrm{D}}{\mathrm{D}t}\left(\frac{v_i v_i}{2}\right) = -\rho g v_3 - (p v_i)_{,i} + (2\mu v_i d_{ij})_{,j} - 2\mu d_{ij} d_{ij}, \tag{2.11.14}$$

where each term has a simple physical interpretation :

$$\rho_0 \frac{\mathrm{D}}{\mathrm{D}t}\left(\frac{v_i v_i}{2}\right) = \text{Rate of change of kinetic energy of a fluid particle} \tag{2.11.15}$$

$$-\rho_0 g v_3 = \text{Rate of change of potential energy of a fluid particle} \tag{2.11.16}$$

$$-(p v_i)_{,i} = \text{Rate of working on the fluid particle by the pressure} \tag{2.11.17}$$

$$(2\mu v_i d_{ij})_{,j} = \text{Rate of working of the viscous stresses on the fluid particle by the neighbouring particles} \tag{2.11.18}$$

$$2\mu d_{ij} d_{ij} = \phi$$
$$= \text{Dissipation of mechanical energy to heat by internal friction.} \tag{2.11.19}$$

It should be noted that the pressure and viscous stresses rates of working only redistribute the kinetic energy (divergence terms), whereas the potential energy and dissipation terms form sources and sinks of energy. Together with the equation of state (2.11.2), these equations form a complete set of the variables $\{v_i\}$, ρ, p, θ and S.

At a water solid boundary, the fluid adheres to the boundary so that at the bottom at any stationary solid boundary the flow velocity must be zero:

$$v_i = 0. \tag{2.11.20}$$

The thermal and salinity conditions may fall into two categories:

$$\frac{\partial \theta}{\partial n} = q_\theta(t), \tag{2.11.21}$$

$$\frac{\partial S}{\partial n} = q_S(t), \tag{2.11.22}$$

or

$$\theta = \theta_b, \tag{2.11.23}$$

$$S = S_b, \tag{2.11.24}$$

or may involve prescribing a combination of the above conditions.

2.12. THE VORTICITY EQUATION

In §1.9 we defined the vorticity of the motion as the curl of the velocity vector and showed that it is proportional to the angular velocity of the principal axes of the rate of deformation tensor. In simple terms the vorticity is a property of the motion that measures the spin or angular velocity of the fluid at any point. Its importance comes from the fact that, in a fluid undergoing incompressible motion, pressure perturbations are transmitted instantly throughout the whole fluid domain whereas vorticity spreads slowly into the fluid domain by diffusion from boundary regions where it is generated. In a stratified fluid such as found in a thermally stratified lake, salt stratified estuaries or thermally stratified meteorological boundary layers, the vorticity, as will be seen below, may also be generated internally, but even for such a case vorticity offers powerful insight into the motion.

An equation for the vorticity may be derived by taking the curl of (2.3.12). In order to facilitate this calculation it is advantageous to note the identity:

$$v_\alpha v_{k,\alpha} = e_{k\beta\delta}\zeta_\beta v_\delta + \left(\frac{v_\beta v_\beta}{2}\right)_{,k}, \tag{2.12.1}$$

where

$$\zeta_\beta = e_{\beta ij} v_{j,i}, \tag{2.12.2}$$

is the β component of the vorticity vector, $\boldsymbol{\zeta}$.

Dividing the momentum equation (2.3.12) by ρ, rearranging the indices and substituting (2.12.1) yields

$$\frac{\partial v_k}{\partial t} + e_{k\beta\delta}\zeta_\beta v_\delta + \left(\frac{v_\beta v_\beta}{2}\right)_{,k} + 2e_{k\alpha\beta}\Omega_\alpha v_\beta = -g x_{3,k} - \frac{p_{,k}}{\rho} + \nu(\theta, \vartheta) v_{k,jj} \tag{2.12.3}$$

where $\nu = \mu/\rho$ is the kinematic viscosity. The curl of (2.12.3) becomes:

$$\frac{\partial}{\partial t}(\zeta_i) + e_{ijk}e_{k\beta\delta}(\zeta_\beta v_\delta) + e_{ijk}\left(\frac{v_\beta v_\beta}{2}\right)_{,kj} + e_{ijk}\Omega_\alpha v_{\beta,j} = -e_{ijk}\left(\frac{p_{,k}}{\rho}\right)_{,j} + \nu\zeta_{i,jj}, \tag{2.12.4}$$

where we have ignored the variations of the kinematic viscosity, ν, when differentiating with respect of x_k.

The terms in (2.12.4) may be rearranged and simplified by noting the identity:

$$\begin{aligned}(\zeta_j + 2\Omega_j)v_{i,j} &= (\zeta_j + 2\Omega_j)\ d_{ij} - \frac{1}{2}e_{ij\alpha}\zeta_\alpha(\zeta_j + 2\Omega_j) \\ &= (\zeta_j + 2\Omega_j)d_{ij} - e_{ij\alpha}\zeta_\alpha\Omega_j\end{aligned} \tag{2.12.5}$$

so that (2.12.4) becomes:

$$\frac{D\zeta_i}{Dt} = (\zeta_j + 2\Omega_j)\ d_{ij} - e_{ijk}\Omega_j\zeta_k + e_{ijk}\frac{1}{\rho^2}\rho_{,j}p_{,k} + \nu\zeta_{i,kk}. \tag{2.12.6}$$

The physical interpretation of each term in (2.12.6) is now given:

$\frac{D\zeta_i}{Dt}$: This is the rate of change of the *i*th component of the vorticity of a fluid particle as it moves along its flow path.

$\zeta_j d_{ij}$: This is the change of the *i*th component of vorticity due to the deformation of the fluid. In §1 we saw that d_{ij} describes the extension and compression along the principal axis of the motion. Thus this term represents the change in the *i*th component of vorticity due to fluid particle changing shape. First, if the fluid particle becomes longer and thinner and conserve angular momentum requires the particle to spin faster. Second, angular momentum changes when vortex lines are tipped.

These two separate actions may conveniently be illustrated by example. Consider first the stretching influence. Let us assume a simple flow field composed of a stagnation flow from the x_3 axis toward the origin and out along the x_1 axis and add to this a simple shear flow with vorticity in the x_1 direction. This flow does not satisfy the flow equations, but may be viewed as a linear approximation to the actual flow as shown in Fig. 2.12.1. This flow may be described by the equations:

$$v_1 = \alpha x_1, \tag{2.12.7}$$

$$v_2 = 0, \tag{2.12.8}$$

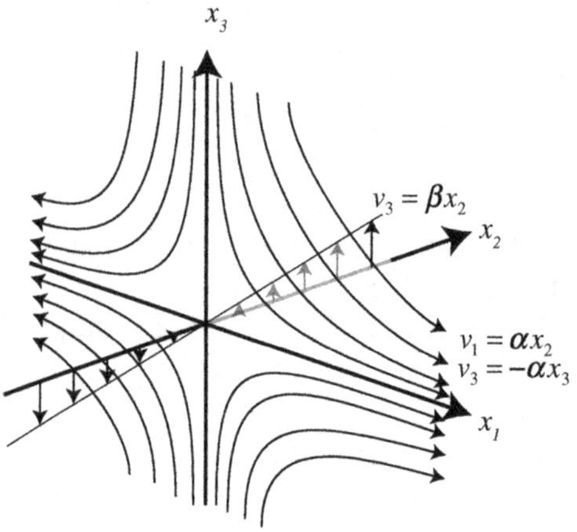

Figure 2.12.1 Schematic of a stagnation flow in the x_1x_3 plane combined with a simple shear in the x_2x_3 plane.

$$v_3 = -\alpha x_3 + \beta x_2. \tag{2.12.9}$$

The stagnation flow has zero vorticity thus,

$$\{\zeta_i\} = \{\beta, 0, 0\}, \tag{2.12.10}$$

and

$$d_{ij} = \begin{bmatrix} \alpha & 0 & 0 \\ 0 & 0 & \frac{\beta}{2} \\ 0 & \frac{\beta}{2} & -\alpha \end{bmatrix}, \tag{2.12.11}$$

so that

$$\{\zeta_i d_{ij}\} = \{\alpha\beta, 0, 0\}, \tag{2.12.12}$$

which clearly shows that the vorticity component ζ_i, due to the simple shear, is intensified by the stretching action, in the x_1 direction, of the stagnation flow (2.12.9).

The tipping effect may be illustrated similarly, with a combination of two simple shear flows, as illustrated in Fig. 2.12.2. This combination is chosen as both components have vorticity and both tip the other's vorticity vector. Such a flow is described by:

$$v_1 = \alpha x_3, \tag{2.12.13}$$

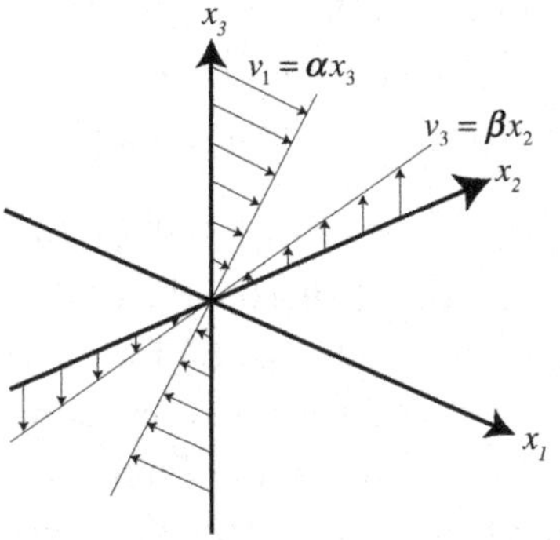

Figure 2.12.2 Schematic of two shear flows, one in the x_2x_3 plane and one in the x_1x_3 plane.

$$v_2 = 0, \tag{2.12.14}$$

$$v_3 = \beta x_2. \tag{2.12.15}$$

This flow satisfies conservation of volume (2.1.1) and has a vorticity vector:

$$\{\zeta_i\} = \{\beta, \alpha, 0\}, \tag{2.12.16}$$

and a deformation tensor

$$d_{ij} = \begin{bmatrix} 0 & 0 & \frac{\alpha}{2} \\ 0 & 0 & \frac{\beta}{2} \\ \frac{\alpha}{2} & \frac{\beta}{2} & 0 \end{bmatrix}, \tag{2.12.17}$$

so that the vorticity source term becomes

$$\{\zeta_i d_{ij}\} = \{0, 0, \alpha\beta\}. \tag{2.12.18}$$

This shows that a flow that has vorticity in the x_1 and x_2 direction will generate vorticity in the x_3 direction due to conservation of angular momentum. This means that the flow described by (2.12.13)–(2.12.15) will quickly develop vorticity in the third direction.

$2\Omega_j d_{ij} - e_{ijk}\Omega_j\zeta_k$: This term is a special case of the tipping effect discussed above; the earth's rotation turns the horizontal vorticity vector generating vorticity in the other horizontal direction. To see this consider a lake situated on the surface of the earth and a local coordinate system with the vertical coordinate extending upwards. Then,

$$\{\Omega_j\} = \{0, 0, \Omega \sin \varphi\}, \tag{2.12.19}$$

where Ω is the angular velocity of the earth and φ is the latitude of the locality of the lake. Given this, the effect of the earth's rotation on the rate of change of vorticity is given by $\{2\Omega d_{i3} \sin \varphi\}$ for $i = 1,2,3$. Now lakes are, in general, much wider and longer than they are deep and further the great majority of lakes are stratified to some degree. Both offer constraints to vertical motions and, in order to understand the effect of the earth's rotation on the motion in a lake, it is sufficient to consider a simple shear flow example:

$$v_1 = \alpha x_2 + \beta x_3, \tag{2.12.20}$$

$$v_2 = 0, \tag{2.12.21}$$

$$v_3 = 0, \tag{2.12.22}$$

where α and β are the shear rates in the x_2 and x_3 direction, respectively.

The deformation tensor corresponding to (2.12.20)–(2.12.22) is given by:

$$d_{ij} = \begin{bmatrix} 0 & \frac{\alpha}{2} & \frac{\beta}{2} \\ \frac{\alpha}{2} & 0 & 0 \\ \frac{\beta}{2} & 0 & 0 \end{bmatrix}, \tag{2.12.23}$$

and thus the contribution to the rate of change of local vorticity is:

$$2\Omega_j d_{ij} - e_{ijk}\Omega_j \zeta_k = \{\Omega\beta \sin \phi, 0, 0\}, \tag{2.12.24}$$

indicating that the earth's rotation combines with the ζ_2 component of vorticity to generate ζ_1 vorticity. This is a most important result and shows that a surface current, with a vertical shear, flowing in the x_1 direction in the southern hemisphere will veer to the left at the surface, setting up a shear flow with a non-zero ζ_1 component of vorticity. It is important to stress that as soon as ζ_1 vorticity is generated it will interact with the earth's rotation to modify the original ζ_2 vorticity.

If there is a shoreline to the left of the current, then the flow will form a boundary current along the shoreline. As we shall see later, this is a common phenomenon in large lakes. On the other hand, if the flow is induced by a wind stress acting in the x_1 direction in the middle of the lake, then the flow will start with only a ζ_2 vorticity component, but will quickly develop a ζ_1 component leading to a spiral like flow. Such a spiral flow was first discovered by Ekman (1905) and the veering in generally is called the Ekman shear.

$e_{ijk}\frac{1}{\rho^2}\rho_{,j}p_{,k}$: This term is called the *baroclinic generation of vorticity* and captures two distinct effects. These are best illustrated by dividing the density and the pressure into background and fluctuating parts:

$$\rho = \rho_0(1 - (\varepsilon h)x_3^* + (\varepsilon\delta)\rho'), \qquad (2.12.25)$$

and

$$p = \rho_0 gh(-x_3^* + \frac{1}{2}(\varepsilon h)x_3^{*2} - (\varepsilon\delta)p'). \qquad (2.12.26)$$

where ρ_0 is the background density, ε is the background density gradient $\frac{1}{\rho_0}\frac{d\rho_e}{dx_3}$, ρ_e is the background density field $\rho_0(1 - \varepsilon x_3)$, ρ' and p' are the dimensionless density and pressure fluctuations, h is the vertical scale of the fluid domain, δ is the vertical scale of the motion ($\delta \ll h$) and x_3^* is the non-dimensional height x_3/h.

These expressions may now be substituted into the baroclinic gravitation term to yield:

$$\frac{1}{\rho^2}e_{ijk}\rho_{,j}p_{,k} = g\varepsilon\left(\frac{\delta}{h}\right)\left(e_{ij3}\rho'_{,j} - (\varepsilon h)e_{i3k}p'_{,k} + O\left((\varepsilon h)\left(\frac{\delta}{h}\right)\right)\right). \qquad (2.12.27)$$

Consider the first term on the *RHS* of (2.12.27). This term is a production term as the vorticity itself does not appear, but only the horizontal density gradient; the vertical density gradient is not active because, when $j = 3$, the symbol e_{ij3} is zero. When $j = 1$, the indices i must equal 2 for the term to be non-zero. Similarly, when $j = 2$ then the indices i must be *1*. This indicates that ζ_1 vorticity is generated by a density gradient in the x_2 direction and analogously ζ_2 vorticity is generated by a density gradient in the x_1 direction. Thus vorticity is generated whenever the isopycnals are tilted by the motion away from their quiescent horizontal state; it is intuitively obvious that whenever isopycnals are tilted, gravity acts to restore them to a horizontal position by generating a moment to counter the tilt. This moment in turn generates vorticity, called the *baroclinic generation*, of a sign so that a velocity field is established which counters the initial motion. Stratified fluids, thus by their nature, behave in a fundamentally different way to a homogeneous fluid; the motion always possesses vorticity, there is a tendency for vertical motions to be suppressed and there is the possibility of internal wave motions.

The second term in (2.12.27) is smaller in magnitude $O(\varepsilon h)$ relative to the first. Once again it is only non-zero when the indices k is either 1 or 2, but now it is the pressure gradient of the motion that is the driving force for the generation mechanisms. This term once again has a simple physical interpretation as the momentum experienced by the fluid whenever a horizontal pressure gradient accelerates the fluid. Since the pressure force acts through the center of volume of a fluid particle, but the inertial resistance acts through the center of mass that, in a stratified fluid, is below the center of volume, the combination of these forces induces a rotation.

If we consider a typical, severe stratification in a lake where say the temperature changes from 15 °C to 30° in 10 m across the thermocline, then ε has a magnitude of about $3 \times 10^{-4}\ \mathrm{m}^{-1}$ and thus (εh) has a magnitude of about 3×10^{-3} which is extremely small and the term may thus be neglected. This is called the Boussinesq approximation (see also §2.11) and, as the reader may show, allows the density ρ in front of the acceleration term $\dfrac{Dv_i}{Dt}$ in (2.3.12) to be replaced by ρ_0.

$\nu\zeta_{i,kk}$: This term is identical to the diffusion term in the temperature equation (2.9.7) and is responsible for redistributing the vorticity; particles that have a large vorticity (large angular momentum) transfer some of this to neighboring particles by viscous friction (exchange of molecules that act to produce viscosity). If we take a Lagrangian control volume and integrate this term then, by Gauss' theorem:

$$\int\limits_{V(t)} \nu\zeta_{i,kk}\,\mathrm{d}V = \nu \int\limits_{S(t)} \zeta_{i,k}\hat{n}_k\,\mathrm{d}S, \tag{2.12.28}$$

so that the net contribution is equal to the flux of vorticity across the surface S; no vorticity is generated or annihilated by this mechanism within the control volume V. This result also provides great insight about the role of the boundary as a source of vorticity. To see this consider a fluid adjacent to a flat horizontal plate that is impulsively started from rest. Further, in order to isolate the effect of the boundary from the internal source of vorticity due to the effects of the stratification or rotation, we shall assume the fluid is homogeneous. If the fluid (water) is initially at rest and the plate is located at $x_3 = 0$, then the boundary condition at $x_3 = 0$ is given by:

$$\begin{aligned} v_1 &= U \quad t \geq 0 \\ &= 0 \quad t < 0, \end{aligned} \tag{2.12.29}$$

$$v_2 = 0, \tag{2.12.30}$$

$$v_3 = 0. \tag{2.12.31}$$

If we evaluate the *RHS* of (2.12.28) at the plate then:

$$\nu \int_S \zeta_{i,k} \hat{n}_k \mathrm{d}S = \nu \int_S v_{i,33} \mathrm{d}S. \tag{2.12.32}$$

From the momentum equation (2.3.12) it follows immediately at $x_3 = 0$:

$$\nu v_{i,33} = U\delta(t) + \frac{1}{\rho_0} p_{,1}, \tag{2.12.33}$$

where $\delta(t)$ is the Dirac delta function. Thus the flux of vorticity at the plate is given by:

$$\nu \int_S \zeta_{i,k} \hat{n}_k \, dS = \int_S \left(U\delta(t) + \frac{1}{\rho_0} p_{,1} \right) \mathrm{d}S. \tag{2.12.34}$$

Thus vorticity is introduced at a solid wall whenever the wall (or the fluid) is moved impulsively and whenever there is a pressure gradient along the wall. The sign of the introduced vorticity may be conveniently visualized by imagining a spherical fluid particle rolling "down" the pressure gradient as shown in Fig. 2.12.3.

In summary, the vorticity equation (2.12.6) thus states that vorticity of a fluid particle moving with the fluid is modified to conserve angular momentum (stretching, tipping and exchange with the earth's angular momentum), by production whenever isopycnals are disturbed from a "natural" horizontal position, redistributed throughout the flow domain by molecular diffusion and lastly by vorticity introduced at boundaries where there is a non-zero pressure gradient.

Before we leave the vorticity equation we shall derive one further important property of the vorticity and introduce the concept of potential vorticity.

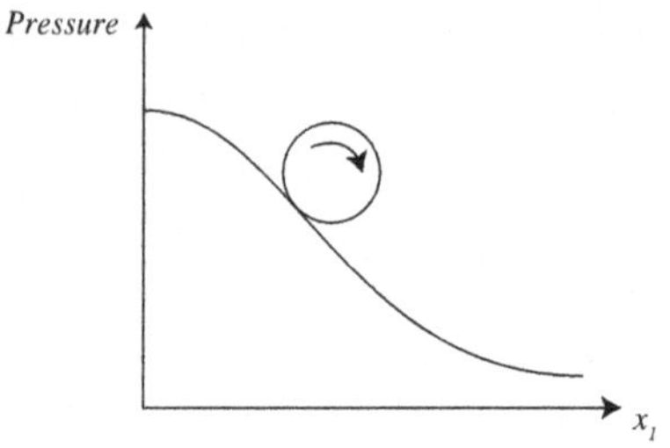

Figure 2.12.3 Vorticity generation at a wall.

Let

$$\omega_i = \zeta_i + 2\Omega_i, \tag{2.12.35}$$

which is called the absolute vorticity. Substituting (2.12.35) into the vorticity equation (2.12.6) yields an equation for the absolute vorticity ω_i:

$$\frac{D\omega_i}{Dt} = \omega_j v_{i,j} - e_{ij3}\frac{g}{\rho_0}\rho_{,j} + \nu\omega_{i,kk}, \tag{2.12.36}$$

where we have also assumed the Boussinesq approximation.

We can now multiply (2.12.36) by $\frac{\rho_{,i}}{\rho}$ and sum over i. After some rearrangement and noting that:

$$\frac{D}{Dt}\left(\frac{\rho_{,i}}{\rho}\right) = -\rho_{,j}v_{j,i}, \tag{2.12.37}$$

we get:

$$\frac{D}{Dt}\left(\frac{\omega_i\rho_{,i}}{\rho}\right) = \frac{\nu\rho_{,i}\omega_{i,kk}}{\rho}. \tag{2.12.38}$$

Now for a motion where the diffusion of vorticity is small we have:

$$\frac{D}{Dt}\left(\frac{\omega_i\rho_{,i}}{\rho}\right) = 0, \tag{2.12.39}$$

or in other words $\frac{\omega_i\rho_{,i}}{\rho}$ is conserved as a fluid particles moves along a particle path. This quantity is called the *potential vorticity*, a quantity that has considerable relevance in coastal ocean flows.

2.13. INTRODUCTION TO SOME STATISTICS CONCEPTS

The motion of water in a lake is a combination of laminar and turbulent flows; the stronger the thermal stratification the more ubiquitous the laminar flow regions become. However, in river flows, in the surface layers of the ocean or lakes, in the benthic and meteorological boundary layer and in isolated interior patches of stratified standing waters, the water often undergoes turbulent motions. In order to discuss such turbulent motions it is necessary to give, at least a rudimentary, intuitive introduction to some statistical concepts needed to describe turbulent motions. No attempt is made at rigor, but hopefully the reader will gain sufficient detail to be able to comfortably manipulate the logic underlying the derivations in §2.14 and §2.15.

The most important concept is that of a random variable. This is a quantity that has a value and a probability that this particular value is realized in a particular experiment. Thus there are four distinct characteristics attached to this concept; a variable value, a unit of the variable, a probability of occurrence and the action to produce the value at a time t. The easiest way to visualize a random variable is to think of a domain that contains all the possible realizations and an action that draws a realization from the domain. The action is not too dissimilar to buying a lottery ticket; all the tickets are printed, the action is the purchase of the ticket and the variable value is the ticket number.

For example suppose we consider the velocity component v_1 at a particular point $\{x_i\}$ and time t. If the flow is laminar, then each time we repeat the identical experiment, the value v_1 will be identical. However, if the flow is turbulent, each new experiment will yield a different value of v_1; the values are random. However, the randomness is determined by a characteristic of the flow variables called the probability function. This arises from carrying out a large number of experiments, each experiment providing a realization of the variable. Suppose we count the number of times a velocity value v_1 falls between $v_1 - \frac{dv_1}{2}$ and $v_1 + \frac{dv_1}{2}$ and plot the frequency of occurrence against v_1. This will form a frequency distribution. If we now divide the frequency of occurrence by N, the total number of experiments, let N approach infinity and progressively reduce the bin size dv_1 as N increases, then our frequency curve approaches a limit that is called the probability density function $P_{x,t}(v_1)$, where $x = \{x_1, x_2, x_3\}$ is the vector position. This function has the property that:

$$\textbf{\textit{The probability that } } \boldsymbol{v_1} \textbf{\textit{ lies between }} \boldsymbol{v_0 - du_1/2} \textbf{\textit{ and }} \boldsymbol{v_0 + du_1/2} \\ \textbf{\textit{at the point }} \boldsymbol{x} \textbf{\textit{ and time }} \boldsymbol{t} \textbf{\textit{ is equal to }} \boldsymbol{P_{x,t}(v_1)dv_1}. \qquad (2.13.1)$$

Together, the variable v_1 and the probability density function $P_{\underset{\sim}{x},t}(v_1)$ form a random variable. The reader will be familiar, from an introductory statistics course, with special examples of probability density functions; the normal, Poisson and the log normal distributions. We shall introduce these distributions as the need arises.

The most important point to remember is that v_1 is the random variable at the time t and position $x = \{x_1, x_2, x_3\}$, the statistical behaviur of which is completely determined by $P_{x,t}(v_1)$. The probability density function is a deterministic function that describes the group or population from which the variable v_1 is drawn and the experiment is the action.

To proceed further we must define such a probability function for each point $\boldsymbol{x} = \{x_1,x_2,x_3\}$ and each time t. It is, however, easy to see that the velocity at one point, say $\boldsymbol{x}^{(1)} = \{x_1^{(1)}, x_2^{(1)}, x_3^{(1)}\}$, is not necessarily independent of the velocity at say, $\boldsymbol{x}^{(2)} = \{x_1^{(2)}, x_2^{(2)}, x_3^{(2)}\}$ if the distance between $\boldsymbol{x}^{(1)}$ and $\boldsymbol{x}^{(2)}$ is small; there is always a certain degree of coherence between neighboring points in the flow.

Thus our probability function must reflect this interdependence so that if $v_1^{(i)}(t_j)$ is the velocity component in the x_1 direction at $\boldsymbol{x}^{(i)}$ and at time t_j we must define a function:

$$P_{(\boldsymbol{x}^{(1)},\, \boldsymbol{x}^{(2)},\, \boldsymbol{x}^{(3)},\ldots,\, \boldsymbol{x}^{(M)}, t_1, t_2, t_3, \ldots, t_N)}\left(v_1^{(i)}(t_j)\right)$$

$$= \textbf{\textit{Probability that the velocity }} v_1^{(i)}(t_j) \textbf{\textit{ at }} x^{(i)} \textbf{\textit{ and }} t_j \textbf{\textit{ is between}}$$

$$v_1^{(i)}(t_j) - \frac{dv_1}{2} \textbf{\textit{ and }} v_1^{(i)}(t_j) + \frac{dv_1}{2} \quad \textbf{\textit{for all } i \textit{ and } j.} \tag{2.13.2}$$

In principle we must define such a function for all possible combinations of position and time. Obviously, in practice, this is a very difficult, if not impossible, task and usually great simplifications are required. The quantity

$$P_{(\boldsymbol{x}^{(1)},\, \boldsymbol{x}^{(2)},\, \boldsymbol{x}^{(3)},\ldots,\boldsymbol{x}^{(M)}, t_1, t_2, t_3, \ldots, t_N)}\left(v^{(i)}(t_j)\right),$$

is called the joint probability density function. In this book we will not make explicit use of the joint probability function, but rather only of some of the resulting properties. When the joint probability density function is independent of the time variable and depends only on the differences in time then, we say the process is *statistically stationary.* Similarly if the probability density function depends only on the difference in the position, then we say the *process is homogeneous.*

Consider now the concept of the mean or average of a random variable u, which could be a component of the velocity, the pressure or any other flow property. Suppose we fix our attention on a single point $\boldsymbol{x}$ and a particular time t, so that the joint probability distribution of (2.13.2) reduces to the simple function of the type (2.13.1). Now we may visualize a repetition of our experiment where we generate many values of the variable u. Let us call these $u^{(i)}$, so that the mean of our variables becomes

$$\overline{u}^{(N)} = \frac{1}{(N-1)} \sum_{i=1}^{N} u^{(i)}. \tag{2.13.3}$$

Suppose we partition the variable $u^{(i)}$ into bins of width Δu. The above sum then becomes

$$\overline{u}^{(N)} = \frac{1}{(n-1)} \sum_{j=1}^{M} \overline{u}^{(j)} n^{(j)}, \tag{2.13.4}$$

where M is the number of bins, $n^{(j)}$ is the number of values of $u^{(i)}$ that fall into bin j and $\overline{\boldsymbol{u}}^{(j)}$ is the central value for that bin. Now the quotient $\dfrac{n^{(i)}}{(n-1)}$ is the likelihood or probability that a value of the variable $u^{(i)}$ falls into our bin j, so that we may rewrite (2.13.4) in the form:

$$\overline{u}^{(N)} = \sum_{j=1}^{M} \overline{u}^{(j)} \, p_{(x,t)} \left(u^{(j)}\right) \Delta u. \tag{2.13.5}$$

In the limit, the bin size Δu shrinks to zero and $N \to \infty$, $\overline{u}^{(N)} \to \overline{u}$, where

$$\overline{u} = \int_{-\infty}^{+\infty} u p_{(x,t)} \left(u\right) du. \tag{2.13.6}$$

In other words $\overline{u}$ is the average of the variable weighted by the probability function. This average is called the *ensemble mean*, in order to emphasize that it is the result of averaging the ensemble of experiments and we shall designate it with an overbar. It should be noted that $\overline{u}$ is a deterministic quantity and is a characteristic of the population or domain from which the random variable u is drawn.

Similarly, we can define the ensemble variance for our variable by:

$$\sigma_u^2 = \int_{-\infty}^{+\infty} (u - \overline{u})^2 \, p_{(x,t)}(u) \, du, \tag{2.13.7}$$

which is just the ensemble mean of the variable $(u - \overline{u})^2$. Since we are finding the mean of a squared quantity, the answer is always positive. The variance is a convenient measure of the range of the variability of the random variable u and it is common to use the square root as a measure of this variability that has the same units as the variable itself; this is called the *standard deviation.*

This concept may be generalized to moments of any order n, such that:

$$\sigma_u^n = \int_{-\infty}^{+\infty} (u - \overline{u})^n p_{(x,t)}(u) du, \tag{2.13.8}$$

Now suppose we free the variable t and consider the velocity at the point $\boldsymbol{x}$ as a function of time. Such random variables are called *time series*. The ensemble mean $\overline{u}(\boldsymbol{x}, t)$ is well-defined and is now a function of time as well as space:

$$\overline{u}(\boldsymbol{x}, t) = \int_{-\infty}^{+\infty} u p_{(\boldsymbol{x},t)}(u) du. \tag{2.13.9}$$

If the quantity $\overline{u}(\boldsymbol{x}, t)$ is constant with time, then we say that the turbulent flow is mean stationary (*the time series is mean stationary*). If both the mean $\overline{\boldsymbol{u}}(\boldsymbol{x})$ and the variance σ_u^2 are constant with time we call the signal a *second order stationary process.*

An alternative method of calculating a mean and variance of the velocity at the point $\boldsymbol{x}$ for a second order stationary time series is through the time integrals:

$$\tilde{u}(\boldsymbol{x}, t) = \frac{1}{T} \int_{t}^{t+T} u(\boldsymbol{x}, t') dt', \tag{2.13.10}$$

$$\tilde{\sigma}_u^2(\boldsymbol{x}, t) = \frac{1}{T} \int_{t}^{t+T} (u(\boldsymbol{x}, t') - \tilde{u}(\boldsymbol{x}, t'))^2 dt', \tag{2.13.11}$$

where $u(\boldsymbol{x},t')$ is a particular realization from one particular experiment. It is important to note that $\tilde{u}(\boldsymbol{x}, t')$ is a statistical or random quantity and has a different value for each experiment and for each value of T and t. However, from a practical point of view (2.13.10) only requires the execution of a single experiment run over a long time T, whereas (2.13.9) requires the impossible task of executing a very large number of identical experiments. The formulations (2.13.10) and (2.13.11) are thus much preferable and it is important to ask when, and in what sense, is (2.13.10) an estimate of (2.13.9) and (2.13.11) of (2.13.7)?

It is intuitively clear that the larger T the more averaging is carried out by the integration in (2.13.10) and so it is reasonable to consider the limit $T \to \infty$; this can, of course, only be done for a stationary process. In order for (2.13.10) and (2.13.11) to be compared in the limit $T \to \infty$ we must introduce the concept of *equality in the square mean*. We say $\overline{u}(\boldsymbol{x}, t)$ and $\tilde{u}(\boldsymbol{x}, t)$ are equal in square mean if,

$$\lim_{T \to \infty} \int_0^T \left(\overline{u}(\boldsymbol{x}, t) - \tilde{u}(\boldsymbol{x}, t; T)\right)^2 p_{(T)}\left(\tilde{u}\right) \mathrm{d}\tilde{u} = 0, \qquad (2.13.12)$$

where $p_{(T)}(\tilde{u})$ is the probability density function of the random variable $\tilde{u}(\boldsymbol{x}, t)$. This says that the time mean $\tilde{u}$ approach $\overline{u}$ such that the mean and variance of $(\tilde{u} - \overline{u})$ approach zero as $T \to \infty$. Such a velocity field is called *ergodic*. The reader is referred to books on statistical time series to see what the requirements are for a process to be ergodic. However, in fluid mechanics it is generally assumed that, if a flow field is stationary then (2.13.10) may effectively be used as an estimate for (2.13.9). This does, however, not mean that every realization of $\tilde{u}$ is equal to $\overline{u}$, quite the contrary, $\tilde{u}$ is a random variable with a mean that approaches $\overline{u}$ and a variance that tends to zero as $T \to \infty$. The above discussion viewed the flow field as evolving in time and, for ease of description, $\tilde{u}(\boldsymbol{x}, t)$, $\boldsymbol{x}$ fixed, is called a *time series*. Equally, we may look at the flow field as a function of $\boldsymbol{x}$ at a fixed time. We may form a similar mean as in (2.13.10) except that any of the $\{x_1,x_2,x_3\}$ are interchanged with t. However, for such a definition to be meaningful, the flow must be homogeneous in $\{x_1,x_2,x_3\}$. In the strictest sense this means that the joint probability distribution is independent of position and depends only on the distance between two points. As with time, it normally suffices to ensure that the mean and variance of a particular variable are independent of $\{x_1,x_2,x_3\}$.

Thus for example, for a flow that is homogeneous in the x_1 direction, it is possible to define the mean:

$$\widehat{u}(x_1, x_2, x_3, t) = \frac{1}{X} \int_{x_1}^{x_1+X} u(x_1', x_2, x_3, t) \mathrm{d}x_1', \tag{2.13.13}$$

and the expression for the variance follows in a similar way. Clearly, the above definition is not restricted to the x_1 coordinate axes and may be generalized to two and three axes. For the case where the flow is homogeneous in both the x_1 and x_2 directions we can form an area average:

$$\left[u(x_1, x_2, x_3, t)\right] = \frac{1}{XY} \int_{x_1}^{x_1+X} \int_{x_2}^{x_2+Y} u(x_1', x_2', x_3, t) \mathrm{d}x_1' \mathrm{d}x_2'. \tag{2.13.14}$$

A very common example of this average is the discharge velocity in a pipe flow defined as:

$$\left[u(\xi, t)\right] = \frac{1}{A} \int_0^{2\pi} \int_0^R u(r, \theta, t)\, r \,\mathrm{d}r \,\mathrm{d}\theta = \frac{Q}{A}, \tag{2.13.15}$$

where A is the area of the pipe, R is the radius and Q is the discharge through the pipe and ξ is the longitudinal coordinate. A further estimate of the mean and variance may be obtained if the flow is homogeneous in all three coordinate variables. Then we can define the volume average:

$$\hat{u}(x_1, x_2, x_3, t) = \frac{1}{V} \int_V u(x_1', x_2', x_3', t) \mathrm{d}V', \tag{2.13.16}$$

where V is a volume over which the average is taken. Clearly all these averages may be defined for both Lagrangian or Eulerian lines, areas or volumes. The advantage of time and space averages over ensemble averages is that they are mentally easily visualized as representing averages in the normal arithmetic sense. Many flows may be locally homogeneous or stationary, but are not homogeneous or stationary in a global sense. An important question therefore arises. Is it possible to use space and time averages to evaluate a meaningful estimate of the ensemble means from a restricted volume V or a restricted range T? To answer this question, consider the time average (2.13.10). The question may be restated as follows: at what time t_2 does the velocity $u^{(2)}$ become independent of the velocity $u^{(1)}$ at t_1. It seems reasonable that we should get a good statistical estimate for our average, provided we extend our averaging time sufficiently far so that the velocities become independent. Consider the integral given by (2.13.10), but defined over t_1 to t_2:

$$\tilde{u}(\boldsymbol{x}, (t_2 - t_1)) = \frac{1}{t_2 - t_1} \int_{t_1}^{t_2} u(\boldsymbol{x}, t')\mathrm{d}t', \tag{2.13.17}$$

where we have emphasized the ergodicity of the variable by writing $\tilde{u}$ as a function of $t_2 - t_1$ and not of t_2 and t_1 separately. Now, if $u^{(2)}$ becomes uncorrelated to $u^{(1)}$, when $t_2 - t_1$ exceeds a certain value, then extending the integral beyond t_2 improves the estimate only marginally and so there is no advantage in extending the integral beyond the point where the correlation ends.

Mathematically, the correlation between $u^{(1)}$ and $u^{(2)}$ may be expressed by the covariance function:

$$\begin{aligned}\overline{C}_u(\boldsymbol{x}, \tau) = \int_{-\infty}^{+\infty} & (u(\boldsymbol{x}, t_1) - \overline{u}(\boldsymbol{x}, t_1))(u(x, t_2) \\ & - \overline{u}(\boldsymbol{x}, t_2))p_{t_1 t_2}(u_1, u_2)\mathrm{d}u_1\mathrm{d}u_2,\end{aligned} \tag{2.13.18}$$

where $p_{t_1 t_2}(u_1, u_2)$ is the joint probability density function for the velocities at the two times t_1 and t_2. The covariance is thus the ensemble mean of the product of u at t_1 and u at t_2, and is written as $\overline{u_1 u_2}$. Since it is the covariance between the same variable, at different times, it is called the *autocovariance*. Thus, once the variables become uncorrelated, the mean of their product is zero which is the same as saying that, if the value of $u^{(1)}$ at t_1 is positive, then the value at t_2 may be equally likely to be positive or negative and of a magnitude independent of the magnitude of $u^{(1)}$ at t_1. The contribution to the average thus becomes increasingly less significant as the record is extended.

Again we may estimate the covariance with a time integral:

$$\tilde{C}_u(\boldsymbol{x}, \tau; t, T) = \frac{1}{T} \int_0^T u(\boldsymbol{x}, t)u(\boldsymbol{x}, t + \tau)\mathrm{d}t. \tag{2.13.19}$$

From the above discussion it follows that:

$$\tilde{C}(\boldsymbol{x},0) = \tilde{\sigma}_u^2 \tag{2.13.20}$$

and for a time series that becomes uncorrelated at large times,

$$\tilde{C}(\boldsymbol{x},0) \to 0 \text{ as } \tau \to \infty. \tag{2.13.21}$$

With this knowledge we can define a time scale for the particular variable u such that:

$$T_c = \int_0^\infty \frac{\tilde{C}_u(\boldsymbol{x},\tau)}{\tilde{C}_u(\boldsymbol{x},0)} d\tau, \tag{2.13.22}$$

which is a measure of the time interval $(t_2 - t_1)$ required for the variable $u^{(2)}$ to become uncorrelated with $u^{(1)}$.

The non-dimensional function, occurring in the integral (2.13.22), is called the autocorrelation function:

$$\tilde{R}(u) = \frac{\tilde{C}_u(\boldsymbol{x},\tau)}{\tilde{C}_u(\boldsymbol{x},0)}. \tag{2.13.23}$$

Thus as a rule of thumb it is only necessary to extend the integral in (2.13.10) to about T_c; extending it beyond this does not add much except some statistical confidence. If the turbulent field $u(\boldsymbol{x},t)$ is such that it has a time scale, T_c, that is very much shorter than the time scale defining the variation of the mean velocity $\overline{u}$, then we may use (2.13.10) as a local estimate of $\overline{u}$. Similar formulations follow for the spatial correlation functions.

A concept that is used very often in analyzing the energy content of a signal such as velocity, displacement or pressure is the power spectral density. Suppose $u(\boldsymbol{x},t)$ is again such a time series and suppose further that $u(\boldsymbol{x},t)$ is real-valued and second order stationary with a time-independent zero mean.

We define the power spectral density, $p_u(f)$, of the random variable $u(\boldsymbol{x},t)$ as the Fourier transform of covariance function $\overline{C}_u(\boldsymbol{x},\tau)$:

$$p_u(f) = \int_{-\infty}^{+\infty} \overline{C}_u(\boldsymbol{x},\tau) e^{-2\pi i f \tau} d\tau. \tag{2.13.24}$$

From Fourier theory, it follows immediately that:

$$\overline{C}_u(\boldsymbol{x},\tau) = \int_{-\infty}^{+\infty} p_u(f)e^{2\pi i f\tau}\mathrm{d}f, \tag{2.13.25}$$

since $p_u(f)$ and $\overline{C}_u(\boldsymbol{x},\tau)$ form a Fourier pair.

The reason why $p_u(f)$ is called the power spectral density follows conveniently from the following argument. Suppose $u(\boldsymbol{x},t)$ is a velocity component at point $\boldsymbol{x}$ and time t, then from (2.13.19):

$$\frac{1}{2}\rho\overline{C}_u(\boldsymbol{x},0) = \frac{1}{2}\rho \int_{-\infty}^{+\infty} (u-\overline{u})^2 p_{(\boldsymbol{x},t)}(u)\mathrm{d}u. \tag{2.13.26}$$

This is the weighted mean of the kinetic energy contained in the velocity fluctuations $(u-\overline{u})$. This is equivalent to the kinetic energy of the turbulent fluctuations (TKE). If we expand the squared quantity we get:

$$\frac{1}{2}\rho\overline{C}_u(\boldsymbol{x},0) = \frac{1}{2}\rho \int_{-\infty}^{+\infty} u^2 p_{(\boldsymbol{x},t)}(u)\,\mathrm{d}u - \frac{1}{2}\rho\overline{u}^2 \tag{2.13.27}$$

$$= \text{Total kinetic energy} - \text{mean kinetic energy}$$

Thus we may write,

$$\mathrm{KE} = \mathrm{TKE} + \mathrm{MKE} \tag{2.13.28}$$

It is thus clear that $p_u(f)\mathrm{d}f$ is the TKE of that part of the signal that has a frequency content between $f-\frac{\mathrm{d}f}{2}$ and $f+\frac{\mathrm{d}f}{2}$; hence the descriptor power spectral density.

It is customary to compute the power spectral density directly from the time series and avoid computing the covariance function. The reason for this is that there are available fast algorithms for computing the Fourier transform called FFT (Fast Fourier Transform). To complete this section of statistical concepts we shall now give an intuitive argument to show how this may be done.

Suppose $u(\boldsymbol{x},t)$ is ergodic with respect to the covariance function; then we may estimate the covariance from the time average estimator:

$$\tilde{C}_u(\boldsymbol{x},t) = \lim_{x\to\infty}\frac{1}{T}\int_0^T u(\boldsymbol{x},t')u(\boldsymbol{x},t'+\tau)\mathrm{d}t', \tag{2.13.29}$$

where lim is the limit in least square mean (2.13.12).

Now let

$$v_T(\boldsymbol{x}, t) = u(\boldsymbol{x}, t) : \quad 0 \leq t \leq T$$
$$= 0 : \quad \text{otherwise} \tag{2.13.30}$$

and

$$V_T(f) = \int_{-\infty}^{+\infty} v_T(\boldsymbol{x}, \boldsymbol{t}) e^{-2\pi i f t} \mathrm{d}t. \tag{2.13.31}$$

Then we may write:

$$\widetilde{C}_u(\boldsymbol{x}, t) = \lim_{x \to \infty} \frac{1}{T} \int_{-\infty}^{+\infty} v_T(\boldsymbol{x}, t') v_T(\boldsymbol{x}, t' + \tau) \mathrm{d}t' \tag{2.13.32}$$

But by (2.13.31) we may write:

$$v_T(\boldsymbol{x}, \boldsymbol{t}) = \int_{-\infty}^{+\infty} V_T(f) e^{2\pi i f t} \mathrm{d}f, \tag{2.13.33}$$

and substituting this into (2.13.32) yields:

$$\widetilde{C}_u(\boldsymbol{x}, \tau) = \lim_{T \to \infty} \frac{1}{T} \int_{-\infty}^{+\infty} v_T(\boldsymbol{x}, t') \int_{-\infty}^{+\infty} V_T(f) e^{2\pi i f (t' + \tau)} \mathrm{d}t'. \tag{2.13.34}$$

Expanding the exponent, leads to:

$$\widetilde{C}_u(\boldsymbol{x}, \tau) = \lim_{T \to \infty} \frac{1}{T} \int_{-\infty}^{+\infty} V_T(\boldsymbol{x}, f) \overline{V}_T(f) e^{2\pi i f \tau} \mathrm{d}f, \tag{2.13.35}$$

where $\overline{\boldsymbol{V}}_T(f)$ is the complex conjugate of $V_T(f)$.

$$\widetilde{C}_u(\boldsymbol{x}, \tau) = \lim_{T \to \infty} \frac{1}{T} \int_{-\infty}^{+\infty} \left| V_T^2(f) \right| e^{2\pi i f \tau} \mathrm{d}f, \tag{2.13.36}$$

that suggests that by (2.13.25):

$$p_u(f) = \frac{1}{T} \left| V_T(\boldsymbol{x}, f)^2 \right|, \tag{2.13.37}$$

and thus

$$p_u(f) = \lim_{T\to\infty} \frac{1}{T}\left| \int_0^T u_T(\boldsymbol{x}, t)e^{-2\pi i\pi ft}\mathrm{d}t \right|^2. \quad (2.13.38)$$

Hence, at least intuitively, the power spectral density may be computed directly from the time series $v_T(\boldsymbol{x},t)$. The computational techniques to carry this out are discussed in books on time series.

2.14. MEAN MOMENTUM AND TRANSPORT EQUATIONS FOR TURBULENT FLOW

The momentum equation, presented in §2.3, is valid for both laminar and turbulent flow but it is convenient, when dealing with turbulent flow to separate the mean flow from the turbulent fluctuations and to separately formulate a momentum equation for the mean and turbulent flow component. A turbulent flow field exhibits random variations in all the flow components; these may be viewed as random variables with well-defined means, as so we may write:

$$v_i = \overline{v}_i + v_i', \quad (2.14.1)$$

$$p_i = \overline{p}_i + p_i', \quad (2.14.2)$$

$$\rho_i = \overline{\rho}_i + \rho_i', \quad (2.14.3)$$

where the overbar designates the ensemble mean and the primes the variations from the mean. Equations (2.14.1)–(2.14.3) may now be substituted into the Boussinesq approximation of the momentum equation (2.11.10) and after taking the ensemble mean of the resulting equation becomes

$$\rho_0 \frac{\overline{\mathrm{D}}\overline{v}_i}{\mathrm{D}t} = 2e_{ijk}\rho_0\Omega_j\overline{v}_k = -g\overline{\rho}\delta_{3i} - \overline{p}_i + \left(\mu\overline{v}_{i,j} - \rho_0\overline{v_i'v_i'}\right)_j, \quad (2.14.4)$$

where the operator:

$$\frac{\overline{\mathrm{D}}}{\mathrm{D}t} = \frac{\partial}{\partial t} + \overline{v}_j(\cdots). \quad (2.14.5)$$

If we compare (2.14.4) with (2.11.10) we see that the two equations are identical except that (2.14.4) has the additional stress term,

$\rho_0\overline{v_i'v_j'}$; this is the covariance of the ith and jth velocity components. The physical interpretation of this term is clear; it is the flux of ith component of momentum moved by the jth velocity component. If the covariance is non-zero then this term will either increase or decrease the ith component of the mean momentum at that point and may thus be interpreted as a stress, additional to the viscous stress $\mu\overline{v}_{i,j}$, changing the momentum of the fluid particle. This term is called the Reynolds stress.

Equation (2.14.4) reveals the action of any turbulent fluctuations, but the Reynolds stresses cannot be evaluated without a knowledge of the joint probability density function of the components of the velocity field. Specification of this probability density function has so far alluded researchers, but many models have been advanced that relate the Reynolds stress to properties of the mean flow and thus provide once again a closed set of equations which may then solved, at least in principle; this is called the turbulence closure problem.

The transport equations (2.11.11) and (2.11.12) may be decomposed in a similar way. To be specific, let's consider the temperature equation (2.11.11) and, in addition to the decompositions (2.14.1)–(2.14.3), we also write:

$$\theta = \overline{\theta} + \theta'. \tag{2.14.6}$$

Substituting these into (2.11.11) and taking the mean leads to the equation:

$$\frac{\overline{\mathrm{D}}\,\overline{\theta}}{\mathrm{D}t} = \left(\kappa\theta_{,i} - (\overline{\theta' v_i'})\right)_{,i}. \tag{2.14.7}$$

Once again this equation is similar to the original equation (2.11.11) except that a correlation term $(\overline{\theta' v_i'})$ now enters the equation, to account for the heat flux resulting from the turbulent fluctuations.

Last, we derive the equivalent mean mechanical energy equations by multiplying the mean momentum equation (2.11.10) by $\overline{v}_i$ and summing over i. This yields an equation for the mean kinetic energy $\overline{v}_i\overline{v}_i/2$:

$$\rho_0\frac{\overline{\mathrm{D}}}{\mathrm{D}t}\left(\frac{\overline{v}_i\overline{v}_i}{2}\right) = -g\overline{\rho}\,\overline{v}_3 - (\overline{v}_i\overline{p})_{,i} + \mu\left(\frac{\overline{v}_i\overline{v}_i}{2}\right)_{,jj} - \mu(\overline{v}_{i,j}\overline{v}_{i,j})$$

$$-(\overline{v}_i\rho_0\overline{v_i'v_j'})_{,j} + \rho_0\overline{v_i'v_j'}\,\overline{v}_{i,j}, \tag{2.14.8}$$

where all but the last two terms have the same form and thus the same physical interpretation as in (2.11.10). The interpretation of the last two terms is as follows:

$(\overline{v}_i \rho_0 \overline{v'_i v'_j})_{,j}$: This is the net rate of working of the Reynolds Stresses on a fluid particle by the mean flow. Since it is a divergence, this term only redistributes kinetic energy and it is not a source or sink of energy.

$\rho_0 \overline{v'_j v'_j} \overline{v}_{i,j}$: This is an exchange term between the mean and turbulent velocity fields (see below §2.15) and most often acts as the main source of turbulent kinetic energy and usually represents a sink in (2.14.8).

2.15. TURBULENT KINETIC ENERGY AND TRANSPORT EQUATIONS

A similar procedure can now be employed to derive an equation for the kinetic energy of the turbulent fluctuations $(\rho_0 \frac{\overline{v'_i v'_i}}{2})$. This is most easily achieved by decomposing (2.11.13) then taking the mean of the resulting equation and finally subtracting (2.14.8) from this result. This yields the equation:

$$\rho_0 \frac{\overline{\mathrm{D}}}{\mathrm{D}t}\left(\frac{\overline{v'_i v'_i}}{2}\right) = -g\overline{\rho' v'_3} - (\overline{v'_i p'})_{,i} + \mu\left(\frac{\overline{v'_i v'_i}}{2}\right)_{,jj} - \mu(\overline{v'_{i,j} v'_{i,j}})$$

$$-\rho_0 \overline{v'_i v'_j}\overline{v}_{i,j} + \rho_0\left(\frac{\overline{v'_j v'_i v'_i}}{2}\right)_{,j}. \quad (2.15.1)$$

The interpretation of each term is as follows:

$\rho_0 \frac{\overline{\mathrm{D}}}{\mathrm{D}t}\left(\frac{\overline{v_i v_i}}{2}\right)$: Rate of change of turbulent kinetic energy of a particle advected by the mean flow.

$-g\overline{\rho' v'_3}$: Rate of change of turbulent potential energy. This term may be a sink when there is active turbulent mixing or a source when the water column re-stratifies after an active event. When divided by the term is called the buoyancy flux.

$-(\overline{v'_i p'})_{,i}$: Rate of work of the turbulent pressure fluctuations.

$\mu\left(\frac{\overline{v'_i v'_i}}{2}\right)_{,jj}$: Rate of diffusion of turbulent kinetic energy.

$-\mu(\overline{v'_{i,j} v'_{i,j}})$: Rate of dissipation of turbulent kinetic energy often written as $-\ \rho 0\varepsilon$, where ε is the dissipation per unit mass.

$-\rho_0\overline{v_i'v_j'}\overline{v}_{i,j}$: Transfer of mean kinetic energy to the turbulent kinetic budget. This is the source of turbulent kinetic energy in a shear flow.

$\rho_0\left(\dfrac{\overline{v_j'v_i'v_i'}}{2}\right)_{,j}$: Transport of turbulent kinetic energy by the turbulent fluctuations themselves.

Once again many of these terms are divergences and merely redistribute the turbulent kinetic energy from one location to another. Only the terms $-g\overline{\rho' v_3'}$ and $-\rho_0\overline{v_i'v_j'}\overline{v}_{i,j}$ are actual sources (or sinks) of turbulent kinetic energy.

The remaining equation of importance is the transport of the variance of the turbulent temperature or salt fields. This can once again be derived by, decomposing the state variable as in (2.14.1)–(2.14.4), substituting this into the temperature equation (as an example) and taking the mean. This yields the mean temperature equation (2.14.7). This mean equation is then subtracted from the original decomposition equation in order to arrive at an equation for the turbulent temperature fluctuations:

$$\frac{\partial\theta'}{\partial t} + \overline{v}_i\theta'_{,i} + v_i'\overline{\theta}_{,i} + v_i'\theta'_{,i} = \kappa\theta'_{,ii}. \tag{2.15.2}$$

Equation (2.15.2) is now multiplied by θ' and the mean is taken of the resulting equation to yield the required results:

$$\frac{\overline{D}\overline{\theta'^2}}{Dt} = -\overline{v_i'\theta'}\overline{\theta}_{,i} - \overline{v_i'\theta'^2_{,i}} - \kappa\left(\overline{\theta'^2}\right)_{,ii} - \kappa\left(\overline{\theta'^2_{,i}}\right). \tag{2.15.3}$$

The terms in this equation have the following physical meaning:

$\left(\dfrac{\overline{D}\overline{\theta'^2}}{Dt}\right)$: Transport of the temperature variance by the mean flow.

$-\overline{v_i'\theta'}\overline{\theta}_{,i}$: Production of temperature variance by the turbulent velocity fluctuations disturbing the mean temperature gradient. This action is ain to stirring.

$\overline{v_i'\theta'^2_{,i}}$: Transport of the temperature variance by the turbulent field.

$-\kappa\left(\overline{\theta'^2}\right)_{,ii}$: Diffusion of the temperature variance.

$-\kappa\left(\overline{\theta'^2_{,i}}\right)$: Destruction of the temperature variance by molecular smoothing.

2.16. SCALING THE EQUATION OF MOTION: LIMITING CASES

In §1.4, we saw that scaling involves two concepts. First, in order to estimate the magnitude of a particular force field we must choose the relevant scales for each physical quantity. Second, we use this scale to non-dimensionalize all the variables yielding non-dimensional variables of order one. We now apply scaling to the conservation equations set up in §2.11; in general we shall assume that the length scale of the motion is L, the velocity scale is U and the time scale is T. Now each term in (2.11.10) is a force per unit volume and may be scaled as follows:

$$\textit{Unsteady inertia}: \quad \rho \frac{\partial v_i}{\partial t} = O\left(\rho_0 \frac{U}{T}\right) = M^{(1)} \tag{2.16.1}$$

$$\textit{Advective inertial}: \quad \rho v_j \frac{\partial v_i}{\partial x_j} = O\left(\rho_0 \frac{U^2}{L}\right) = M^{(2)} \tag{2.16.2}$$

$$\textit{Pressure}: \frac{\partial p}{\partial x_i} = O\left(\frac{\Delta p}{L}\right) = M^{(3)} \tag{2.16.3}$$

$$\textit{Surface buoyancy}: \rho g = O(\rho_0 g) = M^{(4)} \tag{2.16.4}$$

$$\textit{Internal buoyancy}: \quad \rho' g = O(\Delta \rho g) = M^{(5)} \tag{2.16.5}$$

$$\textit{Coriolis}: \quad \rho f v_i = O(\rho_0 f U) = M^{(6)} \tag{2.16.6}$$

$$\textit{Viscous drag}: \quad \mu \frac{\partial^2 v_j}{\partial x_i^2} = O\left(\mu \frac{U}{L^2}\right) = M^{(7)} \tag{2.16.7}$$

$$\textit{Compressibility}: \quad \frac{E_v}{\rho} \frac{\partial \rho}{\partial x_i} = O\left(\frac{E_v}{L} \frac{\Delta \rho}{\rho_0}\right) = M^{(8)} \tag{2.16.8}$$

$$\textit{Surface tension}: \frac{\sigma}{\eta} \frac{\partial^2 \eta}{\partial x_i^2} = O\left(\frac{\sigma}{L^2}\right) = M^{(9)}, \tag{2.16.9}$$

where ρ' is the density perturbation due to the motion of the fluid, $\Delta\rho$ is the density difference across the fluid domain, L is the length scale, Δp is pressure differential across the fluid domain, σ is the surface tension force, η is the displacement of the surface undergoing a curvature of $\frac{\partial^2 \eta}{\partial x_i^2}$.

To assess whether a particular force type is important in a particular flow domain, relative to another, we may simply form the ratio of the above scales, $M^{(i)}$, $i = 1 - 8$, to the particular force we are interested in; such ratios are non-dimensional parameters yielding the relative importance of individual forces. In the above scaling of the forces, it was assumed that the three major scales (L, U, T), called the outer scales are known *a priori*. Further, in using these scales to scale the forces the implicit assumption is that the flow follows these scales. Flows that do adhere to these outer scales and which the forces do scale with these outer scales are called outer flows.

Often, embedded in the outer flow are flow force balances that are internal to the outer flow and are governed by a different force balance that is made possible by the flow assuming different scales of motion. By way of introduction of this subtle concept, consider a homogeneous fluid flowing over a flat plate. If the length of the plate is L, the viscosity of fluid is ν and the velocity of the fluid from the plate is U, then from above, the three forces that could influence the outer flow are the viscous force with magnitude $M^{(7)}$, the advective inertia force of magnitude $M^{(2)}$ and the pressure force $M^{(3)}$. The ratio:

$$\frac{M^{(2)}}{M^{(7)}} = \frac{UL}{\nu}, \tag{2.16.10}$$

and when this is very large, indicates that in the outer flow advective inertia is much larger than the viscous forces, and the two cannot balance. This leaves the pressure to balance the advective inertia forces; such a flow is called inviscid. However, we immediately see that such an outer flow does not exert any drag force on the plate, because the viscous forces have been estimated to be small. Common experience, however, indicates that flows do exert a viscous drag on a flat plate. So where have we gone wrong? The answer lies in assuming that the flow scales with the length scale L. If, instead of fixing the length scale as that implied by the domain size, L, we "float" the length scale and call it δ and then ask whether $M^{(7)}$ can balance $M^{(2)}$, then we see that this is the same as asking whether it is possible that:

$$\mu \frac{\partial^2 v_j}{\partial x_i^2} \sim \rho v_j \frac{\partial v_i}{\partial x_j}. \tag{2.16.11}$$

Now if we let $v_1 \sim U$; $x_1 \sim L$ and $x_3 \sim \delta$, then we see that (2.16.11) is possible if there is a thin layer next to the plate of thickness:

$$\frac{\delta}{L} \sim \left(\frac{\nu}{UL}\right)^{1/2} = \frac{1}{Re^{1/2}}. \tag{2.16.12}$$

So we see that this is an example where there is an outflow where $M^{(3)} \sim M^{(2)}$ and an inner flow in which $M^{(7)} \sim M^{(2)}$, but with different scales; this is a very common feature of fluid flows.

The origin of commonly used non-dimensional parameters follows directly from a force comparison in the so-called outer flow; in other words, scales that follow form the domain in scales:

1. *A viscous, homogeneous, fluid flowing through small domains.*

The two major force types relevant for such flow configurations are inertia and viscous forces and the ratio describing the flow behavior will be:

$$\frac{M^{(2)}}{M^{(7)}} = \frac{UL}{\nu} = Re, \tag{2.16.13}$$

where ν is the coefficient of kinematics viscosity (μ/ρ_0) and Re is called the Reynolds number. When the Re is small, viscous forces will dominate and when Re is large inertia forces dominate.

Clearly, these two forces cannot be the only forces acting since the value of Re can be varied by changing the velocity U through the domain, the size of the domain L and or the viscosity ν of the fluid; these are all externally controlled parameters. As we have seen in §1.11, a fluid moves under the action of different forces and the momentum equations (1.11.5) must be satisfied at all times. Thus, when either inertia forces become smaller or larger than viscous forces, a third force must enter the balance. In steady flows, the pressure terms in (1.11.5) increase or decrease to balance the momentum equation.

For instance, suppose the velocity of the fluid is increased through the domain so that Re becomes increasingly larger. To balance the now larger inertia force, pressure forces must be of the same order as inertia; viscous forces then play a more minor role. In this situation,

$$M^{(3)} = M^{(2)}, \tag{2.16.14}$$

or

$$\Delta p = \rho_0 U^2, \tag{2.16.15}$$

that, as we already know from §1.11, represents the stagnation pressure.

We thus see that, at $Re \sim 1$, the fluid motion is such that viscous forces balance inertia forces. Such flows are called viscous flows and, as discussed in chapter 4, the motion of the fluid is a solution to the full Navier–Stokes equations (2.11.13). By contrast as Re increases to larger values, the viscous force terms on the right hand side of (2.11.13) become increasingly less important and the flow is a solution of (2.11.13) without the viscous terms on the right hand side of (2.11.13). Such flows are called inviscid flows and equation (2.11.13) with the viscous terms removed:

$$\rho\left(\frac{\partial v_i}{\partial t}\right) + \rho v_j\left(\frac{\partial v_i}{\partial x_j}\right) = -\frac{\partial p}{\partial x_i} - \rho g\delta_{i3}, \tag{2.16.16}$$

are called the Euler equations.

For the Euler equations to be valid, further discussed in chapter 3, we have assumed that *Re* is large and that the relevant scales for velocity and length are U and L, respectively. This may not be uniformly true throughout the flow domain, as shown above. Also, (2.11.13) with the viscous terms retained, has second order derivatives of the velocity whereas (2.16.16) is a first order partial differential equation in the velocity $\{v_i\}$. In physical terms, this means that in inviscid flow, having no viscous forces, the fluid can slip past solid boundaries, flowing tangentially along the walls; in such cases, a boundary layer forms, with a small flow length scale adjacent to the solid surface as discussed above.

2. *Fluids with small viscosity moving with a rapidly changing velocity field in a large domain.*

In such flows the motion is obviously unsteady and the two main terms in (2.11.10) are the unsteady inertia forces of magnitude $M^{(1)}$ and the advective inertia forces with magnitude $M^{(2)}$. The ratio

$$\frac{M^{(1)}}{M^{(2)}} = \frac{LT^{-1}}{U} = S_t, \tag{2.16.17}$$

where S_t is called the Strouhol number. If the Strouhol number is small then unsteady inertia is negligible and the flow may be thought of as being steady. In such flows the inertial forces must, once again, be balanced by one of the other forces in (2.11.10).

Once again the scaling may not be uniformly valid as in the case when the overall flow field may be steady, small S_t, but locally the flow may be highly variable. A simple example of such a non-uniformity is the steady flow round a circular cylinder at moderate values of the Reynolds number. This flow is discussed in some depth in §4.5, where it is seen that, even though the free stream velocity U of the fluid impinging on the circular cylinder is steady, the flow in the wake is made up of alternating eddies being shed from the rear of the cylinder.

3. *Fluids with small viscosity moving in a domain with a free surface.*
Many examples come to mind in this category of flow such as open channel flow (§5) and water waves (§6). For definiteness, suppose we consider surface waves that have a periodic motion, then the two dominant forces are surface buoyancy and unsteady inertia such that the force ratio becomes

$$\frac{M^{(1)}}{M^{(4)}} = \frac{U}{gT} = 1. \tag{2.16.18}$$

Two things must be pointed out. First, the buoyancy for $M^{(4)}$ acts through the pressure field $M^{(3)}$ that then, balances the unsteady inertia $M^{(1)}$, so in addition to (2.16.18) we have that the ratio

$$\frac{M^{(3)}}{M^{(1)}} = \frac{\Delta p T}{\rho_0 UL}, \tag{2.16.19}$$

must be of order one, providing a relationship between the pressure magnitude and the velocity associated with the wave field.

Second, the velocity scale U is not known *a priori* and is not imposed through external boundary forces. We may resolve this unknown by noting that if a represents the wave amplitude and we maintain geometry similarity then "a" scales with H, the depth of the water:

$$U = \frac{a}{T} \sim \frac{H}{T} \Rightarrow T \sim \frac{H}{U}. \tag{2.16.20}$$

Substituting the last scale from (2.16.20) into (2.16.18) yields

$$F_r = \frac{U^2}{gh} = 1, \tag{2.16.21}$$

where force ratio:

$$F_r = \frac{U^2}{gh}, \tag{2.16.22}$$

is called the *Froude Number.*

4. *Fluids with small viscosity and a density stratification in the vertical.*
Such flows are collectively known as stratified flows. For definiteness suppose the fluid column has a density given by:

$$\rho = \rho_0 + \rho_e(x_3) + \rho'(x_1, x_2, x_3, t), \tag{2.16.23}$$

where ρ_0 is the average bulk density, $\rho_e(x_3)$ is the density associated with the equilibrium stratification and $\rho'(\underset{\sim}{x}, t)$ is the density perturbation due to the motion. Define:

$$N^2 = -\frac{g}{\rho_0}\frac{d\rho_e}{dx_3}, \tag{2.16.24}$$

that has units of s^{-2}. The parameter N is called the *buoyancy frequency.*

From the definition of N^2 and the assumption that:

$$\rho' \sim \frac{d\rho_e}{dx_3}L, \tag{2.16.25}$$

it follows that:

$$\rho' \sim \frac{\rho_0}{g}N^2 L. \tag{2.16.26}$$

The ratio:

$$\frac{M^{(3)}}{M^{(6)}} = \frac{U^2}{g'L} = \frac{U^2}{N^2L^2} = F_i^2, \tag{2.16.27}$$

where $g' = g\frac{\Delta\rho}{\rho_0}$ and F_i is called the *internal Froude number.* Once again if F_i is small, this would indicate that the induced inertia forces are much smaller than the buoyancy forces imposed by the stratification in the fluid. However, this is based on the assumption that the motion scales with L. It is clear from (2.16.27) that if $F_i \ll 1$ then it is possible that there is a small scale δ embedded in the flow domain where:

$$\delta = \frac{U}{N}, \tag{2.16.28}$$

a velocity concentration exists with a buoyancy–inertia balance dominating the momentum balance.

5. *A viscous fluids with a density stratification moving slowly under the action of buoyancy.* Such flows are common when density gradients become large and when molecular viscosity acts due to a very small vertical length scale. When viscous forces balance buoyancy forces we have the ratio of importance being given by:

$$\frac{M^{(6)}}{M^{(1)}} = \frac{N^2L^3}{\nu U}, \tag{2.16.29}$$

where both L and U are imposed by external forcing. Normally, even for small domain scales and small velocities this ratio is always very large, indicating that buoyancy forces nearly always overwhelm viscous forces, when motions take place on the outer scale dimension. However, there are two important flow types where the force ratio (2.16.29) becomes very important. First, suppose the motion is very slow so that inertia forces balance viscous forces, then from (2.16.10) it follows:

$$U = \frac{\nu}{L}. \tag{2.16.30}$$

Substituting (2.16.30) from (2.16.29) yields:

$$\frac{M^{(6)}}{M^{(1)}} = \frac{N^2 L^4}{\nu^2} = Gr, \tag{2.16.31}$$

where Gr is called the *Grashof number.* The magnitude of the Grashof number indicated the relative magnitude of the buoyancy forces to the viscous forces, when there is a balance of viscous and inertia forces.

If the fluid is strongly diffusive, then the density anomalies are now no longer governed by (2.16.25), but rather diffusion abates the anomaly as fast as it forms. In such cases, (2.16.30) must be modified to read:

$$U = \frac{\kappa}{L}, \tag{2.16.32}$$

where κ is the diffusivity then:

$$\frac{M^{(5)}}{M^{(7)}} = \frac{N^2 L^4}{\nu\kappa} = Ra, \tag{2.16.33}$$

where Ra is called the *Rayleigh number.*

The second flow type occurs when the stratification is very strong, preventing large vertical excursions of the flow; essentially the flow is driven by a small tilt on the density interfaces, called isopycnals, and the motion is restrained by friction at a small vertical scale. Such flows require the scales in (2.16.26) to be modified and are best scaled using the vorticity equation (2.12.6). Assuming baroclinic vorticity generation balances the diffusion of vorticity:

$$\frac{1}{\rho_0^2}\frac{\partial \rho'}{\partial x_3}\frac{\partial p_e}{\partial x_3} \sim \nu \frac{\partial^3 v_1}{\partial x_3^3}, \tag{2.16.34}$$

and from (2.16.23) and (2.16.16) we have:

$$\frac{\partial p_e}{\partial x_3} \sim \rho_e g \sim \rho_0 N^2 \delta, \tag{2.16.35}$$

given that the vertical excursion of the isopycnals is δ over a horizontal distance L. The solution we are seeking is characterized by thin velocity intrusions, being maintained by the horizontal pressure gradient set up but the tilted isopycnals. Given the flow in these velocity concentrations is slow, the density anomaly will be set by advection, moving the tilted isotherms and vertical mass diffusion bringing the changes thus brought about, to rest. Hence from §2.9 and §2.10:

$$v_3 \frac{\partial p_e}{\partial x_3} \sim \kappa_\rho \frac{\partial^2 \rho'}{\partial x_3^2}. \tag{2.16.36}$$

Scaling these simplified balances (2.16.34), (2.16.35), and (2.16.36) leads to a scale for the vertical length scale of the horizontal motion:

$$\frac{\delta}{L} \sim \left(\frac{\nu\kappa}{N^4 L^4}\right)^{\frac{1}{6}}. \tag{2.16.37}$$

In such intrusions buoyancy generated vorticity balances vertical diffusion of vorticity; the strength of the density anomaly being the result of horizontal advection moving the horizontal density gradient and a vertical gradient being established that is sufficiently strong to remove further accumulation of mass by horizontal advection ensuring steady state. Such a multiple interplay is common in stratified flows and we shall give an examples of such flows in §3. A simple rearrangement of (2.16.37) shows that:

$$\delta \sim \left(\frac{\nu\kappa}{N^2}\right)^{\frac{1}{6}} L^{\frac{1}{3}} \tag{2.16.38}$$

showing that such layers thicken as $L^{\frac{1}{3}}$. We shall see that, in a sink flow, the fluid moves toward the sink with the layer progressively thinning. But, a small momentum source moving horizontally through the fluid, will form a forward wake, before again growing as $L^{\frac{1}{3}}$.

The same analysis may be repeated, but now vorticity generation being balanced by the non-linear advection of vorticity:

$$v_3 \frac{\partial^2 v_1}{\partial x_3^2} \sim \frac{1}{\rho^2} \frac{\partial \rho'}{\partial x_1} \rho_0 g, \tag{2.16.39}$$

$$v_3 \frac{\partial \rho_e}{\partial x_3} \sim v_1 \frac{\partial \rho'}{\partial x_1}, \qquad (2.16.40)$$

$$\frac{\partial \rho_e}{\partial x_3} \sim \rho' g. \qquad (2.16.41)$$

Scaling these equations yields

$$\delta \sim \frac{U^2}{N^2} \sim \left(\frac{q}{N}\right)^{\frac{1}{2}}. \qquad (2.16.42)$$

The layer thickness is independent of distance and baroclinic vorticity generation balances the horizontal transport of vorticity.

From this follows a further horizontal length scale, the scale at which (2.16.42) is equal to (2.16.38):

$$L_c = \left(\frac{q^3}{N\nu\kappa}\right)^{\frac{1}{2}}, \qquad (2.16.43)$$

where q is the discharge in the velocity intrusion per unit width.

One last fundamental scale exists for flow in a stratified fluid. Suppose once again the flow has established a thin intrusion and is governed by a buoyancy generation of vorticity being balanced by horizontal advection; this leads to an intrusion thickness given by (2.16.42). However, now suppose q is not prescribed as a boundary condition via a sink or jet, but rather it is an outcome of the flow itself. How can we fix q? In this case, where the motion we are dealing with has a unit aspect ratio, so that $\delta \sim L$, then including a further balance with diffusion of vorticity:

$$\frac{\partial}{\partial x}\left(v_1 \frac{\partial v_1}{\partial x_1}\right) = \nu \frac{\partial^3 v_1}{\partial x_3^3}. \qquad (2.16.44)$$

From this it follows immediately that $q \sim \nu$, so that (2.16.42) becomes

$$\delta_p = \left(\frac{\nu}{N}\right)^{\frac{1}{2}}, \qquad (2.16.45)$$

which is a scale of great importance in turbulence in a stratified flow as is seen in §7.

6. *High-speed flow of a compressible fluid such as air.*

This flow regime is outside the scope of what is presented in this book, but it is instructive to derive the relevant force ratio so that we can determine when the incompressibility assumption may be made. Since we are dealing with a high-speed flow the two forces likely to balance are the inertia force and the pressure due to compressibility

$$\frac{M^{(2)}}{M^{(8)}} = \frac{U^2}{C^2}\frac{\rho_0}{\Delta\rho} = \frac{U^2}{C^2} = M_c^2, \tag{2.16.46}$$

where we have assumed that $\Delta\rho \sim \rho_0$ since the compression of the fluid is large and M_c is called the *Mach number.* The parameter

$$c = \sqrt{\frac{E_\vartheta}{\rho_0}} \tag{2.16.47}$$

was shown in §1.3 to be the speed of sound in the fluid.

All fluids are compressible to some degree, but unless M_c approaches unity, the compressibility may be neglected and conservation of mass simply becomes conservation of volume as discussed in §1.10.

7. *Inviscid fluids moving through a small domain with a free surface.*
Whenever an interface exists between two disparate fluids there is the possibility that a thin sheet is formed with elastic properties; these forces are called surface tension and they are a strong function of the impurities residing in the interface. The most common case is surface tension at the free surface of water; air being the second fluid.

When a train of waves disturbs such an interface, stretching of the interface is resisted by the surface tension. For such flows, unsteady inertia balances surface tension and the ratio becomes

$$\frac{M^{(9)}}{M^{(1)}} = \rho_0 \frac{U^2 L}{\sigma} = W, \tag{2.16.48}$$

where we have assumed that $T = \dfrac{L}{U}$ and where W is called the Weber Number. From the definition, it follows that surface tension becomes important for W of order one or smaller.

8. *Large scale geophysical flows.*
Flows that are found in large lakes, coastal seas or the deep ocean are all influenced by the earth's rotation through the Coriolis acceleration term included in (2.3.12). When multiplied by the density, this becomes the Coriolis force, the magnitude of which scales as $M^{(6)}$. Now, since the fluid must possess a velocity for it to experience a Coriolis force and this force is perpendicular to the velocity (see (2.3.12), an opposing pressure gradient

must be present to balance the Coriolis force, hence the relevant ratio for a free surface sustained flow

$$\frac{M^{(3)}}{M^{(6)}} = \frac{\Delta\rho}{\rho_0 f L U}, \tag{2.16.49}$$

where U is the velocity scale for the velocity perpendicular to the pressure gradient given by $\frac{\Delta\rho}{L}$. The pressure differential may be estimated by assuming:

$$M^{(2)} = M^{(3)}, \tag{2.16.50}$$

or

$$\Delta\rho = \rho U^2, \tag{2.16.51}$$

that may be substituted into (2.16.49) to give:

$$\frac{M^{(3)}}{M^{(6)}} = \frac{U}{fL} = R_o, \tag{2.16.52}$$

where R_o is called the Rossby Number. When there is a balance between $M^{(3)}$ and $M^{(6)}$, then

$$L = \frac{U}{f}, \tag{2.16.53}$$

which is the length scale, perpendicular to U, over which U varies: it is called the *Rossby radius of deformation.*

If the pressure follows from a deflection of the free surface, then we may also say:

$$M^{(3)} = M^{(4)}, \tag{2.16.54}$$

so that:

$$\Delta p = \rho_0 g \delta, \tag{2.16.55}$$

where δ is the surface displacement scale. Substituting (2.16.55) into (2.16.49) and assuming a balance implies:

$$\frac{\delta}{L} = \frac{fU}{g}, \tag{2.16.56}$$

which is the surface slope estimate across a geostrophic current; a current that is maintained by a balance between the Coriolis force and the transverse pressure gradient force. When the geostrophic current is supported by the deflection of a density interface, across which there is a density anomaly of $\Delta\rho$, then (1.17.56) becomes:

$$\frac{\delta}{L} = \frac{fU}{g'} \tag{2.16.57}$$

where $g' = g\frac{\Delta\rho}{\rho}$.

Lastly, consider a wave deflecting the same density interface, the fluid being bounded by a vertical wall, but otherwise being semi-infinite. Suppose the wave is traveling along the vertical wall (representing a coast), then it is possible to have two separate force balances. First, transverse the wall, we balance the pressure gradient $M^{(3)}$ with the transverse Coriolis force $M^{(6)}$. This implies:

$$L = \frac{\Delta p}{\rho f U}, \tag{2.16.58}$$

where the pressure differential in the transverse direction is given by (2.16.55) so that (2.16.57) yields:

$$L = \frac{g'\delta}{fU}. \tag{2.16.59}$$

Second, in the alongshore direction, the normal wave force balance holds $M^{(3)} = M^{(1)}$, so that:

$$U = \frac{\Delta p T}{\rho_0 L_x}, \tag{2.16.60}$$

where T is the period of the wave and L_x the alongshore horizontal scale. Combining (2.16.60), (2.16.57) and (2.16.55) yields

$$L = \frac{C}{f}, \tag{2.16.61}$$

where $C = \frac{L_x}{T}$ is the alongshore wave speed. Such a wave is called a Kelvin wave.

REFERENCES

Aris, R., 1962. Vectors, Tensors, and the Basic Equations of Fluid Mechanics. Courier Dover.

Ekman, V.W., 1905. On the influence of the earth's rotation on ocean-currents. Arkiv För Matematic, Astronomi Och Fysik 2 (11), 1–53.

CHAPTER 3

Some Exact Solutions

Contents

In this chapter I present some exact solutions of the conservation equations presented in §2 with the objective of familiarizing the reader with some simple solution techniques. Problems where the streamlines are parallel are linear and are thus amenable to solution. However, such flows are also directly relevant to our focus as environmental problems most commonly occur in domains that are much shallower than they are long or wide; flows in estuaries, lakes, rivers and coastal seas. The first 6 sections are standard solution, but in sections 8 and 9 we add a more environmental flavor.

3.1. FUNDAMENTAL SCALES AND PROCESSES

In this section we bring together some fundamental solutions of the conservation equations (2.11.1), (2.11.10)–(2.11.12) presented in §2.11 in order to illustrate some fundamental processes operating in a fluid flow and their associated scale. The difficulty in finding exact solutions, or rather any solution, is that the advective acceleration terms $v_j v_{i,j}$ in (2.11.10) are non-linear, invalidating all the tool of classical analysis such as transform techniques. With the widespread use of numerical solutions techniques, this is no longer such a handicap since many problems can now be solved numerically, but numerical solutions have two major drawbacks. First, as with analytic solutions, the numerical solutions also break down for the more challenging problems involving instabilities and turbulence, and so are not a universal tool. Second, numerical solutions usually shed little light on

Environmental Fluid Dynamics
ISBN 978-0-12-088571-8, DOI: 10.1016/B978-0-12-088571-8.00003-6

the underlying physics of the problem. The understanding gained from an evaluation of the analytic solutions of a range of problems is essential if an engineer, biologist or chemist would like to develop a degree of intuition for fluid flow problem and not just possess a black-box solution set of skills. An intuitive understanding of a problem is important when assessing the effect of changes in configuration or flow conditions, an essential skill in engineering design.

Exact solutions are possible under two circumstances. First, when the flow is parallel, then $v_j v_{i,j} = 0$, as can easily be verified, and the equations of motion become linear admitting solutions via classical analysis. Second, a very small number of special solutions have been found where only part of the problem is solved, albeit the important part in the spirit of §2.16. We shall use a selection of the first case to illuminate some important scales of motion and illustrate how advances in fluid dynamics depend greatly on a judicial use of a number of techniques, each adding to a piece of the understanding and together yielding a solution to the whole problem.

3.2. PLANE COUETTE FLOW

Consider the flow of an incompressible fluid between two parallel infinite flat plates as shown in Fig. 3.2.1. The bottom plate is stationary and the top plate is assumed to move, within its own plane, with a speed U in the x_1 direction. Further, suppose that the plate has been moving for some time so that we can assume that the motion is steady.

The origin of the axes (x_1, x_2, x_3) is arbitrary since the plates are infinitely long; there is no beginning or end to the plates. This means that the streamlines must be horizontal and parallel to the plates; any other flow

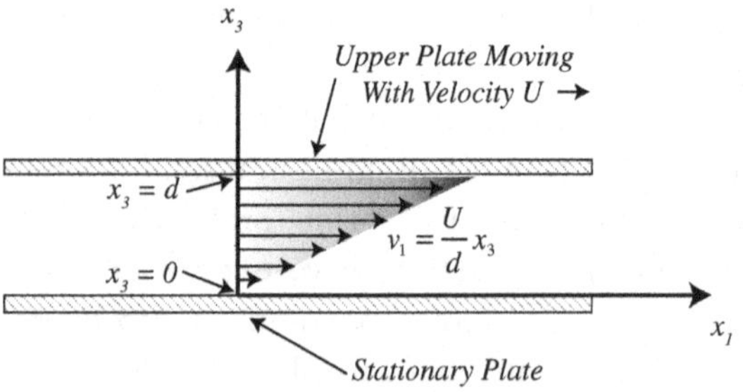

Figure 3.2.1 Schematic of simple shear flow.

configuration would require a horizontal length scale imposed by the boundaries. Thus, the flow must have the property that:

$$v_2 = v_3 = 0. \tag{3.2.1}$$

Given (3.2.1) the conservation of volume (2.11.1) implies,

$$v_1 = v_1(x_3). \tag{3.2.2}$$

Substituting (3.2.1) and (3.2.2) into the momentum equations (2.11.10) leads to:

$$0 = -p_{,1} + \mu v_{1,33}, \tag{3.2.3}$$

$$0 = -p_{,3} - \rho_0 g. \tag{3.2.4}$$

Equation (3.2.4) expresses that the pressure p is hydrostatic. Simple integration leads to:

$$p = -\rho_0 g x_3 + f(x_1), \tag{3.2.5}$$

where $f(x_1)$ is an arbitrary function of x_1 arising from the integration and the first term in (3.2.5) is the hydrostatic pressure distribution. Now once again, since the plates are parallel and infinitely long, and since it is assumed that there is no net horizontal pressure gradient,

$$f(x_1) = p_0. \tag{3.2.6}$$

Substituting (3.2.6) into (3.2.3) yields:

$$v_{1,33} = 0, \tag{3.2.7}$$

that upon integration becomes:

$$v_1 = Ax_3 + B, \tag{3.2.8}$$

where A and B are integration constants. The values of these coefficients may be determined from the no slip boundary condition at $x_3 = 0$ and $x_3 = d$ so that:

$$v_1(x_3) = \frac{U}{d} x_3, \tag{3.2.9}$$

which is the desired solution. We see immediately from (3.2.9) that the velocity field has only one scale, d.

Let us now compute some of the properties of this solution. First consider the stresses

$$\tau_{ij} = \begin{pmatrix} -\rho_0 g x_3 + p_0 & 0 & \dfrac{\mu U}{d} \\ 0 & -\rho_0 g x_3 + p_0 & 0 \\ \dfrac{\mu U}{d} & 0 & -\rho_0 g x_3 + p_0 \end{pmatrix}, \tag{3.2.10}$$

so that the stress on the lower plate is given by:

$$\tau_{3j} = \begin{pmatrix} \dfrac{\mu U}{d} \\ 0 \\ -p_0 \end{pmatrix}, \tag{3.2.11}$$

indicating that p_0 is the pressure on the bottom plate.

The vorticity of the flow may be computed directly from (3.2.9) such that,

$$\zeta_i = e_{ijk}\, v_{k,j} = \begin{pmatrix} 0 \\ \dfrac{U}{d} \\ 0 \end{pmatrix}, \tag{3.2.12}$$

implying a constant vorticity throughout the flow field with zero flux out of either plate; all the vorticity was introduced on initiation of the motion of the plate.

3.3. PLANE POISEUILLE FLOW

Consider again the same geometric configuration as discussed in §3.2, but now we assume a pressure is imposed at the left end ($x_1 = -\infty$) of the duct and also we shall assume that both plates are stationary as shown in Fig. 3.3.1. By the same logic as above in §3.2

$$v_2 = v_3 = 0, \tag{3.3.1}$$

$$v_1 = v_1(x_3), \tag{3.3.2}$$

$$0 = -p_{,1} + \rho_0 \nu_{1,33}, \tag{3.3.3}$$

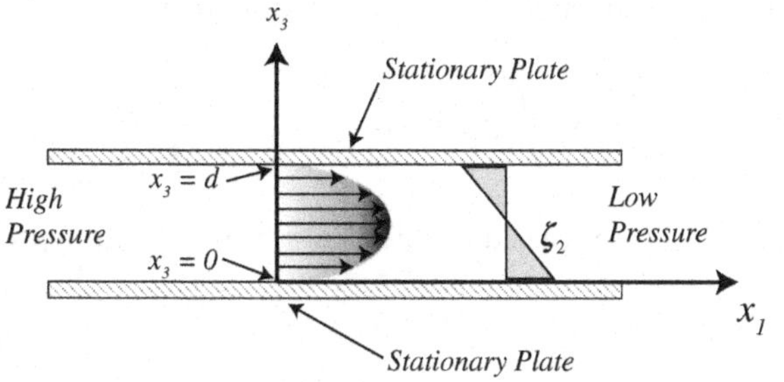

Figure 3.3.1 Schematic of flow through a two-dimensional duct.

$$0 = -p_{,3} - \rho_0 g. \tag{3.3.4}$$

Integrating (3.3.4) with respect to x_3 yields:

$$p = -\rho_0 g x_3 + f(x_1). \tag{3.3.5}$$

Substituting (3.3.5) into (3.3.3) leads to:

$$\frac{\mathrm{d}f(x_1)}{\mathrm{d}x_1} = \frac{\mathrm{d}^2 v_1(x_3)}{\mathrm{d}x_3^2}, \tag{3.3.6}$$

where the ordinary derivative is used as f and v_1 are only functions of one variable. Now the left hand sides of (3.3.6) depend on x_1 and the right hand side on x_3, since the two terms are equal, this is only possible if both sides are equal to a constant, β. Thus

$$p = -\rho_0 g x_3 + \beta x_1 + p_0, \tag{3.3.7}$$

where p_0 is a constant of integration and

$$v_1 = -\frac{\beta d^2}{2\mu}\eta(1-\eta), \tag{3.3.8}$$

where we have used the no slip boundary conditions at $x_3 = 0$ and $x_3 = d$ to evaluate the constants of integration and where

$$\eta = \frac{x_3}{d}, \tag{3.3.9}$$

is the non-dimensional vertical coordinate.

The volume flux through the duct, per unit width:

$$q = \int_0^d v_1(x_3)\mathrm{d}x_3 = -\frac{\beta d^3}{12\mu}. \tag{3.3.10}$$

The minus sign indicates that the fluid flows from the high-pressure end to the low-pressure end.

The stress tensor

$$\tau_{ij} = -p(\theta,\vartheta)\delta_{ij} + \mu(\theta,\vartheta)\{v_{i,j}+v_{j,i}\}$$
$$= \begin{pmatrix} +\rho_0 g x_3 - \beta x_1 - p_0 & 0 & \beta\left(x_3 - \frac{d}{2}\right) \\ 0 & +\rho_0 g x_3 - \beta x_1 - p_0 & 0 \\ \beta\left(x_3 - \frac{d}{2}\right) & 0 & +\rho_0 g x_3 - \beta x_1 - p_0 \end{pmatrix}, \tag{3.3.11}$$

and thus the stress on the bottom plate is given by:

$$\tau_{3j} = \begin{pmatrix} \frac{-\beta d}{2} \\ 0 \\ \beta x_1 - p_0 \end{pmatrix}, \tag{3.3.12}$$

indicating a simple balance between the pressure gradient force in the x_1 direction and the opposing shear stresses exerted by the parallel plates.

The vorticity vector is easily found by substituting the expression for the velocity (3.3.8) into the formula for the vorticity:

$$\zeta_i = e_{i\alpha\beta}v_{\beta,\alpha}$$
$$= \begin{pmatrix} 0 \\ -\left(\frac{\partial v_1}{\partial x_3}\right) \\ 0 \end{pmatrix} \tag{3.3.13}$$
$$= \begin{pmatrix} 0 \\ \frac{\beta}{\mu}\left(x_3 - \frac{d}{2}\right) \\ 0 \end{pmatrix}.$$

The vorticity profile is shown in Fig. 3.3.1 where it is seen that ζ_i is zero at $x_3 = \frac{d}{2}$ and has a linear distribution, identical to the shear stress distribution. This is a beautiful illustration of the discussion on vorticity generation given in §2.12. In the present flow, there is a pressure gradient from left to right ($\beta < 0$), so that positive vorticity is introduced at the lower boundary and negative at the upper boundary (since $\underset{\sim}{\hat{n}}$ is directed downwards). The solution to the flow is that distribution which allows the positive vorticity to diffuse from the lower boundary to the upper boundary and the negative vorticity from the top to the bottom. This is analogous to solving the temperature distribution between a hot and cold plate; a linear distribution is the solution.

Once again we may inspect (3.3.13) in order to ascertain the scale of the flow. From (3.3.8) it is clear that the velocity profile is parabolic with a length scale of d. The velocity scale follows also directly from (3.3.8) or (3.3.9) and is given by:

$$v_1 \sim \frac{\beta d^2}{\mu}, \tag{3.3.14}$$

and using (3.3.13) indicates that:

$$\varsigma_2 \sim \frac{U}{d}, \tag{3.3.15}$$

confirming that the length scale in the vertical is d, even though the velocity distribution is no longer linear in x_3.

3.4. SUPERPOSITION OF PLANE FLOWS

The great simplification afforded by linear systems may also be illustrated by showing that if v_1 is a solution to the plane Couette flow and if $-u_1$ is a solution to the plane Poiseuille flow then:

$$w_1 = v_1 + u_1 = U\eta - \frac{\beta d^2}{2\mu}\eta(1-\eta), \tag{3.4.1}$$

is the solution to the superposed configuration shown in Fig. 3.4.1 and where η is again the non-dimensional vertical coordinate (3.3.9).

The net discharge per unit width through the duct is given by:

$$q = \frac{Ud}{2} - \frac{\beta d^3}{12\mu}. \tag{3.4.2}$$

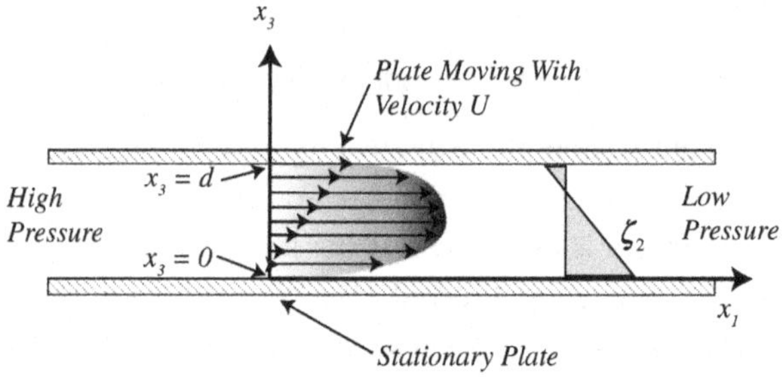

Figure 3.4.1 Combination of Couette and Poiseuille flow.

Clearly we can see from (3.4.2) that the discharge depends on both the velocity U of the upper plate and the pressure arising from the high pressure formed at the center of the gap; there is a competition between the flow driven by the moving upper plate and that induced by the pressure gradient β.

Consider now the flow of a fluid sandwiched between two plates as shown in Fig. 3.4.2. On the left-hand side the gap, between the plates, has a width d_1, and on the right the gap width is d_2. The total length of the plates is L with a very small step ($d_1 \approx d_2$) occurring at mid-length. Now suppose the top plate moves from left to right with a velocity U and the gap size is very much smaller than the length of the configuration.

Under these circumstances we may assume that the entrance flow at A, the adjustment flow at B and the exit flow at C are all confined to their respective regions, and that over the majority of the plates we have a combination of plane Couette and Poiseuille flow so that the solution takes

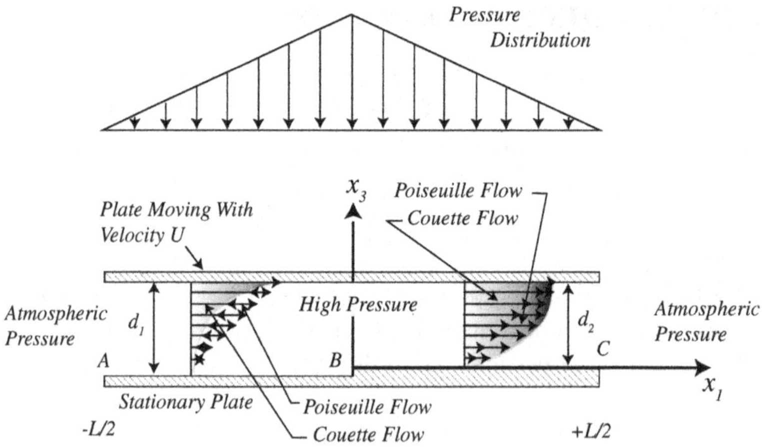

Figure 3.4.2 Build up of pressure due to small step change in duct gap size.

the form (3.4.1). To conserve volume flux, the flows in the right and left-hand ducts must be such as to have the same discharge; this may be achieved by allowing the pressure at B to rise until the discharges are equal:

$$q = \frac{Ud_1}{2} - \frac{\beta d_1^3}{12\mu} = \frac{Ud_2}{2} - \frac{\beta d_2^3}{12\mu}, \tag{3.4.3}$$

where we have assumed for simplicity that:

$$\beta_1 = -\beta_2 = \beta, \tag{3.4.4}$$

since the two halves of the total duct are of equal length $L/2$. Rearranging (3.4.3) yields an expression for β such that:

$$\beta = 6\mu U\left(\frac{d_1 - d_2}{d_1^3 + d_2^3}\right). \tag{3.4.5}$$

The total thrust on the upper plate is given by:

$$T = \int_{-L/2}^{L/2} \tau_{33} dx_1 \tag{3.4.6}$$

Substituting the solution (3.4.1), and using (3.4.5) yields:

$$T = \frac{3}{2}\mu UL^2\left(\frac{d_1 - d_2}{d_1^3 + d_2^3}\right) \tag{3.4.7}$$

Clearly, the higher the upper plate velocity U, the greater the thrust the upper plate can support.

The process of simplifying a complex problem to configurations where solutions are manageable is a necessary skill for the fluid mechanician and used in every problem; the art is to simplify without losing the essence of the problem.

3.5. UNSTEADY PARALLEL FLOW: DIFFUSION OF VORTICITY

A problem of great importance is the motion induced in a fluid overlying a flat plate that is suddenly set into motion; this simple problem allows a complete solution to the full Navier–Stokes equations and so provides deep insight into the way vorticity is generated at boundaries and then diffuses into the main domain, setting the fluid into motion.

The configuration is shown in Fig. 3.5.1 and it is assumed that the plate is of infinite extent, the fluid is of semi-infinite extent and the plate is impulsively set into motion in its own plane, with a velocity U.

Given these assumptions it is possible to make the same simplifications of parallel flow as in §3.2 and seek a solution with the properties:

$$v_2 = v_3 = 0, \qquad (3.5.1)$$

but now we write:

$$v_1 = v_1(x_3, t), \qquad (3.5.2)$$

as the flow is unsteady. With these assumptions the momentum equations (2.11.10) become for $i = 1,2$:

$$\frac{\partial v_1}{\partial t} = \nu \frac{\partial^2 v_1}{\partial x_3^2} - \frac{\partial p}{\partial x_1}, \qquad (3.5.3)$$

$$0 = -\frac{\partial p}{\partial x_3} - \rho_0 g. \qquad (3.5.4)$$

Integrating (3.5.4) with respect to x_3 leads to the result:

$$p = -\rho_0 g x_3 + f(x_1, t), \qquad (3.5.5)$$

but similarly to the Couette flow problem, discussed above, we may assume:

$$f(x_1, t) = p_0 \qquad (3.5.6)$$

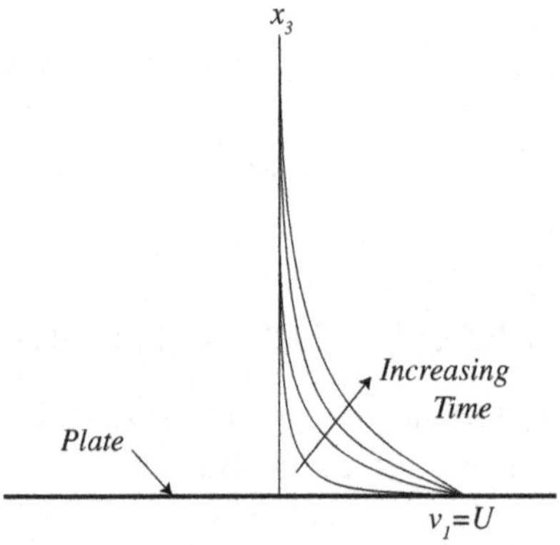

Figure 3.5.1 Schematic of the development of the horizontal velocity field above a flat plate that is abruptly set into motion.

where p_0 is the pressure at the origin. Allowing the pressure p_0 to vary with time does not alter the problem since the fluid is incompressible, and a time varying p_0 would merely superimpose a uniform over-pressure, that does not lead to a motion.

Equation (3.5.3) therefore reduces to:

$$\frac{\partial v_1}{\partial t} = \nu \frac{\partial^2 v_1}{\partial x_3^2} \tag{3.5.7}$$

This is a linear partial differential equation and is amenable to transform techniques. However, in order to develop our insight it is more instructive to use simple-dimensional analysis to find a solution to (3.5.7).

In general, the velocity v_1 can only depend on certain variables and so we may write it in the form:

$$v_1 = v_1(U, x_3, t, \nu). \tag{3.5.8}$$

However, (3.5.7) and the boundary condition at $x_3 = 0$ (the plate) are linear. This is seen by noting that, if v_1 and u_1 are each solutions to the present problem with V and U as the velocity of the plate, then:

$$u_1(x_3, t) + v_1(x_3, t) \tag{3.5.9}$$

is also a solution, but with the plate now moving with a velocity $U + V$.

Thus (3.5.8) may be simplified to read:

$$\frac{v_1}{V} = v_1(x_3, t, \nu), \tag{3.5.10}$$

since the velocity field must depend linearly on the velocity of the plate.

The left-hand side of (3.5.10) is non-dimensional thus, as discussed in §1.2, the right-hand side of (3.5.10) can only be a function of a non-dimensional variable. We have three independent variables x_3, t and ν, and two dimensions, length and time. Hence, we can only form one dimensionless group

$$\eta = \frac{x_3}{(\nu t)^{\frac{1}{2}}}. \tag{3.5.11}$$

We shall call η the self-similarity variable as it allows changes in space and time to be related and thus introduces a self-similarity to the solution. Equation (3.5.10) may then be written:

$$\frac{v_1}{V} = f(\eta), \tag{3.5.12}$$

where $f(\eta)$ is a function obtained by substituting (3.5.12) into (3.5.7) which yields a simple ordinary differential equation:

$$\frac{d^2 f}{d\eta^2} + \frac{\eta}{2}\frac{df}{d\eta} = 0. \tag{3.5.13}$$

We have therefore, through the introduction of the self-similarity variable η, reduced the partial differential equation (3.5.7) for v_1 to the second order ordinary differential equation (3.5.13) in η. This has been possible because there are a family of curves, defined by the value of c, such that:

$$x_3 = c(\nu t)^{\frac{1}{2}}, \tag{3.5.14}$$

in the x_3, ν, t space, along which the solution is constant.

In order to solve (3.5.13) let

$$g = \frac{df}{d\eta}, \tag{3.5.15}$$

so that,

$$\frac{dg}{d\eta} + \frac{\eta}{2} g = 0, \tag{3.5.16}$$

or

$$\frac{1}{g(\eta)}\frac{dg}{d\eta} = -\frac{\eta}{2}. \tag{3.5.17}$$

Integrating (3.5.17) yields:

$$g = g_0\, e^{-\frac{\eta^2}{4}}, \tag{3.5.18}$$

where g_0 is the constant of integration. A further integration allows the determination of the original function f such that:

$$f(\eta) = g_0 \int_0^{\eta} e^{-\frac{\zeta^2}{2}} d\zeta + g_1, \tag{3.5.19}$$

where g_1 is a further integration constant and ζ is the variable of integration.

Remembering that the special function $erf(\eta)$ is given by:

$$erf(\eta) = \left(\frac{2}{\pi}\right)^{1/2} \int_0^{\eta} e^{-t^2}\, dt, \tag{3.5.20}$$

we can rewrite (3.5.19) in the form:

$$v_1 = U\left(1 - erf\left(\frac{x_3}{(4\ \nu t)^{1/2}}\right)\right) = U\left(1 - erf\left(\frac{\eta}{2}\right)\right). \tag{3.5.21}$$

The solution, shown schematically in Fig. 3.5.1, may thus be represented by a single function, when v_1 is plotted against η as shown in Fig. 3.5.2. We can define a velocity layer thickness by defining a layer at the boundary in which the velocity v_1 has decreased to a certain fraction of the velocity of the plate U, by fixing the value of the similarly variable say η_b. This defines a layer thickness:

$$\delta = x_3 = \eta_b(4\nu t)^{\frac{1}{2}}. \tag{3.5.22}$$

If we chose $\eta_b = 3.5$, then the layer thickness is defined by the point where the velocity has dropped to 1% of U.

Equation (3.5.7) is a simple diffusion equation for the velocity v_1 that was already discussed in §2.10 for the temperature field in a fluid; the exact analog being a plate suddenly heated and then kept at a constant temperature; the heat progressively diffuses into the medium, raising its temperature. In the present case, the momentum diffuses into the fluid.

If we note that the vorticity component in the x_2 direction is the only non-zero component of vorticity that is given by:

$$\zeta_2 = \frac{\partial v_1}{\partial x_3}, \tag{3.5.23}$$

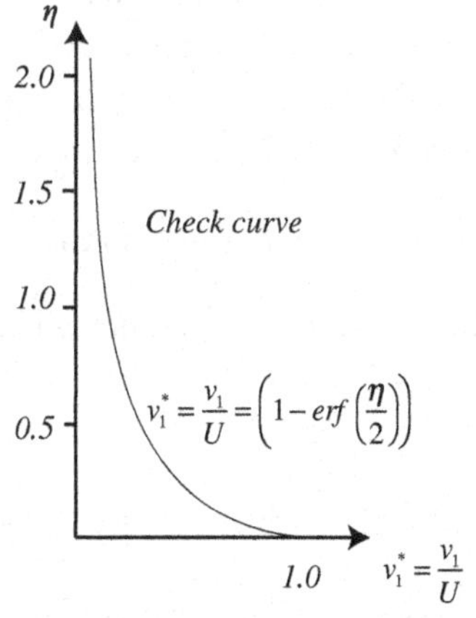

Figure 3.5.2 Non-dimensional velocity profile for the flow above an impulsively started flat plate.

then by differentiating (3.5.7) with respect to x_3 yields:

$$\frac{\partial \zeta_2}{\partial t} = \nu \frac{\partial^2 \zeta_2}{\partial x_3^2}, \tag{3.5.24}$$

showing that the vorticity also satisfies a simple diffusion equation. Naturally, this result follows also directly from the vorticity equation (2.12.6). Now as the plate is abruptly set into motion on the initiation of the problem:

$$\frac{v_1}{U} = 2\{1 - H(x_3)\}; \quad x_3 \geq 0, \tag{3.5.25}$$

where $H(x_3)$ is the Heaviside step function with the property:

$$H(x_3) = 1; \ x_3 > 0, \tag{3.5.26a}$$

$$H(x_3) = \frac{1}{2}; \ x_3 = 0, \tag{3.5.26b}$$

$$H(x_3) = 0; \ x_3 < 0. \tag{3.5.26c}$$

In other words the motion of the plate introduces a very sharp shear layer immediately above the plate when it is set in motion.

The vorticity associated with this shear layer is given by the derivative of (3.5.25), so that at $t = 0$:

$$\zeta_2 = -2U\delta(x_3), \tag{3.5.27}$$

where δ is the Dirac delta function. This allows us to give a very simple interpretation in terms of vorticity of the solution to our problem. On initiation of the motion of the plate, an infinitely thin shear layer is set up at the plate and this shear layer contains an amount of vorticity equal to $-U$ (in the upper half $x_3 > 0$). As time progresses, this vorticity diffuses out to infinity. In the present configuration the time rate of change of the vorticity $\frac{\partial \zeta_2}{\partial t}$ is balanced only by the diffusion $\nu \frac{\partial^2 \zeta_2}{\partial x_3^2}$, so that the vorticity in the total fluid domain, $x_3 \geq 0$, should be conserved and remain equal to $-U$. This is confirmed by noting that:

$$\int_0^\infty \zeta_2 \mathrm{d}x_3 = \int_0^\infty \frac{\partial v_1}{\partial x_3} \mathrm{d}x_3 = -U. \tag{3.5.28}$$

Advection of vorticity plays no role in this example, not because it is absent, but rather because the vorticity is uniform in the variable x_1 (for a fixed x_3 and t) so that the net gain or loss of vorticity at a point is zero. This example has far reaching consequences because (3.5.2) provides a simple length scale for a diffusive boundary layer and we shall often refer to this result.

3.6. OSCILLATING BOUNDARY LAYER

Suppose now that, in the above problem, the plate oscillates rather than being set into motion impulsively. An oscillatory motion of the plate will introduce vorticity, at the wall, with a periodically alternating sign. We shall now show that, as this alternating vorticity diffuses away from the plate, the bands of opposing vorticity diffuse into each other averaging out to zero as we move away from the plate; this is an illustration of how diffusion acts to spread vorticity, which then may average out to zero. The configuration is identical to that discussed in §3.5, but now the plate is oscillating in its own plane so that the boundary condition at $x_3 = 0$ becomes:

$$v_1(x_1, 0, t) = U \cos \omega t. \tag{3.6.1}$$

As the flow is again horizontal and parallel, the equations of motion remain the same as (3.5.3) and (3.5.4). We can anticipate the solution from what has been derived above and the discussion of the vorticity dynamics; the simple start up problem the boundary layer was established by the diffusion of vorticity away from the flat plate. This led to a thickness:

$$\delta \sim (\nu t)^{\frac{1}{2}}, \tag{3.6.2}$$

where t is the elapsed time. In the case where the plate oscillates with a frequency ω, the injection of vorticity at the plate periodically reverses sign. Hence, vorticity of one sign periodically negates vorticity of the other sign and we may expect, from purely dimensional grounds, the vorticity layer will be confined to a thickness:

$$\delta \sim \left(\frac{\nu}{\omega}\right)^{\frac{1}{2}}. \tag{3.6.3}$$

This scale is of central importance in stratified flow turbulence. The solution to (3.5.7) with the boundary condition (3.6.1) may be found by seeking a periodic solution in the form:

$$v_1(x_1, x_3, t) = Real\big(f(x_3)e^{i\omega t}\big). \tag{3.6.4}$$

Substituting (3.6.4) into (3.5.7) leads to an equation for the unknown function $f(x_3)$:

$$\frac{d^2 f}{dx_3^2} - \frac{i\omega}{\nu} f = 0 \tag{3.6.5}$$

The solution to (3.6.5) is easily obtained by noting:

$$(i)^{\frac{1}{2}} = \pm \frac{(1+i)}{\sqrt{2}} \tag{3.6.6}$$

leading to:

$$v_1^* = \frac{v_1(x_1, x_3, t)}{U} = e^{-\eta} \cos(\omega t - \eta), \tag{3.6.7}$$

where

$$\eta = \frac{x_3}{\left(\dfrac{2\nu}{\omega}\right)^{\frac{1}{2}}}. \tag{3.6.8}$$

The solution (3.6.7) is shown in Fig. 3.6.1 where it is seen that the alternating vorticity source at the plate does indeed diffuse out from the plate, spreading by diffusion to form a boundary layer of thickness (3.6.3).

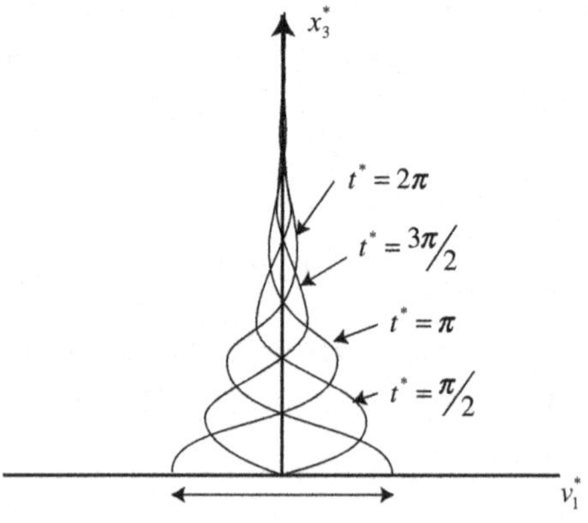

Figure 3.6.1 Velocity distribution induced by oscillating flat plate.

3.7. NATURAL CONVECTION IN A LONG CAVITY

We shall now turn to a problem, the solution of which was obtained only relatively recently and that was motivated by a diverse set of environmental applications, where the depth of the domain is much smaller than the length and where a longitudinal density gradient drives convection; this configuration is generally referred to as the long cavity or long box problem. The density gradient may be due to a temperature gradients, such as are found in embayments of lakes or reservoirs, or due to salinity variations, as are observed in shallow estuaries. For such configurations, we may assume that the aspect ratio, defined as the depth divided by the length:

$$A = \frac{h}{L}, \tag{3.7.1}$$

is very small.

The configuration is shown in Fig. 3.7.1, where we have indicated, by means of arrows, the direction of flow.

If the aspect ratio is small, we may assume that the upper and lower boundaries are flat, horizontal and infinite in extent; variability such as caused by sloping boundaries will be neglected in this first order problem. Further, we shall assume that the configuration is 2D and there are no variations in the x_2 direction and the bottom and top boundaries are fixed and insulated. Other geometries also admit solutions, but here we wish to

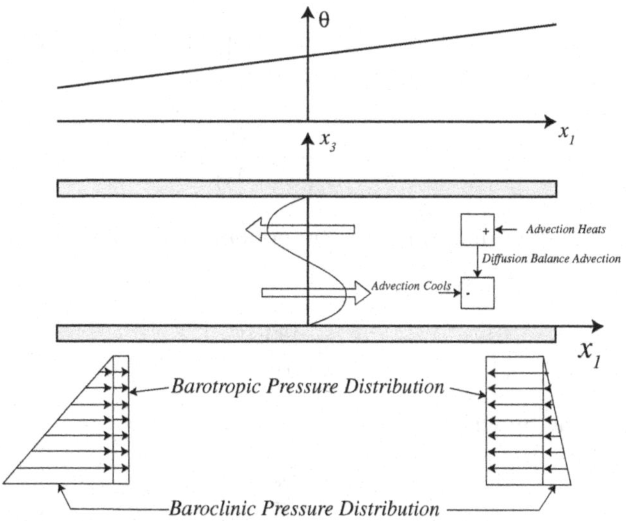

Figure 3.7.1 The long box configuration.

keep the problem as simple as possible in order to elucidate the processes. With these assumptions, we may again write:

$$v_2 = v_3 = 0, \tag{3.7.2}$$

$$v_1 = v_1(x_3), \tag{3.7.3}$$

and thus the equations of motion become as before:

$$0 = -\frac{\partial p}{\partial x_1} + \mu \frac{\partial^2 v_1}{\partial x_3^2}, \tag{3.7.4}$$

$$0 = -\frac{\partial p}{\partial x_3} - g\rho, \tag{3.7.5}$$

but now the density is both a function of the vertical and horizontal coordinates and is governed by the mass diffusion equation (2.11.11) and or (2.11.12):

$$v_1 \frac{\partial \rho}{\partial x_1} = \kappa_\rho \left(\frac{\partial^2 \rho}{\partial x_1^2} + \frac{\partial^2 \rho}{\partial x_3^2} \right), \tag{3.7.6}$$

where κ_ρ is the effective diffusion coefficient, the magnitude depending on the whether the stratification is due to temperature or salinity and whether the flow is laminar or turbulent.

Eliminating the pressure p from (3.7.4) and (3.7.5) by differentiating (3.7.4) with respect to x_3 and (3.7.5) with respect to x_1, and subtracting the two equations yields the vorticity equation:

$$0 = \mu \frac{\partial^3 v_1}{\partial x_3^3} + g \frac{\partial \rho}{\partial x_1}. \tag{3.7.7}$$

If we note that the only non-zero component of vorticity is ζ_2 and this is given by:

$$\zeta_2 = \frac{\partial v_1}{\partial x_3} \tag{3.7.8}$$

then (3.7.7) may be written as:

$$\frac{\mu}{g} \frac{\partial^2 \zeta_2}{\partial x_3^2} + \frac{\partial \rho}{\partial x_1} = 0. \tag{3.7.9}$$

The long box flow is thus simply a flow where vertical diffusion of vorticity is equal, locally, to the baroclinic generation of vorticity, again with the bottom plate being a source of positive vorticity and the upper plate being a sink of vorticity.

Combining (3.7.9) and (3.7.3) shows that:

$$\frac{\partial \rho}{\partial x_1} = f(x_3), \tag{3.7.10}$$

so that:

$$\rho = f(x_3)x_1 + h(x_3), \tag{3.7.11}$$

where $f(x_3)$ and $h(x_3)$ are arbitrary functions, yet to be determined.

We may assume that there are two symmetries to the problem. First, there is no net flow in the x_1 direction, so that:

$$\int_0^h v_1(x_3)\mathrm{d}x_3 = 0, \tag{3.7.12}$$

since the problem is assumed to be part of a closed system. Second, as there is no length scale vertically or horizontally imposed externally by the boundary conditions, the problem must be anti-symmetric about the line $x_3 = \frac{h}{2}$; the velocity and density must be odd functions about the central plane. Another way of expressing this result is to note that the solution must be anti-symmetric with respect to a reversal of the end conditions and also with respect to a reversal in the direction of gravity.

Introducing (3.7.11) into (3.7.6) and integrating the result over the depth of the cavity yields:

$$\int_0^h v_1(x_3)\frac{\partial \rho}{\partial x_1}\mathrm{d}x_3 = 0, \tag{3.7.13}$$

where we have used the boundary conditions:

$$\frac{\partial \rho}{\partial x_3} = 0; x_3 = 0, h. \tag{3.7.14}$$

Now since both x_1 and $\frac{\partial \rho}{\partial x_1}$ are odd function of x_3 about $x_3 = \frac{h}{2}$, (3.7.12) and (3.7.13) together require that:

$$\frac{\partial \rho}{\partial x_1} = C, \tag{3.7.15}$$

where C is a constant; the density gradient in the x_1 direction is constant.

Substituting (3.7.15) into (3.7.7) and integrating using the boundary conditions,

$$v_1 = 0; \;\; x_3 = 0, h, \tag{3.7.16}$$

together with the symmetry requirements, allows the determination of the solution for v_1,

$$v_1(x_3) = -\frac{Cgh^3}{6\mu}\eta\left(\eta - \frac{1}{2}\right)(\eta - 1), \tag{3.7.17}$$

where

$$\eta = \frac{x_3}{h}, \tag{3.7.18}$$

Substituting (3.7.17) back into (3.7.6) and then integrating the results leads to an expression for the density:

$$\rho(x_1, x_3) = \rho_0 + Cx_1 - \frac{C^2gh^3}{24\mu\kappa}\left\{\eta^3\left(\frac{\eta^2}{5} - \frac{\eta}{2} + \frac{1}{3}\right) - \frac{5}{60}\right\}, \tag{3.7.19}$$

where we have chosen the last integration constant so that:

$$\rho = \rho_0; \;\; x_1 = 0; \;\; \eta = \frac{1}{2} \tag{3.7.20}$$

which merely fixes the overall magnitude of the fluid density with a background level of ρ_0.

The solution, given by (3.7.17) and (3.7.19), is shown in Fig. 3.7.2, where it is also compared to experimental data. Many other companion configurations (free surface, net through flow, surface heat flux) have been investigated, but the above adequately illustrates the underlying physics of the solution.

The flow captured by these solutions may be interpreted by noting that the fluid on the right (Fig. 3.7.1) is warmer and thus lighter than the fluid on the left and so experiences a slightly lower higher hydrostatic pressure,

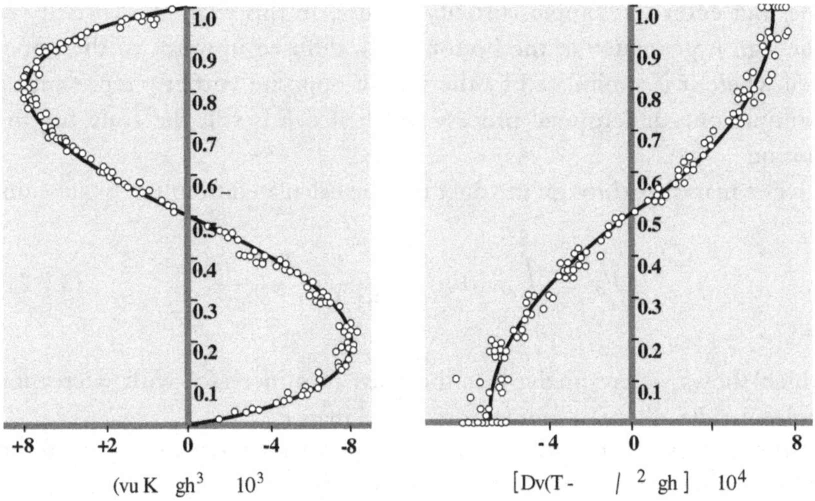

Figure 3.7.2 Theoretical solution compared to experimental data (Imberger 1972).

setting up a pressure gradient from left to right. However, conservation of volume requires a return flow from left at *C* to the right at *D* under the top plate; there is thus a pressure increase from *C* to *D* resulting from the addition of a barotropic pressure gradient, supported by the fixed top plate and the baroclinic pressure gradient induced by the density gradient. This motion, however, advects the background density gradient Cx_1 causing a depletion in density in the top half and an accumulation in the lower half of the duct that must be balanced some way if the flow is to be in steady state (note in Fig. 3.7.1 we show the temperature not the density); vertical diffusion arrests the changes due to horizontal advection, once a sufficient vertical density gradient has been established. The density distribution that satisfies this requirement is, as we have seen, the fifth order polynomial component of (3.7.19).

The vorticity dynamics is clear from the vorticity diffusion equation (3.7.9) and by remembering that at a wall, where there is a pressure gradient, the no slip boundary condition generates a vorticity flux from the wall into the fluid. In our present case, the longitudinal density gradient generates a uniform source of positive vorticity within the fluid column (see Fig. 3.7.2(b)). At the walls, the above described pressure gradient introduces negative vorticity which diffusion spreads into the fluid column; the result is a parabolic distribution of vorticity. It is perhaps interesting to note that the horizontal advection of the fluid does not enter this simple vorticity picture. In this example, vorticity of one sign is generated at the bottom wall, diffused upwards to the upper wall where it is annihilated by the flux of opposite vorticity representing a simple one-dimensional process and hence h is still the scale for the motion.

The mass flux through the duct may be calculated from the expression:

$$F_\rho = \int_0^h \rho v_1 \mathrm{d}x_3 = \frac{1}{51,840}\frac{C^3 g^2 h^9}{\mu^2 \kappa}, \tag{3.7.21}$$

which shows, interestingly, that the mass flux increases with decreasing molecular diffusivity of momentum μ and mass κ_ρ.

Before leaving this section we shall consider an approximation to an important dispersion problem that follows directly from the above solution. Suppose, instead of a linear density distribution in the x_1 direction as above, we have a slowly varying density function as illustrated in Fig. 3.7.3.

For the approximation to remain valid the scale of the density variations L must be long compared to the duct height h or $h/L < 1$.

Consider a mass balance for the small control volume shown in Fig. 3.7.3:

$$\frac{\partial}{\partial t}\left(\int_0^h \rho \mathrm{d}x_3\right) \mathrm{d}x_1 = \frac{\partial F_\rho}{\partial x_1}\mathrm{d}x_1. \tag{3.7.22}$$

Now if we define the mean density:

$$\overline{\rho} = \overline{\rho}(x_1), \tag{3.7.23}$$

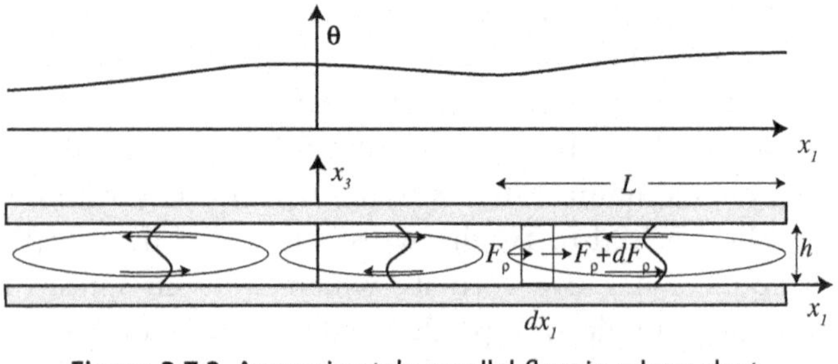

Figure 3.7.3 Approximately parallel flow in a long duct.

as the density averaged over the depth, then we may rewrite (3.7.22) in the form:

$$\frac{\partial \overline{\rho}}{\partial t} + \frac{1}{h}\frac{\partial F_\rho}{\partial x_1} = 0. \tag{3.7.24}$$

Since the density gradient $\frac{\partial \overline{\rho}}{\partial x_1}$ is assumed to vary only slowly, we may assume the long box solution (3.7.21) for the mass flux is applicable locally, so that (3.7.21):

$$\frac{\partial \overline{\rho}}{\partial t} = \frac{3}{51,840}\frac{h^8 g^2}{\mu^2 \kappa} C^2 \frac{\partial C}{\partial x_1}, \tag{3.7.25}$$

but also from (3.7.14)

$$\frac{\partial C}{\partial x_1} = \frac{\partial^2 \overline{\rho}}{\partial x_1^2}, \tag{3.7.26}$$

leading to the equation:

$$\frac{\partial \overline{\rho}}{\partial t} = \kappa_1 \frac{\partial^2 \overline{\rho}}{\partial x_1^2}, \tag{3.7.27}$$

where the longitudinal diffusion coefficient is given by:

$$\kappa_1 = \frac{1}{17,280}\frac{h^8 g^2 C^2}{\mu^2 \kappa}. \tag{3.7.28}$$

If we linearize the equation by assuming κ_1 is a constant, equal to the local mean value, then we see that the depth averaged density disperses longitudinally with an effective diffusion coefficient given by (3.7.28); it is interesting that the rate of dispersion increases with decreasing diffusion $\mu^2\kappa$ in the fluid.

It is also noteworthy that the steady state solution to (3.7.27) is:

$$\frac{\partial \overline{\rho}}{\partial x_1} = C = \mathit{Constant}, \tag{3.7.29}$$

which is the above parallel flow solution.

Further, if $\overline{\rho}$ varies with x_1 in a non-linear fashion as shown in Fig. 3.7.3, then the flow is no longer parallel and the fluid moves in cells as shown in Fig. 3.7.3; however, provided the density $\overline{\rho}$ varies only very slowly then the neglected advective acceleration terms will be small.

We can quantify this proviso as follows. The ratio of the neglected to the retained terms in (3.7.4), for the case where the vertical velocities are not exactly zero, is given by:

$$R = \frac{\rho_0 v_1 \dfrac{\partial v_1}{\partial x_1}}{\mu \dfrac{\partial^2 v_1}{\partial x_3^2}} \tag{3.7.30}$$

Substituting from the solution (3.7.19) and (3.7.4) leads to:

$$R = \frac{gh^5}{\nu^2} \frac{1}{\rho_0} \frac{\partial^2 \overline{\rho}}{\partial x_1^2}, \tag{3.7.31}$$

thus provided:

$$R \ll 1. \tag{3.7.32}$$

the model equation (3.7.27) will capture the dispersion of the mass by the processes of horizontal advection and vertical diffusion. We may recast (3.7.32) a little if we introduce the length scale L characterizing the density variation so that:

$$\frac{\partial^2 \rho}{\partial x^2} \sim \frac{C}{L}, \tag{3.7.33}$$

where C is the density gradient (3.7.29), then (3.7.32) and (3.7.31) become:

$$R = \frac{Uh^2}{L\nu} \ll 1, \tag{3.7.34}$$

where $U = \dfrac{Cgh^3}{\mu}$ is the velocity scale of the motion. The interpretation of (3.7.34) is simply that distance traveled by the fluid in the time $\dfrac{h^2}{\nu}$ (the time the velocity takes to diffuse over the depth) is small compared to the length of the density variations. In other words, the flow has enough lateral room to adjust to the local gradient. We will return to this discussion in the chapter on dispersion. Dispersion is thus a combination of vertical diffusion balancing the horizontal advective distortion; this is called shear dispersion.

3.8. NATURAL CONVECTION IN A SLOPING CAVITY

We now consider the flow in a sloping cavity filled with a stably stratified fluid as shown in Fig. 3.8.1; this is the generalization of §3.7 above. The shallow nature of the long box, forces the flow within the box, to be parallel to the top and bottom boundaries of the cavity and hence it is advantageous to use a coordinate system (ζ_1, ζ_3) embedded in the cavity as shown in Fig. 3.8.1.

We shall assume at the outset that the temperature of the fluid in the cavity is stably stratified as this is the flow of most relevance for environmental flows, but the reader may wish to also derive the solution, by an analogous approach, for the unstable case. Once again we may assume that the transverse velocity is zero, ($u_2 = 0$) and thus:

$$u_1 = u(\zeta_3). \tag{3.8.1}$$

The momentum equations become:

$$0 = -\frac{\partial p}{\partial \zeta_1} - \rho g \sin(\theta) + \mu \frac{\partial^2 u_1}{\partial \zeta_3^2}, \tag{3.8.2}$$

$$0 = -\frac{\partial p}{\partial \zeta_3} - \rho g \cos(\theta), \tag{3.8.3}$$

and the species diffusion equation becomes:

$$u_1 \frac{\partial \rho}{\partial \zeta_1} = \kappa_\rho \left\{ \frac{\partial^2 \rho}{\partial \zeta_1^3} + \frac{\partial^2 \rho}{\partial \zeta_3^3} \right\}. \tag{3.8.4}$$

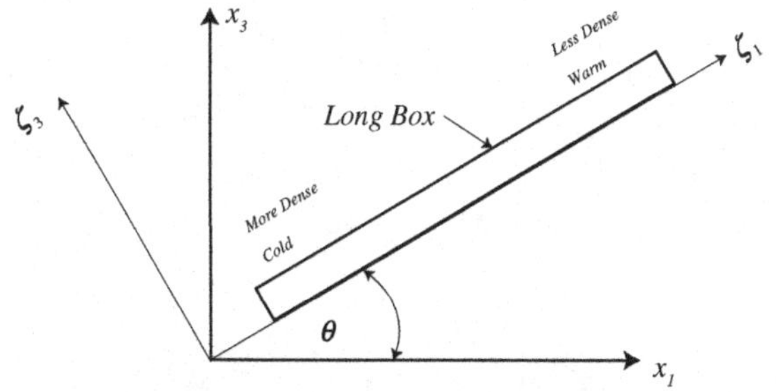

Figure 3.8.1 Flow in a tilted long box, the flow being supported by a stable temperature gradient along the box.

Following §3.7, we eliminate the pressure from (3.8.2) and (3.8.3) by cross-differentiation:

$$0 = -g\sin(\theta)\frac{\partial\rho}{\partial\zeta_3} + g\cos(\theta)\frac{\partial\rho}{\partial\zeta_1} + \mu\frac{\partial^3 u_1}{\partial\zeta_3^3}, \tag{3.8.5}$$

which is the counterpart to (3.7.7 with $\theta = 0$). Once again we make use of the flow symmetry and seek a solution of the form:

$$\rho = -C\zeta_1 + f(\zeta_3), \tag{3.8.6}$$

where C is a linear stable density gradient.

Substituting (3.8.6) into (3.8.5) leads to:

$$0 = -g\sin(\theta)\frac{\partial f}{\partial\zeta_3} - gC\cos(\theta) + \mu\frac{\partial^3 u_1}{\partial\zeta_3^3}. \tag{3.8.7}$$

Similarly, substituting (3.8.6) into (3.8.4) leads to:

$$-u_1(\zeta_3)C = \kappa_\rho\left\{\frac{\partial^2 f(\zeta_3)}{\partial\zeta_3^2}\right\}. \tag{3.8.8}$$

We now eliminate h from (3.8.7) and (3.8.8) by differentiating (3.8.7) with respect to ζ_3:

$$\frac{d^4 u_1(\zeta_3)}{d\zeta_3^4} + 4\beta^4 u_1(\zeta_3) = 0; \quad \beta^4 = \frac{gC}{4\kappa_\rho\mu}\sin(\theta), \tag{3.8.9}$$

that is an equation for the longitudinal velocity $u_1(\zeta_3)$ forming the counterpart to (3.7.7) and where we have used the ordinary derivative as $u_1(\zeta_3)$ is only a function of ζ_3. Equation (3.8.9) is a linear fourth order differential equation and so we seek a solution of the form:

$$u_1(\zeta_3) = \sum_{i=1}^{i=4} A_i e^{\lambda_i \zeta_3}. \tag{3.8.10}$$

This is a solution provided:

$$\lambda_i^4 = -4\beta^4 \tag{3.8.11}$$

so that:

$$\lambda_i = \sqrt{2}\beta e^{i\pi/4};\ \sqrt{2}\beta e^{i3\pi/4};\ \sqrt{2}\beta e^{i5\pi/4};\ \sqrt{2}\beta e^{i7\pi/4}. \tag{3.8.12}$$

These roots are shown diagrammatically in Fig. 3.8.2.

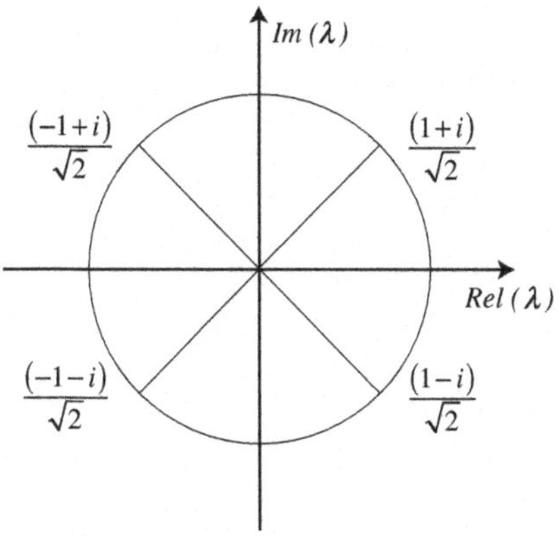

Figure 3.8.2 Fundamental roots of equation (3.8.12).

Substituting this into (3.8.9) leads to the general set of solutions:

$$u_1(\zeta_3) = A_1 e^{\beta\zeta_3} e^{i\beta\zeta_3};\ A_2 e^{\beta\zeta_3} e^{-i\beta\zeta_3};\ A_3 e^{-\beta\zeta_3} e^{i\beta\zeta_3};\ A_4 e^{-\beta\zeta_3} e^{-i\beta\zeta_3} \tag{3.8.14}$$

Now boundary condition at the two plates requires that the velocity there is zero (no slip), the velocity is anti-symmetric about the mid-depth of the cavity and that there is no mass flux at the upper and lower boundaries:

$$\frac{\partial h}{\partial \zeta_3} = 0 \ at \ \zeta_3 = 0, d. \tag{3.8.15}$$

This suggests a solution of the form:

$$u(\zeta_3) = A\{\sin\beta\zeta_3 \sinh\beta(h-\zeta_3) - \sin\beta(h-\zeta_3)\sinh\beta\zeta_3\}, \tag{3.8.16}$$

where A is a constant and h is the cavity depth. Substituting (3.8.16) into (3.8.7) and noting that the mass flux at the two boundaries of the cavity must be zero determines the constant A_1:

$$A = \frac{2\kappa B}{\tan(\theta)(\sin\beta h + \sinh\beta h)}. \tag{3.8.17}$$

Integrating (3.8.7) yields and expression for the density function $h(\zeta_3)$.

The configuration discussed in §3.7 may be retrieved by considering the limit as the slope tends to zero.

CHAPTER 4

Effect of Viscosity

Contents

The objective of this chapter is to introduce the reader to the influence of the fluid viscosity and emphasis is placed on two extremes, first, where viscosity dominate the flow dynamics and second, where viscous effects are confined to a thin boundary layer.

4.1. FLOW AROUND A SPHERE AT LOW REYNOLDS NUMBER

Consider flow about a body (sphere for definiteness) as shown in Fig. 4.1.1 and suppose U is the incoming speed of the fluid, L is the dimension of the body (for the case of a sphere, the diameter of the sphere), ν is the kinematic viscosity of the fluid that is assumed to be homogeneous with a density ρ_0.

For convenience, we repeat here the momentum equations (2.12.10) using the summation notation in order to simplify the writing of the equations:

$$\rho_0\left(\frac{\partial v_i}{\partial t}+v_j\frac{\partial v_i}{\partial x_j}\right) = -\frac{\partial p'}{\partial x_i}+\mu\frac{\partial^2 v_i}{\partial x_j \partial x_j}, \tag{4.1.1}$$

Environmental Fluid Dynamics
ISBN 978-0-12-088571-8, DOI: 10.1016/B978-0-12-088571-8.00004-8

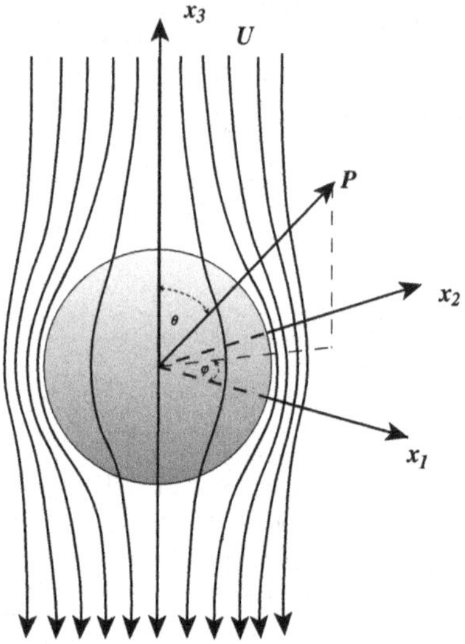

Figure 4.1.1 Flow around a sphere.

where p' is the pressure varion due to the motion, over and above the hydrostatic pressure. Given the scales of the problem, outlined above, we can introduce new non-dimensional variables:

$$v_i^* = \frac{v_i}{U}; \quad p^* = \frac{p'}{P_0}; \quad t^* = \frac{t}{T}; \quad x_i^* = \frac{x_i}{L}, \tag{4.1.2}$$

where P_0 is an unknown pressure scale and T is a time scale, then (4.1.1) becomes:

$$\rho_0 \frac{U}{T}\frac{\partial v_i^*}{\partial t^*} + \frac{\rho_0 U^2}{L} v_j^* \frac{\partial v_i^*}{\partial x_j^*} = -\frac{P_0}{L}\frac{\partial p^*}{\partial x_i^*} + \frac{\mu U}{L^2}\frac{\partial^2 v_i^*}{\partial x_j^* \partial x_j^*}. \tag{4.1.3}$$

For slow flows of a viscous fluid the dominant term in (4.1.3) is the viscous shear stress, the last term in (4.1.3), so we multiply (4.1.3) by $L^2/\mu U$ in order to make the last term O(1):

$$\frac{L^2}{\nu T}\frac{\partial v_i^*}{\partial t^*} + \frac{UL}{\nu} v_j^* \frac{\partial v_i^*}{\partial x_i^*} = \frac{-P_0 L}{\mu U}\frac{\partial p^*}{\partial x_i^*} + \frac{\partial^2 v_i^*}{\partial x_j^* \partial x_j^*}, \tag{4.1.4}$$

Suppose the velocity varies only on a time scale $T = \frac{L}{U}$, that is the advective time scale and the Reynolds number for the flow around the body is given by:

$$\mathrm{Re} = \frac{UL}{\nu} \ll 1, \tag{4.1.5}$$

then if we set:

$$P_0 = \frac{\nu U}{L}, \tag{4.1.6}$$

so that the pressure term becomes of the same magnitude as the viscous term, (4.1.4) becomes:

$$\mathrm{Re}\left(\frac{\partial v_i^*}{\partial t^*} + v_j^* \frac{\partial v_i^*}{\partial x_j^*}\right) = -\frac{\partial p^*}{\partial x_i^*} + \frac{\partial^2 v_i^*}{\partial x_j^* \partial x_j^*}. \tag{4.1.7}$$

Assuming that the above scaling is correct then, if $\mathrm{Re} \ll 1$, the first order equation for the flow is given by:

$$0 = -\frac{\partial p^*}{\partial x_i^*} + \frac{\partial^2 v_i^*}{\partial x_j^* \partial x_j^*}; \quad i = 1, 2, 3. \tag{4.1.8}$$

We may eliminate the pressure gradient term by introducing the vorticity ζ_i:

$$\zeta_i = e_{ijk} \frac{\partial v_k^*}{\partial x_j^*}. \tag{4.1.9}$$

This is most easily done by multiplying (4.1.8) by e_{ijk} and then taking the derivative with respect to x_j^*. This yields:

$$\frac{\partial^2 \zeta_i}{\partial x_j^* \partial x_j^*} = 0. \tag{4.1.10}$$

since

$$e_{ijk} \frac{\partial^2 p^*}{\partial x_k^* \partial x_j^*} = 0. \tag{4.1.11}$$

Equation (4.1.10) has a simple interpretation if we compare it with the diffusion equation (2.9.7); the vorticity is steady with no sources of vorticity within the flow field. Vorticity is generated at the surface of the sphere and diffuses out to infinity. This scaling procedure and the conclusions derived from it about the diffusion processes are valid for any shaped body with a slow viscous fluid moving about it. For the case of a sphere we now derive an analytical solution to (4.1.10).

For the spherical body, we may note that the flow would clearly be independent of the azimuthal coordinate φ so it is advantageous to introduce a spherical polar coordinate system (r, θ, φ) as shown in Fig. 4.1.1. In such a coordinate system, the conservation of mass equation becomes (see Appendix):

$$\frac{\partial v_i^*}{\partial x_i^*} = \frac{1}{r^2}\frac{\partial}{\partial r}(r^2 v_r) + \frac{1}{r\sin\theta}\frac{\partial}{\partial\theta}(v_\theta \sin\theta) = 0, \tag{4.1.12}$$

so that we may introduce a stream function $\psi(r, \theta)$ such that:

$$v_r = \frac{1}{r^2\sin\theta}\frac{\partial\psi}{\partial\theta};\; v_\theta = \frac{-1}{r\sin\theta}\frac{\partial\psi}{\partial r}, \tag{4.1.13}$$

where v_r is the radial component of the velocity and v_θ is the velocity component in the θ direction.

The solution for ψ may now be obtained by noting that the only non-zero component of vorticity is in the azimuthal direction φ:

$$\zeta_\varphi = \frac{1}{r}\left\{\frac{\partial}{\partial r}(rv_\theta) - \frac{\partial v_r}{\partial\theta}\right\}. \tag{4.1.14}$$

Substituting from (4.1.13) into (4.1.14) and then substituting the result into (4.1.10) leads to an equation for ψ:

$$\begin{aligned}&\frac{1}{r^2}\frac{\partial}{\partial r}\left\{r^2\frac{\partial}{\partial r}\left[\frac{1}{r}\left(\frac{\partial}{\partial r}\left(-\frac{1}{\sin\theta}\frac{\partial\phi}{\partial r}\right)\right)-\frac{\partial}{\partial\theta}\left(\frac{1}{r^2\sin\theta}\frac{\partial\phi}{\partial\theta}\right)\right]\right\}\\&+\frac{1}{r^2\sin\theta}\left\{\frac{\partial}{\partial\theta}\left(\sin\theta\left[\frac{1}{r}\left(\frac{\partial^2}{\partial\theta\,\partial r}\left(-\frac{1}{\sin\theta}\frac{\partial\phi}{\partial r}\right)\right.\right.\right.\right.\\&\left.\left.\left.\left.-\frac{\partial^2}{\partial\theta^2}\left(\frac{1}{r^2\sin\theta}\frac{\partial\phi}{\partial\theta}\right)\right)\right]\right)\right\}=0.\end{aligned} \tag{4.1.15}$$

Direct substitution shows the solution to this equation is given by:

$$\varphi = Ur^2\left(\frac{3}{4}\frac{a}{r} - \frac{1}{4}\frac{a^3}{r^3} - \frac{1}{2}\right)\sin^2\theta, \tag{4.1.16}$$

where a is the radius of the sphere. In advanced books on slow viscous flow this solution is interpreted in terms of particular flow singularities. Here, it suffices to note that the velocities from (4.1.13) both decrease as r^{-1}.

The solution (4.1.6) is valid for the special case where the Reynolds number is low enough so that viscous forces balance the pressure forces pushing the fluid past the stationary sphere. A simple example of such a flow is a small silt particle or an algal cell settling in a lake, river or estuary. Clearly such a particle will fall or rise under gravity at a velocity where the gravity force, minus buoyancy, just balances the drag exerted by viscous stress acting on the surface of the particle.

We may use the above solution to calculate this drag force. However, we must remember that we have also retained the pressure force in our simplified flow model, so we must calculate both the viscous drag and the pressure drag on the spherical surface. In order to do this we must first derive an expression for the surface shear stress acting on the surface of the sphere ($r = a$) by noting that the spherical polar equivalent to (2.11.8) is given by:

$$\tau_\theta = \frac{\mu}{r}\frac{\partial v_\theta}{\partial \theta}\bigg|_{r=a}. \tag{4.1.17}$$

The pressure acting on the surface of the sphere may be derived from the spherical polar equivalent of (4.1.8) and then setting $r = a$:

$$p = p_\infty - \frac{3}{2}\frac{\mu U}{r^2}\cos\theta. \tag{4.1.18}$$

The pressure is normal to the surface of the sphere and p_∞ is the pressure at infinity. The total drag F_D exerted by the flow on the sphere is then given by:

$$\begin{aligned} F_D &= 2\pi \int_0^\pi (\tau_\theta r \sin\theta + pr\cos\theta)\,d\theta \\ &= 6\pi a \mu U. \end{aligned} \tag{4.1.19}$$

It is customary to define, what is called the drag coefficient C_D, such that:

$$C_D = \frac{F_D}{\frac{1}{2}\rho_0 U^2 \pi a^2} = \frac{24}{\mathrm{Re}}, \tag{4.1.20}$$

where

$$\mathrm{Re} = \frac{\mathrm{Ua}}{\nu}, \tag{4.1.21}$$

is the Reynolds number of the flow around the sphere.

Example

Consider a spherical silt particle with a radius $a = 3 \times 10^{-5}$ m and a density of $\rho_s = 2430.0$ kg m^{-3} sinking in estuarine water with a temperature of 20 °C and a salinity of 34.1 psu. Find the settling velocity.

Solution

Assume the particle has reached its terminal velocity where the gravity force acts downward and buoyancy and viscous forces act to retard the falling particle. Then from above:

$$\text{Drag force} = 6\pi\mu\text{Ua}$$

$$\text{Particle weight} = \frac{4}{3}\pi a^3 g\rho_s$$

$$\text{Particle buoyance force} = \frac{4}{3}\pi a^3 g\rho_w$$

At terminal velocity there is a balance:

$$U = \frac{4}{3}\frac{\pi a^3 g}{6\pi\mu a}(\rho_s - \rho_w).$$

Water with a temperature of 20 °C and a salinity of 34.1 psu has a density $\rho_w = 1024.1$ kg m^{-3}. Substituting these values into the expression for U, yields:

$$U = 3 \times 10^{-4}\ \text{ms}^{-1}.$$

To ensure consistency of the solution with our assumption of slow viscous flow we need to check the value of the Reynolds number:

$$\text{Re} = \frac{Ua}{\nu} = 0.01.$$

In such problems, it is always advisable to check the Reynolds number to make sure that it is indeed small as assumed in all of the above theoretical development.

4.2. POROUS MEDIA FLOW

When rainfall falls to the ground some of the water runs off into streams, some ponds or is intercepted by the trees and evaporates and some penetrates the ground. The component that penetrates the ground moves through the soil matrix under the influence of gravity and often travels many hundreds of kilometers to ultimately be stored in subterranean basins or in some cases it may vent into a stream or the ocean. The velocities associated with this water movement are usually very small and also the soil pore sizes are small, so that the Reynolds number, based on the pore velocity and pore size is small allowing the same low Reynolds number assumption to be made as in §4.1. Suppose we consider, as shown in Fig. 4.2.1, a cubic control volume V and side areas A, large enough to enclose many sand grains, but smaller than the length scale of the mean flow variations throughout the porous media.

The water will move slowly through the interstices in the soil matrix as shown in Fig. 4.2.1 and when summed over the area A, will lead to a discharge, Q, through the area A. The water velocity, $\{v_i\}$, through the pores will be larger than the discharge velocity:

$$u_d = \frac{Q}{A}, \tag{4.2.1}$$

but the discharge velocity, u_d, may conveniently be used as a scale for the interstitial velocity and with this we may define a representative Reynolds number:

$$\mathrm{Re} = \frac{du_d}{\nu}, \tag{4.2.2}$$

d is the length scale of the soil grains. In what follows, we assume $\mathrm{Re} \ll 1$.

In order to find a general resistance law for the flow u_d, consider first a single grain size as shown in Fig. 4.2.2.

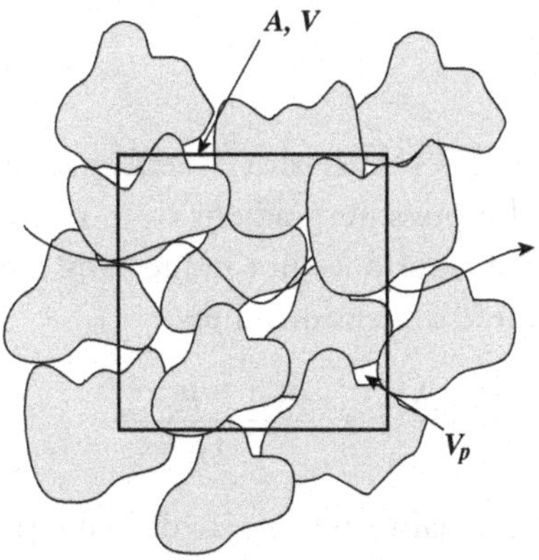

Figure 4.2.1 Schematic of the structure of a porous media such as sand or silt.

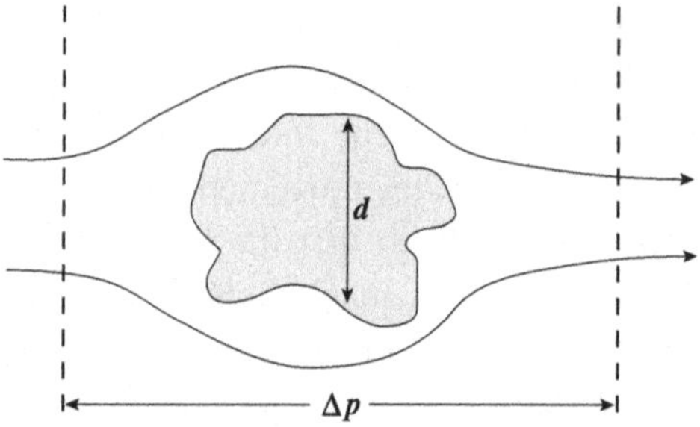

Figure 4.2.2 Flow past a single soil grain in the soil matrix.

In §4.1, we saw that the total pressure and viscous drag force, F_{D} on such a grain is given by:

$$F_{\mathrm{D}} \sim \mu u_d d, \tag{4.2.3}$$

where we have assumed that the pore water velocity $\{v_i\}$ scales with the discharge velocity u_d, d is the grain size and μ is the viscosity of the water moving through the pores. If we assume that the soil matrix is fixed in space then this drag force must be overcome by the pressure gradient pushing the water through the soil matrix. Suppose we let Δp be the pressure differential across the soil volume as shown in Fig. 4.2.1, then in order for the pressure force to balance the viscous drag we must have:

$$\Delta p d^2 \sim \mu u_d d. \tag{4.2.4}$$

This yields an estimate for the pore velocity scale u_d in terms of the applied pressure:

$$u_d \sim \frac{\Delta p d}{\mu}. \tag{4.2.5}$$

Now the pressure differential that is effective in pushing the fluid through the pores is the pressure gradient over and above the hydrostatic pressure gradient, as the latter does not induce any motion. Further, let us generalize (4.2.5) to three dimensions then

$$\frac{\Delta p_i}{d} \rightarrow \frac{\partial(p + \rho g x_3)}{\partial x_i}, \tag{4.2.6}$$

where $(p + \rho g x_3)$ is the dynamic plus the hydrostatic pressure, or put simple the net pressure over and above the hydrostatic pressure. Substituting (4.2.6) into (4.2.5) leads to a scale for the pore discharge velocity u_d in the x_i direction:

$$u_{di} \sim \frac{-d^2}{\mu} \frac{\partial(p + \rho g x_3)}{\partial x_i}. \tag{4.2.7}$$

We now introduce the pressure head h:

$$h = \frac{p + \rho g x_3}{\rho g}. \tag{4.2.8}$$

Substituting (4.2.8) into (4.2.7) leads to a relationship for the local discharge velocity:

$$u_{di} = -\left(\frac{\beta g d^2 A_v}{\nu A}\right) \frac{\partial h}{\partial x_i} = -k \frac{\partial h}{\partial x_i}, \tag{4.2.9}$$

where

$$k = \left(\frac{\beta g d^2 A_v}{\nu A}\right), \tag{4.2.10}$$

is the coefficient of permeability of the porous media and β is a numerical coefficient of O(1). The relationship (4.2.9) is called D'Arcy's (1856) Law in honor of D'Arcy who first derived this relationship. The value of k may be found in many books, Table 4.2.1 shows a few representative values.

In order to obtain a single equation of motion, we must also use the condition that the water flowing through the pores is incompressible. Repeating (2.11.1) here

$$v_{i,i} = 0, \tag{4.2.11}$$

and noting that:

$$u_{di} \sim \frac{A_v}{A} v_i \tag{4.2.12}$$

Table 4.2.1 Typical Values for the Coefficient of Permeability

Soil type	k: Coefficient of permeability m s^{-1}
Clean gravel	10^{-2}
Coarse sand	10^{-2} to 10^{-4}
Fine sand	10^{-4} to 10^{-5}
Silt	10^{-5} to 10^{-7}

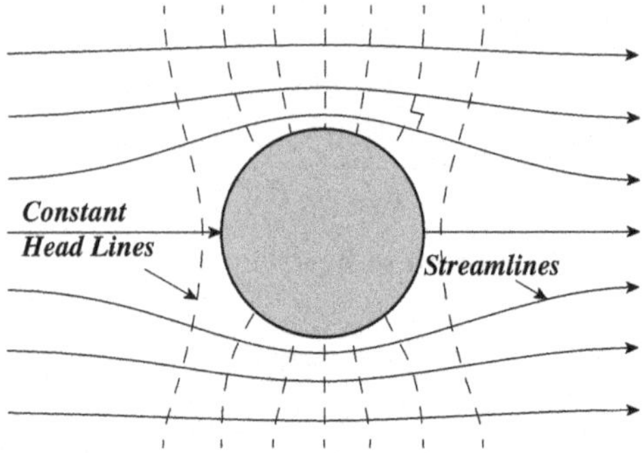

Figure 4.2.3 Flow in porous media around an impervious cylinder.

then provided $\frac{A_v}{A}$ is locally constant (4.2.11) becomes:

$$u_{di,i} = 0. \tag{4.2.13}$$

Combining (4.2.13) with (4.2.9) leads to the single equation for the pressure head h:

$$\frac{\partial^2 h}{\partial x_1^2} + \frac{\partial^2 h}{\partial x_2^2} + \frac{\partial^2 h}{\partial x_3^2} = 0, \tag{4.2.14}$$

which is the Laplace equation. Further, it follows immediately that the vorticity,

$$\zeta_i = e_{ijk}\frac{\partial u_k}{\partial x_j} = -ke_{ijk}\frac{\partial^2 h}{\partial x_k \partial x_j} = 0. \tag{4.2.15}$$

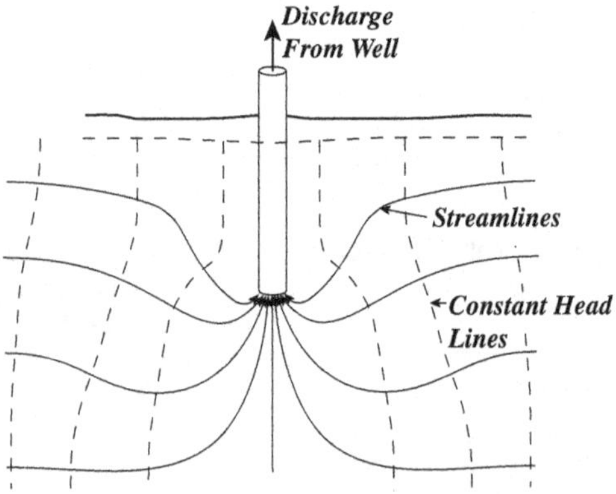

Figure 4.2.4 Flow into a well from which a discharge of Q is being withdrawn.

Flows that have the property that the vorticity is zero are called irrotational; flow through a porous media is one such flow. From (4.2.9) is follows that the velocity is perpendicular to the constant pressure or constant head surface. As an example we have sketched, in Fig. 4.2.3, streamlines for flow around an impervious cylinder embedded in a porous media.

Another very common such flow is flow into a ground water well shown in Fig. 4.2.4.

4.3. BOUNDARY LAYER FLOW

So far we have discussed low Reynolds number flows around a sphere and flow through a porous media. In these two problems, we showed that vorticity is generated at the wall and then diffuses out into the flow domain. In low Reynolds number flows, diffusion dominates advection and so the fluid velocity does not change the vorticity field to any appreciable extent. However, for larger fluid velocities advection progressively influences the vorticity balance. To illustrate the role of advection, we now discuss the very simplest high Reynolds number flow problem; flow over a flat plate as shown in Fig. 4.3.1.

In order to define a Reynolds number, we need to add a well-defined length scale to this problem, because for the case of a semi-infinitely long plate the problem does not possess a length scale. This may be achieved by painting a line on the plate and letting the distance from the leading edge of the plate to the painted line be length L as shown in Fig. 4.3.1.

We can then define a Reynolds number

$$\mathrm{Re} = \frac{UL}{\nu} \tag{4.3.1}$$

In what follows, we shall assume that $\mathrm{Re} \gg 1$ and use this assumption to simplify the equations of motion in order to both be able to solve the

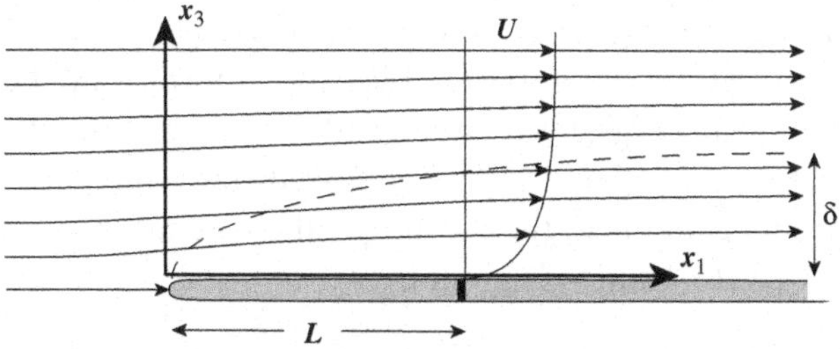

Figure 4.3.1 Parallel flow impinging on a collinear infinitely thin flat plate.

equations, and also to provide a fundamental understanding of the dynamics that cover such large Reynolds number flows. Clearly, with this definition of the length scale L, the discussion following will apply to downstream of our arbitrarily painted line.

However, before we begin with a more formal analysis of the problem shown in Fig. 4.3.1, we first use some simple scaling arguments to elucidate the dynamics of the problem. Inspection of the flow, shown in Fig. 4.3.1, clearly reveals that upstream of the plate, the flow is parallel and uniform; in brief the flow does not know anything about the presence of the plate downstream. The vorticity of this parallel flow is zero; upstream of the plate the flow possess no vorticity. On the other hand, as shown in Fig. 4.3.1, there is clearly vorticity associated with the velocity profile over the plate downstream of the leading edge. In Fig. 4.3.1, we show only the upper half of the plate, but a symmetric velocity profile would form on the underside of the plate. So where is the origin of this vorticity? In §2.12, we showed that in a homogeneous fluid vorticity can only be generated at a solid wall and then only if there is a pressure gradient along the wall. A little distance away from the leading edge of the plate the streamlines are essentially parallel so there is no pressure gradient along the surface of our plate except near the very leading edge.

This means that the vorticity must originate at the leading edge of the plate, where the flow must accommodate the presence of the plate, no matter how thin the plate is, and make its way around the leading edge. Hence the picture that emerges is that vorticity free flow strikes the plate, acquires a small amount of vorticity at the leading edge and then the flow advects this vorticity over the plate. For large Reynolds number flows, the advection is much faster than the rate at which vorticity diffuses away from the plate so the layer, shown in Fig. 4.3.1, will be quite thin. The layer may be defined as the thickness of fluid in which the velocity ranges from zero at the plate to a constant fraction (usually taken as 99%) of the free stream velocity U. Alternatively we may define the layer as that fluid layer in which the vorticity changes from a maximum at the plate to a certain fraction (usually taken as 1%) of that maximum. This layer is called the "boundary layer" and is the layer that is formed by the vorticity acquired at the leading edge, being swept, with a velocity U, over the plate by the fluid and diffusing in the x_3 as it progresses downstream.

With this model in mind, the distance x_1 the vorticity, generated at the leading edge, travels in time t is given by:

$$x_1 \sim Ut \tag{4.3.2}$$

Now in time t the vorticity diffuses, transversely to the plate, a distance δ, given by:

$$\delta = (\nu t)^{\frac{1}{2}}. \tag{4.3.3}$$

Eliminating t from (4.3.2) and (4.3.3) leads to an expression for the boundary layer thickness:

$$\frac{\delta}{x_1} \sim \frac{\nu}{(UL)^{\frac{1}{2}}}\left(\frac{L}{x_1}\right)^{\frac{1}{2}} \sim \frac{1}{\mathrm{Re}^{\frac{1}{2}}}(x_1^*)^{-\frac{1}{2}}, \tag{4.3.4}$$

where

$$x_1^* = \frac{x_1}{L}. \tag{4.3.5}$$

We see from this simple scaling analysis that the thickness of the boundary layer decrease, as the half power, with the Reynolds number and the distance downstream (local Reynolds number). What is the role of this boundary layer? Clearly the flow outside the boundary layer is essentially independent of the fluid viscosity and so is called the "outer flow". On the other hand, the flow inside the boundary layer is where the viscosity of the fluid reduced the velocity of the fluid from the "free stream" velocity U to zero at the plate surface required by the no slip condition; the flow within the boundary layer region is called the inner flow. We would thus expect the equations describing the flow in the outer flow region to be independent of viscosity and the flow in the inner region being a balance between viscous forces and streamwise inertia, or in terms of vorticity a balance between transverse diffusion and streamwise advection of vorticity.

We are now ready to try to solve for the flow over a flat plate as shown in Fig. 4.3.1. First, consider the momentum equations (2.11.10) that are repeated here for convenience in a form applicable to the flat plate problem, where the flow is steady, non-rotating and the fluid is of constant density:

$$v_1\frac{\partial v_1}{\partial x_1} + v_3\frac{\partial v_1}{\partial x_3} = -\frac{1}{\rho_0}\frac{\partial p}{\partial x_1} + \nu\left(\frac{\partial^2 v_1}{\partial x_1^2} + \frac{\partial^2 v_1}{\partial x_3^2}\right), \tag{4.3.6}$$

$$v_1\frac{\partial v_3}{\partial x_1} + v_3\frac{\partial v_3}{\partial x_3} = -\frac{1}{\rho_0}\frac{\partial p}{\partial x_3} - g + \nu\left(\frac{\partial^2 v_3}{\partial x_1^2} + \frac{\partial^2 v_3}{\partial x_3^2}\right). \tag{4.3.7}$$

Now similarly the conservation of mass equation follows from (2.11.3):

$$\frac{\partial v_1}{\partial x_1}+\frac{\partial v_3}{\partial x_3}=0. \qquad (4.3.8)$$

Following the example in §4.2, we scale the equations in order to eliminate small terms and so simplify the equations before we try to solve them. Given our discussion above of the inner and outer flow, we expect different simplifications to apply in each region and so we divide the discussion, commencing with the outer flow region:

$$x_1 \sim L; \quad x_3 \sim L; \quad v_1 \sim U; \quad v_3 \sim U; \quad p \sim U^2, \qquad (4.3.9)$$

and so introduce the following new variables:

$$x_1^* = \frac{x_1}{L}; \quad x_3^* = \frac{x_3}{L}; \quad v_1^* = \frac{v_1}{U}; \quad v_3^* = \frac{v_3}{U}; \quad p^* = \frac{(p-p_h)}{\rho U^2}, \qquad (4.3.10)$$

where $p_h =$ the hydrostatic pressure. This leads to:

$$v_1^*\frac{\partial v_1^*}{\partial x_1^*}+v_3^*\frac{\partial v_1^*}{\partial v_3^*} = -\frac{\partial p^*}{\partial v_1^*}+\frac{1}{\mathrm{Re}}\left(\frac{\partial^2 v_1^*}{\partial x_1^{*2}}+\frac{\partial^2 v_1^*}{\partial x_3^{*2}}\right), \qquad (4.3.11)$$

$$v_1^*\frac{\partial v_3^*}{\partial x_1^*}+v_3^*\frac{\partial v_3^*}{\partial v_3^*} = -\frac{\partial p^*}{\partial v_3^*}+\frac{1}{\mathrm{Re}}\left(\frac{\partial^2 v_3^*}{\partial x_1^{*2}}+\frac{\partial^2 v_3^*}{\partial x_3^{*2}}\right), \qquad (4.3.12)$$

and

$$\frac{\partial v_1^*}{\partial x_1^*}+\frac{\partial v_3^*}{\partial x_3^*}=0. \qquad (4.3.13)$$

Now if $\mathrm{Re} \gg 1$ and the scaling has correctly captured the magnitudes of each term in the equations of motion, then (4.3.11)–(4.3.13) reduce to:

$$v_1^*\frac{\partial v_1^*}{\partial x_1^*}+v_3^*\frac{\partial v_1^*}{\partial v_3^*} = -\frac{\partial p^*}{\partial x_1^*}, \qquad (4.3.14)$$

$$v_1^*\frac{\partial v_3^*}{\partial x_1^*}+v_3^*\frac{\partial v_3^*}{\partial v_3^*} = -\frac{\partial p^*}{\partial x_3^*}, \qquad (4.3.15)$$

$$\frac{\partial v_1^*}{\partial x_1^*}+\frac{\partial v_3^*}{\partial x_3^*}=0. \qquad (4.3.16)$$

Since the scaling has eliminated the viscous terms we have, by default, eliminated the boundary layer and so also the ability of the flow to satisfy the no slip condition at the plate. Hence we need to drop the condition: $v_1^*(x_1, 0) = 0$.

A simple solution satisfying equations (4.3.14)–(4.3.16) is given by:

$$v_1^*(x_1^*, x_3^*) = 1 \Rightarrow v_1(x_1, x_3) = U, \tag{4.3.17}$$

$$p^*(x_1^*, x_3^*) = 0 \Rightarrow p(x_1, x_3) = 0, \tag{4.3.18}$$

$$v_3^*(x_1^*, x_3^*) = 0 \Rightarrow v_3(x_1, x_3) = 0. \tag{4.3.19}$$

This is the outer solution and represents parallel flow past the plate; in the limit of large Reynolds number the boundary layer is infinitesimally thin and the outer flow does not "feel" the plate.

However, we know that this does not satisfy the condition of no slip at the boundary. To satisfy the no slip boundary condition, we must allow for a thin boundary layer adjacent to the plate where viscosity can bring the free stream velocity U to zero at the plate as explained above.

Near the plate we may assume:

$$x_1 \sim L; \quad x_3 \sim \delta; \quad v_1 \sim U; \quad v_3 \sim W, \tag{4.3.20}$$

where W is, as yet, an unknown velocity scale for the vertical velocity v_3. In the boundary layer this may be small, but cannot be zero, as the flow must move away from the plate in order to accommodate the smaller volume flux through the boundary layer, as we move downstream (see Fig. 4.3.1).

We can see this by examining the conservation of volume equation (4.3.8):

$$\frac{\partial v_1^*}{\partial x_1^*} + \frac{WL}{U\delta} \frac{\partial v_3^{**}}{\partial x_3^{**}} = 0, \tag{4.3.21}$$

where

$$x_1^* = \frac{x_1}{L}; \quad x_3^{**} = \frac{x_3}{\delta}; \quad v_1^* = \frac{v_1}{U}; \quad v_3^{**} = \frac{v_3}{W}. \tag{4.3.22}$$

In order that (4.3.21) can remain satisfied, the second term must remain order one, so that:

$$W = \frac{\delta}{L} U. \tag{4.3.23}$$

Once again we let:

$$p^* = \frac{p}{\rho_0 u^2}, \tag{4.3.24}$$

and

$$\delta = \left(\frac{L\nu}{U}\right)^{\frac{1}{2}}, \tag{4.3.25}$$

and noting Re $\gg$ 1, then (4.3.6)–(4.3.8) become:

$$v_1^* \frac{\partial v_1^*}{\partial x_1^*} + v_3^{**} \frac{\partial v_1^*}{\partial x_3^{**}} = -\frac{\partial p^*}{\partial x_1^*} + \frac{\partial^2 v_1^*}{\partial x_3^{**2}}, \tag{4.3.26}$$

$$\frac{\partial p^*}{\partial x_3^{**}} = 0, \tag{4.3.27}$$

$$\frac{\partial v_1^*}{\partial x_1^*} + \frac{\partial v_3^{**}}{\partial x_3^{**}} = 0. \tag{4.3.28}$$

These are the boundary layer equations.

The boundary conditions for this inner problem follow simply from the scaling assumption. At the plate we must have no slip. Now $x_3^{**} \to \infty$ may be approached by letting the Reynolds number Re $\to \infty$ and keeping x_3 fixed, but arbitrarily small. This is seen by noting that:

$$x_3^{**} = \frac{x_3}{L}\frac{L}{\delta} = \frac{x_3 \mathrm{Re}^{\frac{1}{2}}}{L}. \tag{4.3.29}$$

Hence as Re $\to \infty$; x_3 fixed$\Rightarrow x_3^{**} \to \infty$. This means that that the correct outer boundary condition is given by:

$$v_1^{**}(x_1^*, \infty) = v_1^*(x_1^*, 0). \tag{4.3.30}$$

Vorticity Flux

The flux of vorticity across any vertical section is given by:

$$F_\zeta = \int_0^\infty \zeta_2 v_1 \mathrm{d}x_3, \tag{4.3.31}$$

now

$$\zeta_2 = \frac{\partial v_1}{\partial x_3} - \frac{\partial v_3}{\partial x_1} \sim O\left(\frac{u}{\delta}\right) - O\left(\frac{\delta U}{L^2}\right) \sim O\left(\frac{u}{\delta}\right) \quad \text{as } \delta \to 0, \tag{4.3.32}$$

so to first approximation:

$$\zeta_2 = \frac{\partial v_1}{\partial x_3}. \tag{4.3.33}$$

Substituting (4.3.33) into (4.3.31) leads to:

$$F_\zeta = \int_0^\infty v_1 \frac{\partial v_1}{\partial x_3}\, \mathrm{d}x_3 = \left.\frac{v_1^2}{2}\right|_0^\infty = \frac{U^2}{2}. \tag{4.3.34}$$

Hence the flux of vorticity along the plate is constant confirming that the vorticity comes from the leading edge, the assumption which we started with, showing the our arguments are consistent.

The solution to (4.3.26)–(4.3.28) forms one of the most important corner stones of high Reynolds number fluid mechanics.

First we note that (4.3.27) admits a solution in the form of a stream function $\psi(x_1^*, x_3^{**})$ where:

$$v_1^* = \frac{\partial \psi}{\partial x_3^{**}}, \tag{4.3.35}$$

$$v_3^{**} = -\frac{\partial \psi}{\partial x_1^{**}}. \tag{4.3.36}$$

Substituting (4.3.35) and (4.3.36) into (4.3.25) and (4.3.26) leads to an equation for $\psi(x_1^*, x_3^{**})$:

$$\frac{\partial \psi(x_1^*, x_3^{**})}{\partial x_1^*} \frac{\partial \psi(x_1^*, x_3^{**})}{\partial x_3^{**} \partial x_1^*} - \frac{\partial \psi(x_1^*, x_3^{**})}{\partial x_3^{**}} \frac{\partial \psi(x_1^*, x_3^{**})}{\partial x_3^{**} \partial x_3^{**}} = \frac{\partial \psi(x_1^*, x_3^{**})}{\partial x_3^{**}}, \tag{4.3.37}$$

with the boundary conditions:

$$x_3^{**} = 0 \quad \psi = 0; \quad \frac{\partial \psi}{\partial x_3} = 0; \quad \frac{\partial \psi}{\partial x_1} = 0, \tag{4.3.38}$$

and

$$x_3^{**} \to \infty \quad \frac{\partial \psi}{\partial x_3} \to 1. \tag{4.3.39}$$

Table 4.3.1 Blasius Flow Profile

η	f	f'	f''
0.0	0.0	0.0	0.3320573362
1.0	0.1655717258	0.3297800312	0.3230071167
2.0	0.6500243699	0.629765736	0.2667515457
3.0	1.396808231	0.8460444437	0.1613603195
4.0	2.305746418	0.9555182298	0.0642341210
5.0	3.283273665	0.9915419002	0.01590679869

Now from (4.3.29) and (4.3.25), $\delta \sim (x_3^{**})^{\frac{1}{2}}$, hence we look for a solution of the form:

$$\frac{\psi}{\left(x_1^*\right)^{\frac{1}{2}}} = f\left(\frac{x_3^{**}}{\left(x_1^*\right)^{\frac{1}{2}}}\right), \tag{4.3.40}$$

and let

$$\eta = \frac{x_3^{**}}{\left(x_1^*\right)^{\frac{1}{2}}}. \tag{4.3.41}$$

Substituting (4.3.40) into (4.3.37) yields an equation for f:

$$2\frac{d^3 f}{d\eta^3} + f\frac{d^2 f}{d\eta^2} = 0 \tag{4.3.42}$$

$$f(0) = f'(0) = 0; \quad f'(\infty) = 1. \tag{4.3.43}$$

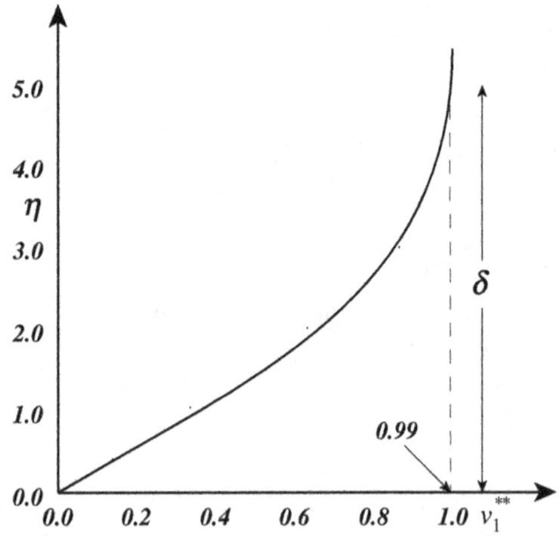

Figure 4.3.2 The Blasius velocity profile of the velocity in a boundary layer adjacent to a flat plate with flow at high Reynolds number.

The solution to (4.3.42), found numerically, is tabulated in Table 4.3.1 and shown in Fig. 4.3.2.

This is called the Blasius profile. If we now define the boundary layer thickness to be the distance at which the velocity has reached 99% of the free stream velocity U, or where the vorticity has decreased to 1% of the value near the plate then, from Table 4.3.1 we see:

$$\frac{\delta}{x_1} = 5.0\frac{1}{(\mathrm{Re}_{x_1})^{\frac{1}{2}}} \tag{4.3.44}$$

Displacement Thickness

The displacement thickness is the thickness that the boundary layer presents to the upstream coming flow due to the velocity deficit as shown in Fig. 4.3.3.

The displacement thickness, Δ, is defined by equating the full flow to the velocity defect flow:

$$\rho_0 U\Delta = \int_0^\infty \rho_0(U - v_1)\mathrm{d}x_3. \tag{4.3.45}$$

Dividing by U leads to:

$$\Delta = \int_0^\infty \left(1 - \frac{v_1}{U}\right)\mathrm{d}x_3. \tag{4.3.46}$$

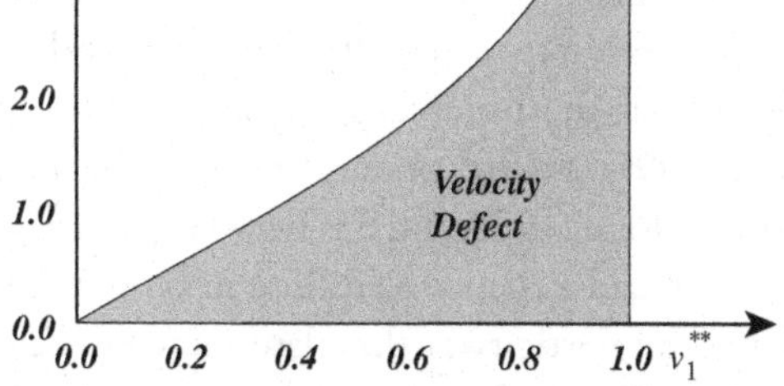

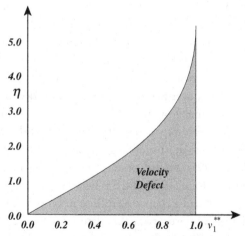

Figure 4.3.3 The shaded area represents the velocity defect due to the boundary layer flow.

Using the numerical solution of (4.3.42) (Table 4.3.1) in (4.3.45) provides the value for the displacement thickness:

$$\frac{\Delta}{x_1} = \frac{1.72}{\text{Re}}\left(\frac{L}{x_1}\right)^{\frac{1}{2}}. \tag{4.3.47}$$

Skin Friction

The viscous drag exerted by the flow on the plate is given by:

$$\tau_w = \mu \left.\frac{\partial v_1}{\partial x_3}\right|_{x_3=0}. \tag{4.3.48}$$

Once again substituting the numerically obtained solution yields:

$$\tau_w = \left(\frac{\mu U}{\delta}\right)\frac{\partial v_1^{**}}{\partial x_3^{**}} = (x_1^*)^{-\frac{1}{2}}\left(\frac{\mu U}{\delta}\right)\left.\frac{\partial^2 f}{\partial^2 \eta}\right|_{x_3=0} = 0.332\mu U\left(\frac{U}{\nu x_1}\right)^{\frac{1}{2}}, \tag{4.3.49}$$

(see Table 4.3.1) that shows that the drag on the plate has a singularity at $x_1 = 0$ and decays as the half power of x_1.

4.4. ADVERSE AND FAVORABLE PRESSURE GRADIENTS

In §4.3, we discussed the boundary layer flow over a flat plate with no pressure gradient. When the outer flow has a non-zero pressure gradient then the vertical momentum equation (4.3.27) implies that this pressure gradient is imposed on the flow in the inner region. Hence for a non-zero pressure gradient by §2.12 vorticity is generated at the plate and diffuses into the flow. Alternatively from a momentum point of view, for a pressure gradient that causes the outer flow to accelerate with downstream distance, the outer edge of the inner flow must also accelerate in order to match the outer flow. This requires a larger velocity jump across the boundary layer. The increased velocity at the outer edge means that the point where the velocity has reached 99% of the outer flow may be further away from the plate because the velocity differential is now larger or it could be closer if the momentum has not had a chance to diffuse into the inner flow. Similarly, in terms of vorticity, the vorticity at the plate increases due to the pressure gradient generating a vorticity flux across the plate. Hence, once again the boundary layer, the region containing 99% of the vorticity, can get larger if

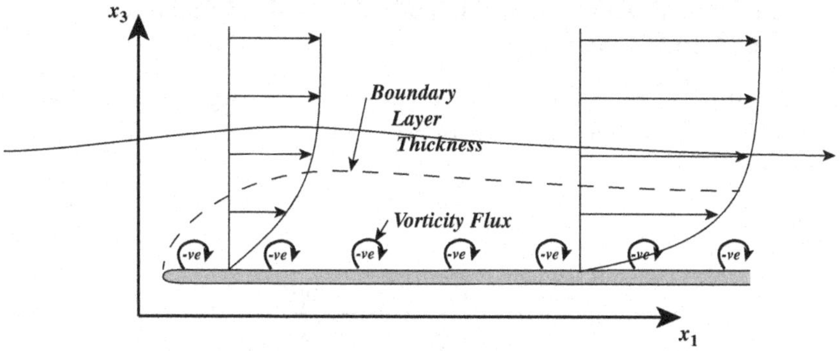

Figure 4.4.1 Boundary layer becomes thinner.

this increase vorticity diffuses away from the plate or it may get thinner if the vorticity at the plate increases faster than it can diffuse away. This scenario is illustrated in Fig. 4.4.1.

When the pressure gradient opposes the outer flow and the outer flow slows with downstream distance then is clear that the vorticity at the plate decreases and so the point where the vorticity is 1% of the value at the plate moves away from the plate; in other words there is a rapid thickening of the boundary layer. In the extreme case where the flow near the plate is reversed the flow will separate from the plate (see Fig. 4.4.1).

1. Favorable $\dfrac{\partial P}{\partial x} < 0 \quad \dfrac{\partial v_1}{\partial x_1} > 0$
2. Adverse Pressure Gradient (Fig. 4.4.2)

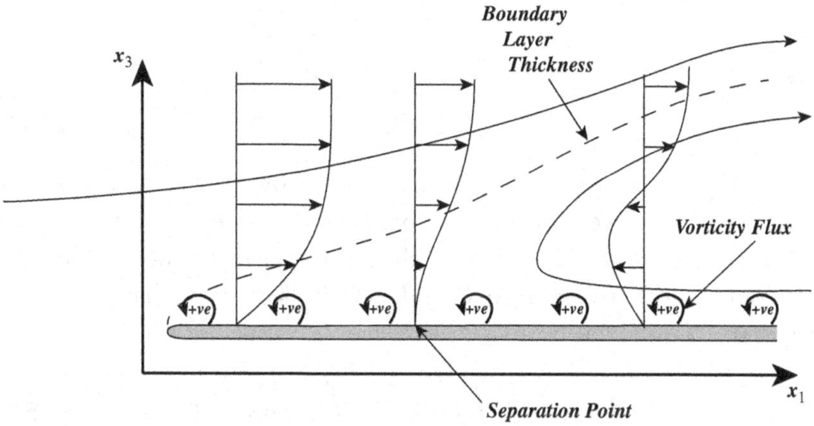

Figure 4.4.2 Boundary layer becomes thicker.

4.5. FLOW AROUND A CYLINDER WITH INCREASING REYNOLDS NUMBER

We saw in §4.4 that when the Reynolds number is large the equations of motion simplify to the Euler equations:

$$\frac{\partial v_i}{\partial x_i} = 0, \tag{4.5.1}$$

$$\frac{\partial v_i}{\partial t} + v_j \frac{\partial v_i}{\partial x_j} = -\frac{1}{\rho_0}\frac{\partial p}{\partial x_i}. \tag{4.5.2}$$

If we take the curl of (4.5.2), we get

$$\frac{D\zeta_i}{Dt} = \zeta_j \frac{\partial v_i}{\partial x_j}. \tag{4.5.3}$$

where ζ_i is the ith component of the vorticity and

$$\frac{D\zeta_i}{Dt} = \frac{\partial v_i}{\partial t} + v_j \frac{\partial v_i}{\partial x_j}, \tag{4.5.4}$$

is the rate of change of vorticity following a fluid particle (see §1.13).

Assuming the flow is again uniform at upstream infinity so that:

$$v_i = (U, -\infty, x_3), \tag{4.5.5}$$

then at infinity $\zeta_i = 0$ for all i, and thus combining this with (4.5.4) we see that:

$$\frac{D\zeta_i}{Dt} = 0, \tag{4.5.6}$$

so that:

$$\zeta_i \equiv 0, \tag{4.5.7}$$

the flow is irrotational everywhere. Such flows are called potential flows for reasons we will now explore. Let P be a fixed point in space and Q a point at x, From Stokes' theorem applied to the line integral along any path between P to Q and then back to P we have in vector notation:

$$\int_P^Q \boldsymbol{v}\cdot\hat{\boldsymbol{t}}\,ds + \int_Q^P \boldsymbol{v}\cdot\hat{\boldsymbol{t}}\,ds = \int_S \zeta\cdot\hat{n}\,dS = 0 \tag{4.5.8}$$

where $\hat{\boldsymbol{t}}$ is the unit tangential vector to the line along which the integration is taken in (4.5.6) and the surface S is that formed by the line PQP and $\hat{\boldsymbol{n}}$ is the unit normal to the surface S. This implies that we can thus define a function $\phi(\underset{\sim}{x})$ such that:

$$\phi(x) = \int_{P}^{Q(x)} \boldsymbol{v} \cdot \hat{\boldsymbol{t}}\, ds, \tag{4.5.9}$$

which is uniquely determined to within an arbitrary constant by the position x, independent of the path P to Q.

This function has the property, from Leibnitz' rule (see any undergraduate book on calculus),

$$v_i = \frac{\partial \phi}{\partial x_i}. \tag{4.5.10}$$

We call ϕ the velocity potential of the flow as the velocity may be derived from a simple gradient operation. Substituting (4.5.10) into the conservation of volume equation (4.5.1) leads to a single linear equation for ϕ:

$$\frac{\partial^2 \phi}{\partial x_1^2} + \frac{\partial^2 \phi}{\partial x_2^2} + \frac{\partial^2 \phi}{\partial x_3^2} = 0, \tag{4.5.11}$$

which is called the Laplace equation.

It is interesting to note, that the solution to the Euler equation, derived under the assumption of large Reynolds number is also a solution to the full Navier Stokes equation. This may be seen by substituting (4.5.10) into the neglected viscous term of the Navier Stokes equations:

$$\mu \frac{\partial}{\partial x_i}\left(\frac{\partial^2 \phi}{\partial x_j \partial x_j}\right) = 0. \tag{4.5.12}$$

This means that any irrotational flow solution is also a solution of the full Navier Stokes equations, including viscosity, but the solution does not satisfy the no slip condition at a wall, there we need a boundary layer.

For the case of 2D flow, we can introduce yet another function called the stream function. To see this note that for any function $\psi(x_1,\ x_3,\ t)$ conservation of volume (4.5.1) is satisfied if we define the velocities though the relationships:

$$v_1 = \frac{\partial \psi(x_1, x_3, t)}{\partial x_3} \quad \text{and} \quad v_3 = -\frac{\partial \psi(x_1, x_3, t)}{\partial x_1}. \tag{4.5.13}$$

Substituting (4.5.13) into (4.5.7):

$$\frac{\partial^2 \psi}{\partial x_1^2} + \frac{\partial^2 \psi}{\partial x_3^2} = 0, \tag{4.5.14}$$

This means that the outer flow of large 2D Reynolds number flow is governed by two linear equations (4.5.11) and (4.5.15). The solution defines the following:

$$\psi = \text{constant} - \text{streamlines}$$

$$\phi = \text{constant} - \text{potential lines}$$

the two being perpendicular to each other as shown in Fig. 4.5.1.

The important feature of (4.5.11) and (4.5.14) is that they are linear equations and we may use standard undergraduate mathematics to find solutions. To be specific, consider the outer flow around a cylinder. The Laplace equation (4.5.11) may be written in cylindrical polars (Appendix):

$$\nabla^2 \phi = \frac{\partial^2 \phi}{\partial x_1^2} + \frac{\partial^2 \phi}{\partial x_3^2} = \frac{1}{r}\frac{\partial}{\partial r}\left(r\frac{\partial \phi}{\partial r}\right) + \frac{1}{r^2}\frac{\partial^2 \phi}{\partial \theta^2} = 0. \tag{4.5.15}$$

Seek a solution of the form of separation of variables:

$$\phi = \phi_1(r)\phi_2(\theta) + \phi_{\text{uniform}}, \tag{4.5.16}$$

where

$$\phi_{\text{uniform}} = +Ur\cos\theta, \tag{4.5.17}$$

describes the velocity potential corresponding to the upstream uniform flow.

Substituting (4.5.16) into (4.5.15) and solving for ϕ_1 and ϕ_2 noting that for the uniform flow:

$$v_r = \frac{\partial \phi}{\partial r} = +U\cos\theta, \tag{4.5.18a}$$

$$v_\theta = \frac{1}{r}\frac{\partial \phi}{\partial \theta} = -U\sin\theta, \tag{4.5.18b}$$

leads to:

$$\frac{1}{r}\frac{\partial}{\partial r}\left(r\phi_2\frac{\partial \phi_1}{\partial r}\right) + \frac{1}{r^2}\phi_1\frac{\partial^2 \phi_2}{\partial \theta^2} = 0. \tag{4.5.19}$$

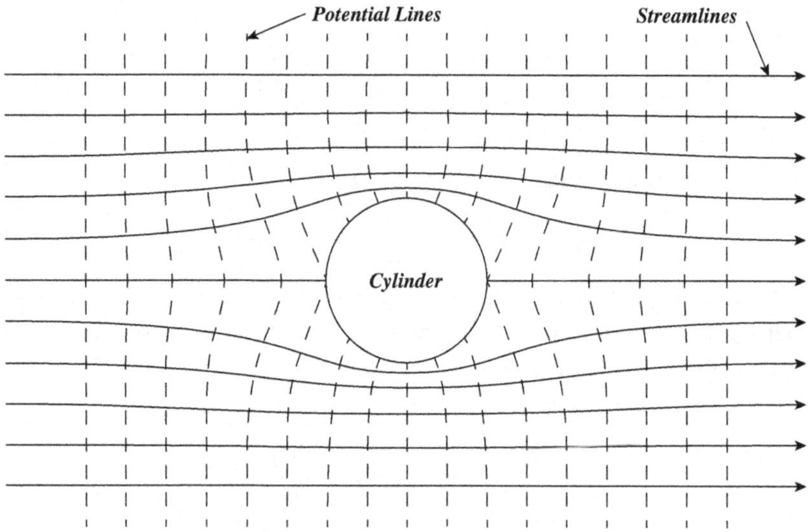

Figure 4.5.1 Potential lines and streamlines.

Separating variables

$$\frac{r}{\theta_1}\frac{\partial}{\partial r}\left(r\frac{\partial \phi_1}{\partial r}\right) = \kappa^2 \quad \text{and} \quad \frac{1}{\phi_{2_1}}\frac{\partial^2 \phi_2}{\partial \theta^2} = -\kappa^2, \tag{4.5.20}$$

$$\phi_2 = C\cos\kappa\theta + D\sin\kappa\theta, \tag{4.5.21}$$

and

$$r^2\frac{\mathrm{d}^2\phi_1}{\mathrm{d}r^2} + r\frac{\mathrm{d}\phi_1}{\mathrm{d}r} - \phi_1\kappa^2 = 0, \tag{4.5.22}$$

that is an equidimensional differential equation, the solution to which is:

$$\phi_1 = Ar^{-\kappa} + Br^{\kappa}. \tag{4.5.23}$$

(Hint let $r = e^t$, $t = \ell nr$)

Now at the cylinder boundary $u_r = 0 \quad r = a \Rightarrow D = 0, \quad \kappa = 1,$ so that:

$$\phi = Ur\cos\theta + \frac{Ua^2}{r}\cos\theta, \tag{4.5.24}$$

$$v_r = \frac{\partial \phi}{\partial r} = U\cos\theta - \frac{Ua^2}{r}\cos\theta, \tag{4.5.25}$$

and

$$v_r = \frac{1}{r}\frac{\partial \phi}{\partial \theta} = -U \sin\theta - \frac{Ua^2}{r^2}\sin\theta. \tag{4.5.26}$$

From this solution we see that when $r = a$,

$$v_r = 0 \quad \text{and} \quad v_\theta = -2U \sin\theta. \tag{4.5.27}$$

This means that there is a stagnation point at $r = a$ and $\theta = 0°$, $180°$; in other words the flow is perfectly symmetrical with respect to upstream and downstream as seen in Fig. 4.5.1.

Now Bernoulli's equation states.

$$\frac{p}{\rho g} = \text{const} - x_3 - \frac{(v_r^2 + v_\theta^2)}{2g} \tag{4.5.28}$$

that implies, by symmetry, that the net pressure force on the cylinder is zero. This contradiction can only be resolved by including the effects of the inner flow. This is difficult to do and we shall content ourselves with a heuristic discussion highlighting some key features. Once again let us define a Reynolds number Re given by:

$$\text{Re} = \frac{UD}{\nu}, \tag{4.5.29}$$

where D is the cylinder diameter. This may be rewritten in the form:

$$\text{Re} = \frac{D^2/\nu}{D/U} = \frac{\text{Diffusion} - \text{time-scale}}{\text{Advection} - \text{time-scale}} \tag{4.5.30}$$

showing that when the Reynolds number becomes large, advection is much faster than diffusion so that the vorticity generated at the cylinder wall is increasingly swept into the wake formed behind the cylinder. Second, inspection of the outer flow solution at the surface of the cylinder (4.5.27) shows that from the front stagnation point to the mid-section the outer flow is accelerating and so the pressure gradient is favorable and the boundary layer (the inner solution) may be expected to remain a very thin layer and has essentially no effect on the outer flow. However, after the flow reaches the widest section of the cylinder at $\theta = 90°$, the flow slows until it reaches the rear stagnation point at which point the velocity is zero and the pressure back to the stagnation pressure. The pressure gradient is thus adverse and we may expect the streamlines to separate at the cylinder vertex or shortly after and being a high Reynolds number flow the vorticity, introduced as the fluid moved over the front face of the cylinder in the favorable pressure gradient region, is swept into the wake. We may therefore describe the streamline pattern for the flow around a cylinder as a function of the Reynolds number:

Re: $0 < \text{Re} \leq 1$ (Fig. 4.5.2).

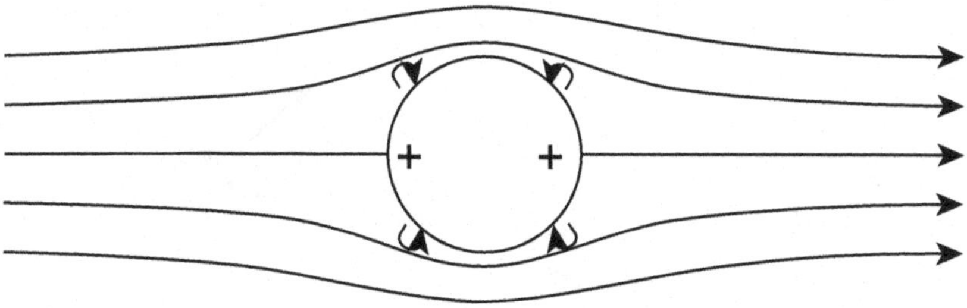

Figure 4.5.2 Diffusion much faster than advection, hence the flow is symmetric.

Re: $1 < \text{Re} \leq 10$ (Fig. 4.5.3).

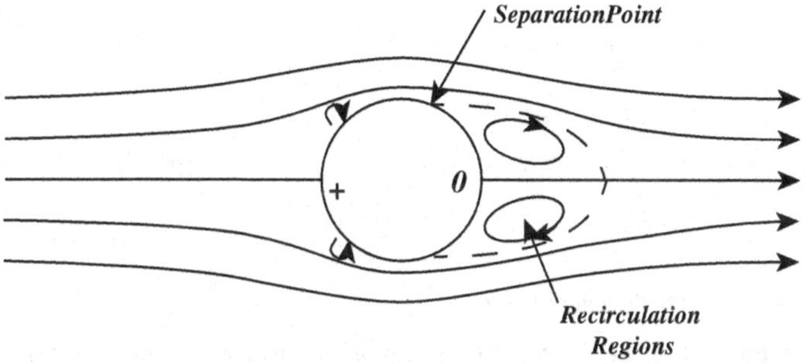

Figure 4.5.3 Wake forms as vorticity is swept back behind the cylinder. Excess vorticity behind the cylinder leads to closed streamlines.

Re: $10 < \text{Re} \leq 60$ (Fig. 4.5.4).

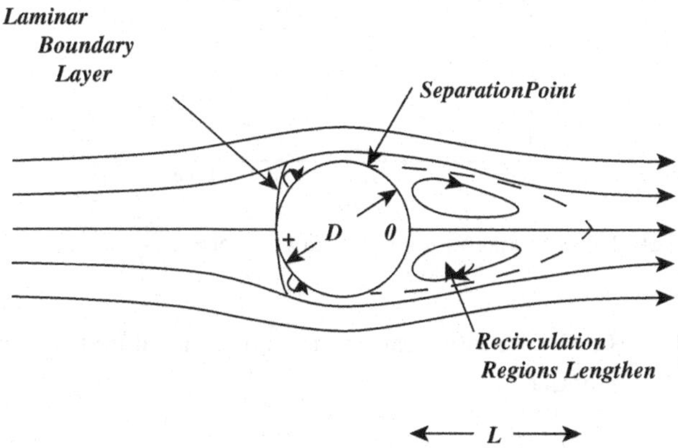

Figure 4.5.4 Advection dominates and boundary forms at the front of the cylinder.

Re: $60 < \text{Re} \leq 100$ (Figs. 4.5.5–4.5.6b).

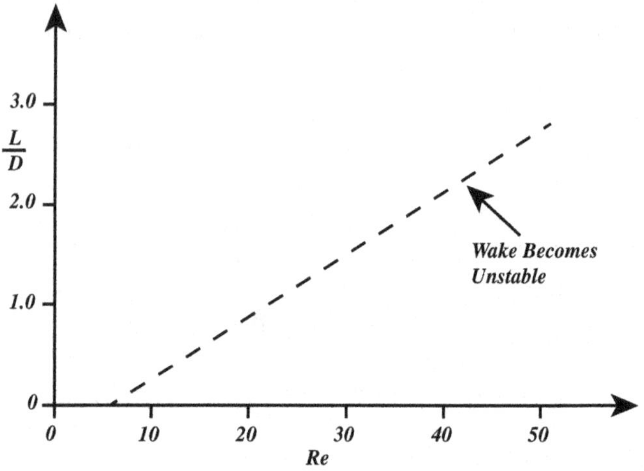

Figure 4.5.5 The ratio of the wake length to the cylinder diameter as a function of Re.

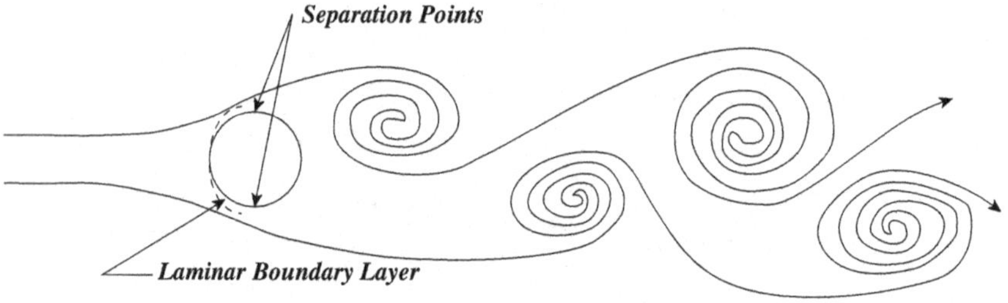

Figure 4.5.6a Eddies begin to shed from alternative sides of the cylinder with a frequency *f* that depends on Re.

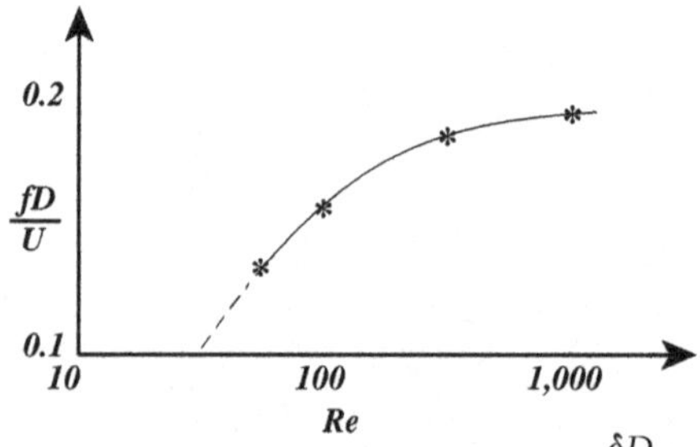

Figure 4.5.6b The Strouhal number as a function of Re. $\frac{\delta D}{U} = 0.198\left(1 - \frac{19.7}{\text{Re}}\right)$.

Re: $10^2 < \text{Re} \leq 10^5$. Flow pattern remains unchanged, but eddy shedding becomes irregular.

Re: $Re > 10^5$ (Fig. 4.5.7).

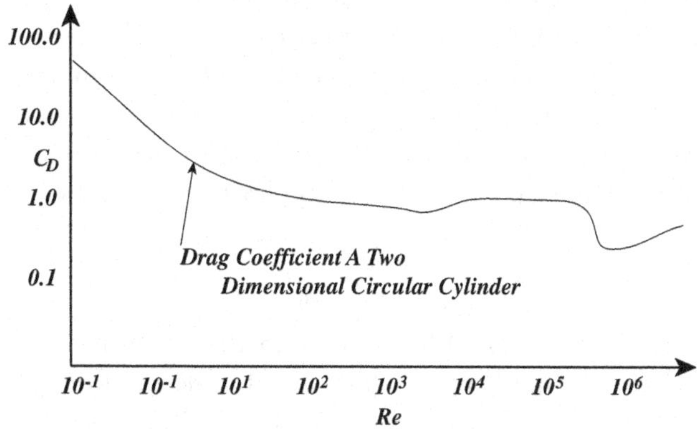

Figure 4.5.7 Boundary layer or font face becomes turbulent. Greater momentum transfer moves separation point to the rear.

In general, based on dimensional reasoning alone, we may write:

$$C_D = \frac{F_D}{\frac{1}{2}\rho U^2 A} = f(\text{Re}), \qquad (4.5.31)$$

where the dependence on the Reynolds number is shown in Fig. 4.5.7 and is completely determined by the behavior of the wake region. In the range $Re > 10^5$ the drag is changing from that dominated by viscous stress on the boundary (see equation (4.1.20) to a pressure induced drag, where the wake prevents the pressure recovery due to the separation of the flow; the drag is simply the result of the forward stagnation pressure point remaining, but the rear side of the cylinder is at or near the pressure in the flow near the maximum width point. In the region $60 < Re < 10^4$, the drag coefficient is essentially constant, indicating that the wake shape does not change over this range. The dip at or near $Re \sim 10^5$ is due to the forward boundary layer becoming turbulent and thus is able to sweep the separation point a little beyond the point of maximum thickness; the low pressure wake area thus contract a little lowering the drag coefficient. The exact Reynolds number that this occurs depends on the wall roughness; the rougher the wall of the cylinder the earlier the forward boundary layer becomes turbulent.

For $Re > 10^5$, the drag coefficient once again rises with increasing Reynolds number. As the separation points move back to the point of width and the pressure recovery in the wake is increasingly less and less.

4.6. FLOW AROUND A CYLINDER WITH CIRCULATION

In order to understand the concept of lift it is useful to consider the flow about a rotating cylinder in a uniform flow with a velocity U. In §4.5, we showed that the governing equation for the outer flow, at large Reynolds number, is the linear Laplace equation. It therefore follows that we may add two solutions and gain a third solution. The solution for uniform flow past a cylinder is given by (4.5.24) and a further solution to (4.5.15) is given by:

$$\phi = \kappa \frac{\theta}{2\pi}. \tag{4.6.1}$$

Simple substitution of (4.6.1) into (4.5.15) confirms that (4.6.1) is indeed a solution to the Laplace equation. Also, noting from (4.5.18) it is clear that at $r = a$

$$v_r = 0, \tag{4.6.2}$$

and

$$v_\theta = \frac{\kappa}{2\pi a}, \tag{4.6.3}$$

so that (4.6.1) is indeed a solution to our flow problem. It is not difficult to show the corresponding stream function is given by:

$$\psi = \frac{-\kappa}{2\pi} \log \frac{r}{a} \tag{4.6.4}$$

The flow given by (4.6.1) and (4.6.4) represents a circulation around the cylinder with a circulation velocity that is given by (4.6.3) at the cylinder and decaying as r^{-1} as we move away from the cylinder, contributing nothing to the uniform flow at $r = \infty$. As noted above, that the flow equations are linear and for outer flows we cannot satisfy the no slip condition at the cylinder boundary a combined solution is given by:

$$\phi = -u \cos\theta \left(r + \frac{a^2}{r} \right) + \frac{\kappa\theta}{2\pi}. \tag{4.6.5}$$

This solution has the property that the velocity at the cylinder boundary is given by:

$$\begin{aligned} v_r &= 0 \\ v_\theta &= \left. \frac{1}{r}\frac{\partial\phi}{\partial\theta} \right|_{ra} = 2U \sin\theta + \frac{\kappa}{2\pi a}. \end{aligned} \tag{4.6.6}$$

We see immediately that the tangential velocity at the cylinder boundary is increased, over that of the uniform flow case, at $\theta = 90°$, but is decreased at $\theta = -90°$, from which the Bernoulli equation would already suggest that the pressure at $\theta = -90°$ is somewhat higher than that at the point $\theta = 90°$; this implies that there is a net force on the cylinder perpendicular to the direction of the uniform flow. Such a perpendicular force is called lift. Circulation about the cylinder thus induces lift, the value of κ determines the speed of the tangential velocity v_θ and the lift force. Further, we notice from (4.6.6) that for rotations larger than a certain value the tangential velocity at the cylinder boundary vanishes at the points given by:

$$\sin\theta = -\frac{\kappa}{2\pi a u_\infty}. \tag{4.6.7}$$

The streamline pattern, as a function of κ, is shown in Fig. 4.6.1, where it is seen that when $\dfrac{\kappa}{aU} = 4\pi$, there is only one stagnation point in the flow, directly underneath the cylinder; clearly there will be a net pressure force upwards even though the flow is from left to right. Further, it is seen that once $\kappa > 4\pi aU$ then the flow circulates uniformly around the cylinder. In such a case we may expect the outer solution to remain valid even when viscosity is present, as there is no adverse pressure gradient remaining at the cylinder.

An expression for the lift may be calculated by noting the pressure is once again given by the Bernoulli equation, so that:

$$p|_{r=a} = p_0 - \frac{1}{2}\rho\left(2u_\infty \sin\theta + \frac{\kappa}{2\pi a}\right)^2 + \rho x_3. \tag{4.6.8}$$

The lift force is therefore given by integrating the vertical component of the pressure around the cylinder boundary:

$$F_{x_3} = -\int_0^{2\pi} P|_{r=a}\, a\sin\theta\, d\theta. \tag{4.6.9}$$

Substituting (4.6.8) into (4.6.9) and carrying out the integration yields:

$$F_{x_3} = \rho U \kappa, \tag{4.6.10}$$

that shows that the lift is directly proportional to the oncoming velocity U and the circulation κ.

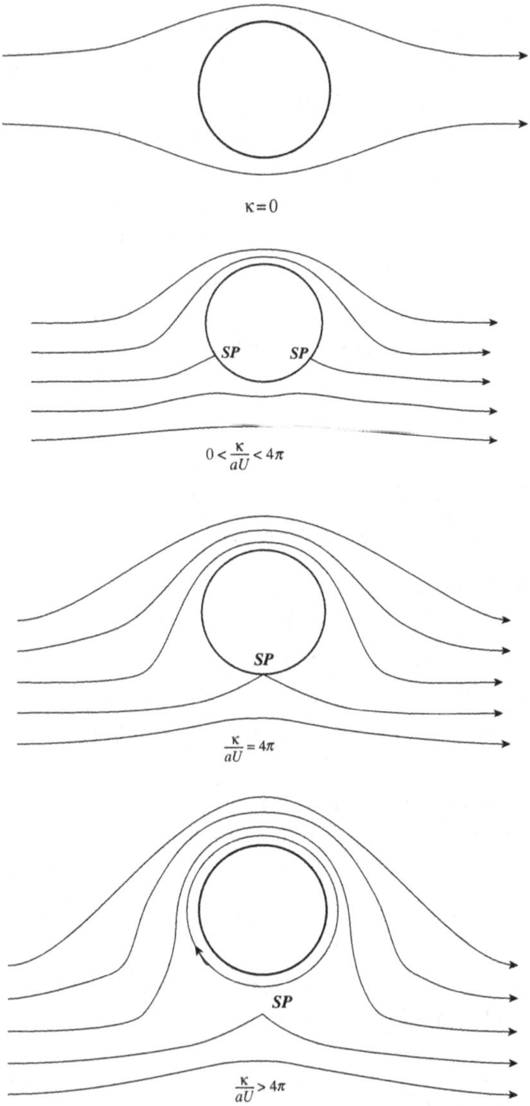

Figure 4.6.1 Schematic of streamlines about a cylinder with circulation.

4.7. FLOW ABOUT A CYLINDER IN AN ACCELERATING FLOW

Consider now the flow about a cylinder in a uniform flow that is varying with time, such that:

$$U = U(t). \tag{4.7.1}$$

Since the fluid is assumed to be incompressible, any pressure signal is transmitted throughout the flow domain infinitely fast, so that the flow will adjust immediately everywhere to a changing upstream velocity, so that the velocity potential derived in §4.5 will still be valid and we may write:

$$\phi = -U(t)\, r\cos\theta - \frac{U(t)a^2\cos\theta}{r}. \tag{4.7.2}$$

The force on the cylinder will now, however, be different because the unsteady inertia term $\frac{\partial u}{\partial t}$ contributes to the force when U is a function of time. The Bernoulli equation for unsteady irrotational flow follows from (2.5.14) and is given by:

$$\frac{p}{\rho} + \frac{\partial\phi}{\partial t} + \frac{1}{2}q^2 + gx_3 = F(t). \tag{4.7.3}$$

At the front of the cylinder there is a stagnation pressure point, so that $p = p_s$; $x_3 = 0,\ q = 0$

$$\frac{p}{\rho} = \frac{\partial\phi}{\partial t} - \frac{1}{2}q^2 - gx_3 + \frac{p_s}{\rho}. \tag{4.7.4}$$

The drag force F_D is given by:

$$F_D = \int_0^{\pi} \frac{p - p_s}{\phi} a\cos\theta\, d\theta. \tag{4.7.5}$$

Substituting the solution (4.7.2) into (4.7.4) leads to:

$$\frac{p - p_s}{\rho} = \left\{ -\frac{\partial U}{\partial t} a\cos\theta + \frac{\partial U}{\partial t}\frac{a^2}{a}\cos\theta \right\} = \frac{1}{2}\{4U^2\sin^2\theta\} - ga\sin\theta. \tag{4.7.6}$$

Substituting (4.7.6) into (4.7.5) and evaluating the integral provides the expression for the drag force in an unsteady uniform flow:

$$F_D = -2\pi a^2 \rho \frac{dU}{dt}. \tag{4.7.7}$$

This may conveniently be written in the form:

$$F_D = (M_0 + M_a)\,\frac{dU}{dt}, \tag{4.7.8}$$

where M_0 is the mass of the fluid displaced by the cylinder and M_a is the extra mass, or added mass, that is influenced by the cylinder in which case the cylinder is equal to M_0. However, in general for bodies of arbitrary shape the added mass must be determined explicitly. The quantity $(M_0 + M_a)$ is called the virtual mass.

The above is all based on the assumption that the outer flow may be used to fully describe the motion around the cylinder. We can check this assumption by estimating the time scale for the establishment of the flow and compare this with the rate of advection.

The time taken to establish the flow around the cylinder when the free stream velocity U is the time it takes for the flow to sweep the vorticity generated in the forward boundary layer around the cylinder which is $O(D/U)$, where D is the diameter of the cylinder. This time must be compared with the time, T, that is characteristic of the flow acceleration, i.e. the time over which, say, the free stream velocity doubles. The ratio of the two time scales is

$$K = \frac{UT}{D}. \tag{4.7.9}$$

This ratio is called the Keulegan Carpenter number and is the ratio of the time scale of acceleration to the time scale it takes to set up flow around a cylinder. If $K < 1$ then the flow free stream velocity changes more quickly that the flow takes to set up so the forces due to the unsteady acceleration as given by (4.7.8) will dominate. On the other hand for cases where $M > 1$, the flow free stream velocity changes slowly compared to the flow set up time and we can expect the forces to be given by (4.5.31).

4.8. FORCES ON A CYLINDER IN A COMBINED FLOW REGIME

In §4.5, §4.6 and §4.7, we considered forces on a cylinder under separate flow conditions. In general, it was shown that if $-x_1$ is the direction from which the free stream velocity of the fluid impinges on a cylinder then the net force $\underset{\sim}{F}$ on the cylinder may be written as:

$$\underset{\sim}{F} = f_D(\text{Re}, K)\hat{\imath}_1 + g_L(\text{Re}, K)\hat{\imath}_3, \tag{4.8.1}$$

where $f_D(\text{Re}, K)$ and $g_L(\text{Re}, K)$ are two unknown functions of the Reynolds number Re and the Keulegan number K, the first being the net drag force and the second being the lift force. We have seen that the drag force is the force that is in-line with the direction of the oncoming flow and the lift force is, by definition, the force perpendicular to the direction of flow. In §4.5 and §4.7, we saw that the drag force is made up of two components, the first is essentially due to the pressure differential set up by the flow separation and the second is due to the force that needs to be exerted when a fluid is accelerated. When $K < 1$ the acceleration force dominates and when $K > 1$ the pressure of form drag dominates. At present no theory exists for the case of a combined oncoming flow, but Morison et al (1950) suggested that a simple addition of the forces would lead to a useful estimate of the force and so he suggested:

$$\underset{\sim}{F} = \left\{\frac{1}{2}C_D(\text{Re}, K)\rho A U^2 + C_M M_v \frac{\partial U}{\partial t}\right\}\hat{\imath}_1 + C_L(\text{Re}, K)\rho\kappa U\hat{\imath}_3, \quad (4.8.2)$$

where C_D (Re, K) is the drag coefficient, C_M is the added mass coefficient, M_v is the virtual mass, A is the cross-sectional area presented by the cylinder to the flow and the other symbols have their normal meaning from the previous sections. It has been found that (4.8.2) is universally valid for all bluff bodies, provided the coefficients are suitable adjusted. For a cylinder of unit length A is equal to the cylinder diameter.

For a cylinder in a flow where the Reynolds number is such that $10^2 < \text{Re} < 10^5$, a good approximation for the values of the drag coefficient $C_D \sim 1.2$–1.4 and added mass coefficient $C_M \sim 2.0$.

4.9. SHALLOW VISCOUS FLOWS

Environmental flows mostly take place in shallow domains as for instance in rivers, estuaries, lakes and coastal seas, as well as the flow of air in the atmospheric boundary layer over a terrestrial domain. When the domain is very shallow we may assume the flow is approximately parallel and then, as shown in §3.4, we may neglect the non-linear inertia terms in the momentum equations and assume the flow is locally parallel and the bottom friction balances the applied surface wind stress and the barotric pressure gradient set up to accommodate volume considerations. In deriving the flow described in §3.4, we had assumed a solid lid moving with a velocity U (see 3.4) together with a pressure gradient. Here we replace these two assumptions and assume the upper surface is a free surface exposed to a surface wind stress $\{\tau_1, \tau_2\}$ and or a net through flow. The coordinate system is as shown in Fig. 4.9.1, the water surface elevation, above the origin is designated by η, the depth d(x_1,d$_2$) is assumed to vary only weakly, then using the same logic as in §3.4, we may derive the horizontal velocity:

$$v_i(x_1, x_2, x_3) = -\frac{gd^2}{2\nu}\frac{\partial \eta}{\partial x_i}\left(1 - \left(\frac{x_3}{\text{d}}\right)^2\right) + \frac{\tau_i d}{\rho_0 \nu}\left(1 + \left(\frac{x_3}{\text{d}}\right)\right) + \text{O}\left(\frac{\eta}{\text{d}}\right);$$

$$i = 1, 2, 3, \quad (4.9.1)$$

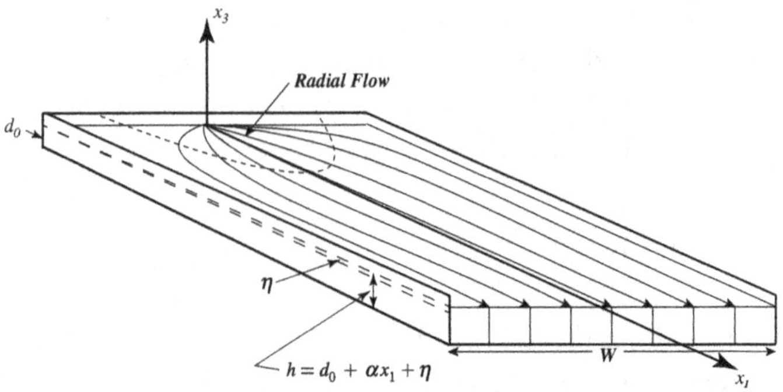

Figure 4.9.1 River inflow to a shallow estuary with a gentle linear increase in depth.

Integrating the velocity over the depth of the water colum leads to the unit volume flux:

$$q_i(x_1, x_2) = \int_{-d}^{\eta} v_i(x_1, x_2, x_3)\mathrm{d}x_3 = -\frac{gd^3}{3\nu}\frac{\partial\eta}{\partial x_i} + \frac{\tau_i d^2}{2\rho_0\nu} + O\left(\frac{\eta}{\mathrm{d}}\right); i = 1, 2, \tag{4.9.2}$$

where we have assumed that the surface displacement is very small compared to the flowing depth.

Integrating the conservation of volume equation, (4.5.1), and assuming the flow at the free surface is parallel to the free surface leads to:

$$\frac{\partial q_1}{\partial x_1} + \frac{\partial q_2}{\partial x_2} = 0. \tag{4.9.3}$$

Given (4.9.3) allows us to introduce a depth averaged stream function $\psi(x_1,x_2)$, such that:

$$q_1 = \frac{\partial\psi(x_1, x_2)}{\partial x_2}, \tag{4.9.4}$$

$$q_2 = -\frac{\partial\psi(x_1, x_2)}{\partial x_1}. \tag{4.9.5}$$

In §4.3 we showed that the flow equations are linear, so the solutions for $i=1$ and 2 may be added to yield a third solution. A disturbance to the flow such as a river inflow (Fig. 4.9.1) or a surface wind stress (Fig 4.9.1) will cause the water surface elevation to be set up and this in turn leads to a flow. To illustrate the solution methodology, we discuss three separate configurations, all important simplifications of common environmental flows.

Case 1: River flowing into a shallow receiving water with constant depth d_0, width $W$and no wind stress. When the domain depth is constant, we may substitute (4.9.4) and (4.9.5) into (4.9.3) to arrive at the Laplace equation for the stream function:

$$\frac{\partial^2 \psi(x_1, x_2)}{\partial x_1^2} + \frac{\partial^2 \psi(x_1, x_2)}{\partial x_2^2} = 0. \qquad (4.9.6)$$

Alternatively by substituting (4.9.2) into (4.9.3) we may show:

$$\frac{\partial^2 \eta(x_1, x_2)}{\partial x_1^2} + \frac{\partial^2 \eta(x_1, x_2)}{\partial x_2^2} = 0 \qquad (4.9.7)$$

The solution to the configuration shown in Fig. 4.9.1 may be derived by separation of variables:

$$\eta(x_1, x_2) = \sum_{n=1}^{\infty} \frac{3Q_1 \boldsymbol{\nu}}{g\pi d^3 n} e^{\frac{-2\pi n}{W} x_1} \cos\frac{2n\pi x_1}{W} - \frac{3Q_1 \boldsymbol{\nu}}{gWd^3} x_1, \qquad (4.9.8)$$

where Q_1 is the inflow discharge The flow is radial near the origin and approaches uniform flow for large x_1.

Case 2: River inflow into a semi-infinte shallow receiving water with depth increasing with distance. Locally, we may assume that the (4.9.2) is still valid, substituted this into (4.9.3), account for the depth variation with respect to x_1 and then changing back to the unit discharge q_1:

$$\frac{\partial^2 q_1}{\partial x_1^2} + \frac{\partial^2 q_1}{\partial x_1^2} - \beta \frac{\partial q_1}{\partial x_1} = 0; \; \beta = \frac{3}{d} \frac{dd}{dx_1}. \qquad (4.9.9)$$

As it stands (4.9.9) is difficult to solve, even in the semi-infinite domain under consideration here.

We make progress, by first assuming to depth increseases exponentially so that β is constant and second by noting that the flow prefers to flow along the path ofleast bottom friction, in other words, along the line of steepest descent, which is along the x_1 axis; the second term in (4.9.9) is thus larger than the first and the first may be neglected. Second, given there is no length scale in the problem, a self similar solution should exist so we see a solution of the form:

$$q_1 = \frac{Q_1 \beta^{\frac{1}{2}}}{2\pi^{\frac{1}{2}} x_1^{\frac{1}{2}}} e^{-\frac{\beta x_2^2}{x_1}}, \qquad (4.9.10)$$

that has the property that the total inflow in the x_1 direction in Q_1. From (4.9.10) is is seen, when the depth increases from the point of inflow, the inflowing stream forms a jet like inflow structure.

Case 3: Infinitely long parallel duct with a longitudinal wind stress and zero net flow. The no net volume flux is imposed as this may be used as a model of

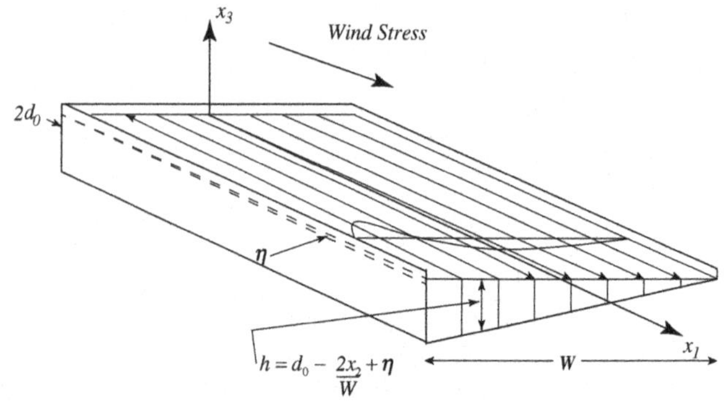

Figure 4.9.2 Topographic circulation indued by a wind stress acting on a triangular shapped shallow basin.

a wind stress acting on a shallow elongated lake; the water surface elevation build up to induce a return flow, negating the flow induced by the wind stress. If the duct is very much longer than wide (Fig. 4.9.2), we may assume that q_2 is zero and thus by (4.9.2) so the surface elevation is horiziontal in the tranbsverse direction. For definiteness we assume that the depth is uniform in the longitudinal direction x_1, and varies in the x_2 direction, given by:

$$d = d_0\left(1 - \frac{2x_2}{W}\right). \tag{4.9.11}$$

Subsituting this into (4.9.2) and requiring that the net volume flux is zero, determines the free surface slope as a function of the applied surface stress:

$$q_1 = \frac{\tau_1 d_0^2}{6\nu\rho_0}\left(1 - \frac{2x_2}{W}\right)^2\left(1 + \frac{4x_2}{W}\right). \tag{4.9.12}$$

This flow is sketched in Fig. 4.9.2 where it is seen that the depth averaged flow is against the wind stress in the deep part of the duct and with the wind stress in the shallows. In a lake with end boundaries, there would be a region where the flow turns to conserve volume and induce the baraotropic reverse flow pressure gradient. Such circulations are called "topographic gyres", very important and frequently occurring featues in shallow lakes.

REFERENCES

Darcy H, 1856. Les Fontaines Publiques de la Ville de Dijon. Victor Dalmont, Paris, p. 647.

Morison, J. R., O'Brien M. P., Johnson J. W., Schaaf S. A, 1950. The force exerted by surface waves on piles. Petroleum Transactions (American Institute of Mining Engineers) 189, 149–154.

CHAPTER 5

Fundamentals of Hydraulics

Contents

We present here some fundamental hydraulic concepts that provide the foundations for the ideas discussed in the next chapter on Environmental Hydraulics. The emphasis here is on elucidating the fundamental underlying concepts in hydraulics, not to provide the civil engineer with a practical text book, rather to help the environmental engineer to understand the principles of jets and plumes, pipe flow and open channel flow. However, once the reader has negotiated and understood these underlying principles, presented here, he or she will find it easier to navigate through a specialist hydraulics book when the need arises.

5.1. SIMPLE JETS

A simple jet is a flow that is set up by a fluid issuing from a nozzle into an ambient body of receiving fluid as shown in Fig. 5.1.1. The word "simple" is intended to imply that the jet is driven solely by the momentum of the fluid issuing from the nozzle, so in this sense it is the flow arising by allowing the diameter of the nozzle to go to zero, but keeping the momentum flux

Environmental Fluid Dynamics
ISBN 978-0-12-088571-8, DOI: 10.1016/B978-0-12-088571-8.00005-X

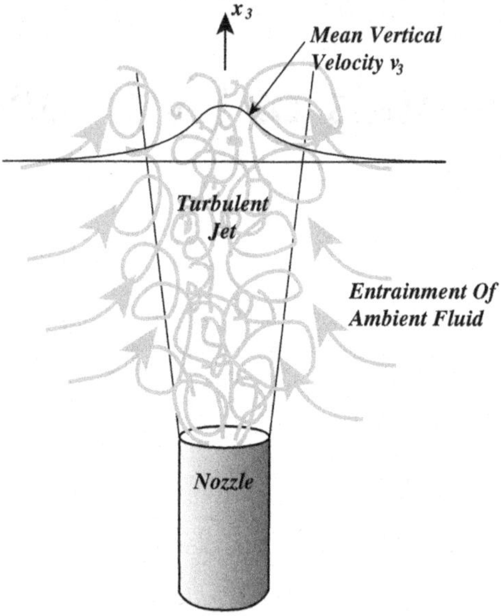

Figure 5.1.1 Schematic of flow induced by a simple jet.

constant. To see that this is indeed possible consider the expression for the volume flux Q:

$$Q = \pi R^2 u_d, \tag{5.1.1}$$

where Q is the nozzle discharge, or volume flux, where R is the nozzle radius and u_d is called the discharge velocity. The kinematic momentum flux is given by:

$$M = \pi R^2 u_d^2. \tag{5.1.2}$$

Thus if we let $R \to 0$; $M = \text{Const.}$, then

$$Q = \pi R^2 \left(\frac{M}{\pi R^2}\right)^{\frac{1}{2}} \sim R \to 0 \text{ as } R \to 0 \tag{5.1.3}$$

In this limiting process, the discharge velocity at the nozzle will approach infinity. Thus in the limiting example, no mass exits at the nozzle and the nozzle becomes simply a source of momentum flux M.

Suppose ν is the kinematic viscosity, then the Reynolds number of the flow is given by:

$$Re = \frac{Ru_d}{\nu}. \tag{5.1.4}$$

When this Reynolds number exceeds about 1000 then both the flow in the pipe and in the jet, formed by the discharge, will be turbulent.

The mean flow properties of such flows may be determined from some simple dimensional analysis and the use of experimental data for the determination of the resulting unknown constants.

The volume flux of the jet at a height of x_3, is given by:

$$\mu = \int_A \overline{v}_3(r, x_3)\mathrm{d}A \quad \mathrm{m}^3\,\mathrm{s}^{-1}, \tag{5.1.5}$$

where $\overline{v}_3(x_3)$ is the mean vertical velocity in the jet at height x_3 and radial distance r; for the case of a circular nozzle the mean flow will be axi-symmetric.

The kinematic momentum flux of any height x_3 is defined by:

$$m = \int_A \overline{v}_3^2(r, x_3)\mathrm{d}A \quad m^4\,s^{-2}. \tag{5.1.6}$$

and again initially at the nozzle

$$m = u_d Q = M \tag{5.1.7}$$

The dilution D of the fluid in the jet is defined by the ratio:

$$D = \frac{\mu}{Q}, \tag{5.1.8}$$

The quantities m and μ are of central importance as they describe the characteristics of the jet motion that are needed in engineering applications. Now these bulk quantities can only depend on the external influences, so for the volume flux we may write, for a particular nozzle configuration, from dimensional reasoning:

$$f(\mu, Q, A, \nu, x_3) = 0, \tag{5.1.9}$$

where f is an unknown function. From dimensional analysis (5.1.9) may be rewritten in the form:

$$f\left(\frac{\mu}{Q}, \frac{x_3}{A^{\frac{1}{2}}}, Re\right) = 0, \tag{5.1.10}$$

where A is the area of the nozzle. For generality for the case of a non-circular nozzle area we define the Reynolds number (5.1.4) as:

$$Re = \frac{A^{\frac{1}{2}} u_d}{\nu}. \tag{5.1.11}$$

Define a length scale:

$$\ell_Q = \sqrt{A}, \tag{5.1.12}$$

then we may note that:

$$\ell_Q = \frac{Q}{M^{1/2}}, \tag{5.1.13}$$

If the Reynolds number is very large then relationship (5.1.10) will be independent of Re and we may write:

$$\frac{\mu}{Q} = f\left(\frac{x_3}{\ell_Q}\right), \tag{5.1.14}$$

The unknown function f in (5.1.14) may be determined for $x_3 \gg \ell_Q$. This is done by noting that $x_3 \rightarrow \infty$ is the same as $Q \rightarrow 0$ in (5.1.14), but when $Q \rightarrow 0$ (5.1.14) should become independent of Q so that we may write:

$$\frac{\mu}{Q} = \alpha\left(\frac{x_3}{\ell_Q}\right) \quad x_3 \gg \ell_Q \text{ or } Q \rightarrow 0, \tag{5.1.15}$$

where α is a constant, yet to be determined.

Many experiments have been performed with the aim of determining the coefficient α (see Fischer et al. 1979), The mean value of α for all the published results is

$$\alpha = 0.25, \tag{5.1.16}$$

so that the volume flux μ becomes:

$$\mu = 0.25 M^{1/2} x_3. \tag{5.1.17}$$

Using similar reasoning it follows:

$$\mathrm{m} = \mathrm{M}, \tag{5.1.18}$$

which states that the momentum flux of the jet, at any height x_3, is constant and equal to the momentum flux at the nozzle. At first sight this seems strange. However, on reflection the result is obvious since the fluid which is entrained into the jet is motionless and so brings no momentum to the jet; hence the momentum flux remains constant, the mass of moving fluid increases and the velocity decreases so that the momentum remains constant.

Similarly, the centerline velocity $\overline{v}_{3m}(x_3)$ may be written as:

$$\overline{v}_{3m} = 7.0\frac{M}{Q}\left(\frac{\ell_Q}{x_3}\right) \quad z \gg \ell_Q \text{ or } Q \to 0, \tag{5.1.19}$$

where the coefficient was obtained from experiments (see Fischer et al., 1979).

Suppose we also have a tracer issuing from the nozzle with an initial concentration C_0 then the mass flux of the tracer in the plume is given by:

$$T = C_0 Q. \tag{5.1.20}$$

The mean concentration averaged across the jet area $[\overline{C}]$ at any height x_3 is given by:

$$[\overline{C}]\mu = T, \tag{5.1.21}$$

so that:

$$\frac{[\overline{C}]}{C_0} = 4.0\frac{\ell_Q}{x_3}, \tag{5.1.22}$$

where the coefficient is once again determined from experimental data (see Fisher et al., 1979).

The maximum mean concentration at the centerline of the jet is given by:

$$\frac{\overline{C}_m}{C_0} = 5.64\frac{\ell_Q}{x_3} \tag{5.1.23}$$

again the coefficient was obtained by fitting experimental data to the general relationship (see Fischer et al., 1979).

The width of the jet b_w may be defined as the radius where the mean velocity is e^{-1} times the maximum mean velocity. Using the same dimensional reasoning and fitting the resulting relationship to experimental data yields:

$$b_w = 0.107x; \tag{5.1.24}$$

the jet expands linearly with distance from the nozzle.

5.2. SIMPLE PLUME

Whenever fluid, of a different density, is injected into an ambient fluid the density difference will contribute a buoyancy force. When the injected fluid is lighter than the ambient, buoyancy will cause the injected fluid to rise and when the injected fluid is heavier than the ambient, buoyancy will cause the injected fluid to fall under gravity. If the ambient receiving fluid is homogeneous, and the injected fluid is, say lighter than the ambient fluid, then as the injected fluid rises it will again entrain ambient fluid into the rising plume. It is important to note that no matter how much ambient fluid is entrained, the plume fluid mixture will retain its buoyancy and so will continue to accelerate upwards, entraining more fluid as it rises. This differentiates a plume from a jet; in a jet, as we saw in §5.1, the momentum remains constant, but in a plume the momentum flux continues to increase.

A plume is energized by the source of buoyancy flux B, where:

$$B = \frac{\Delta\rho}{\rho_0} gQ \quad \mathrm{m^4 s^{-3}}, \tag{5.2.1}$$

where $\Delta\rho$ is the density difference between the issuing and ambient fluid, ρ_0 is a density of the ambient fluid, g is the acceleration due to gravity and Q is again the discharge from the nozzle. Since, as noted above, the ambient fluid brings no new buoyancy, the buoyancy flux β remains constant with height:

$$\beta = \int_A g\frac{\overline{\rho}}{\rho_0}\overline{v}_3 \mathrm{d}A = \frac{\Delta\rho}{\rho} gQ = B. \tag{5.2.2}$$

By comparing (5.1.1), (5.1.2) and (5.2.1) it is seen that it is possible to consider the limit:

$$\frac{\Delta\rho}{\rho_0} Q = Const; \quad Q \to 0, \tag{5.2.3}$$

$$Q \to 0; \quad M \to 0; \quad Q = Constant. \tag{5.2.4}$$

Following the nomenclature of §5.1, we shall call this limit a "simple" plume as only the parameter B enters the problem.

Once again we are interested in the four mean properties: μ the volume flux, m the momentum flux, $\overline{v}_{3m}$ the centerline vertical velocity and the width of the plume b_w.

Consider first the volume flux retaining the initial volume flux Q. We may then write:

$$f(\mu, B, x_3, Q, \nu, \kappa) = 0, \tag{5.2.5}$$

where κ is the diffusivity of the buoyancy agent, usually temperature or salinity.

There are six independent and two dependent variables so that (5.2.5) may be rewritten as:

$$f\left(\frac{\mu}{B^{1/2}x_3^{5/3}}, \frac{\nu}{\kappa}, \frac{Q}{\nu x_3}, \frac{\mu}{Q}\right) = 0. \tag{5.2.6}$$

Once again we may invoke a number of special conditions. First, consider the case where the Reynolds number $Q/\nu x_3 >>> 1$ and where the Prandtl number ν/κ is fixed. Under such conditions:

$$f\left(\frac{\mu}{B^{1/3}x_3^{5/3}}, \frac{\mu}{Q}\right) = 0. \tag{5.2.7}$$

At the nozzle:

$$m = u_d Q = M, \tag{5.2.8}$$

$$\mu = Q. \tag{5.2.9}$$

A simple plume, by definition, is the case where $Q \to 0$, $B =$ constant and $M \to 0$. For such a case we only have a source of buoyancy and Q does not enter (5.2.7). On the other hand, if Q is non-zero then we can form a length scale, equivalent to ℓ_Q defined by (5.1.13):

$$\ell_B = \frac{Q^{3/5}}{B^{1/5}}, \tag{5.2.10}$$

which represents the effective height where the influence of the initial volume flux Q recedes and buoyancy determines the plume behavior. Now if $Q = 0$ then (5.2.7) simply becomes:

$$\mu = 0.15 B^{1/3} x_3^{5/3}, \tag{5.2.11}$$

where the coefficient 0.15 was obtained from fitting experimental data to (5.2.11) (see Fischer et al., 1979).

Similarly, we may write an expression for the maximum mean centerline velocity:

$$\overline{v}_{3m} = 4.7\left(\frac{B}{x_3}\right)^{1/3}, \tag{5.2.12}$$

and the momentum flux:

$$m = 0.35B^{2/3}x_3^{4/3}. \tag{5.2.13}$$

We see for the simple plume, as distinct from the simple jet, buoyancy inherent in the plume continues to accelerate the fluid upward and thus the momentum flux increases with height.

The average across the plume of the mean concentration $[\overline{C}]$ of a tracer with concentration C_0 near the nozzle is again given by:

$$\frac{[\overline{C}]}{C_0} = 6.7\left(\frac{\ell_B}{x_3}\right)^{5/3}, \tag{5.2.14}$$

and the maximum mean concentration

$$\frac{\overline{C}_m}{C_0} = 9.1\left(\frac{\ell_B}{x_3}\right)^{1/3}. \tag{5.2.15}$$

Lastly, if we define the width b_w of the plume as the radius where the mean velocity has decreased to e^{-1} of the mean centerline velocity then:

$$f(b_{\mathrm{w}}, B, x_3) = 0, \tag{5.2.16}$$

so that it is immediately apparent from dimensional analysis that:

$$b_{\mathrm{w}} \sim x_3. \tag{5.2.17}$$

Experimental data may again be used to fit the width and this leads to:

$$b_{\mathrm{w}} = 0.10x_3. \tag{5.2.18}$$

The plume has a further interesting property. By combining (5.2.11) with (5.2.13) we may write:

$$\mu = 0.25m^{1/2}x_3, \tag{5.2.19}$$

which is identical to the equation for the volume flux for a simple jet (5.1.17). Thus if we use the local momentum flux, jets and plumes have the same volume flux relationship. This means that, in the case of a plume, buoyancy imparts momentum to the plume fluid and then momentum entrains ambient fluid via a turbulent breakdown in a way identical to that taking place in a jet; the slight difference in the jet and plume width is the result that in a jet m is a constant and in a plume increases with x_3.

Alternatively, for the case of a plume, we can use (5.2.11) and (5.2.13) to eliminate x_3 and obtain an expression for the volume flux μ in terms of m and $\beta = B$.

$$\mu = 0.57B^{-1/2}m^{5/4}. \qquad (5.2.20)$$

From this expression we see that:

$$\frac{\mu B^{1/2}}{m^{5/4}} = R_p = 0.557 \qquad (5.2.21)$$

The constant R_p is called the plume Richardson number since it is the ratio of the buoyancy to the momentum influences.

5.3. BUOYANT JET

A buoyant jet is a combination of a jet and a plume so that at the nozzle $Q \neq 0$; $M \neq 0$; $B \neq 0$; the water issuing into the ambient receiving water has buoyancy, momentum and a volume flux.

Given the discussion in §5.1 and §5.2 we have two non-zero length scales, ℓ_Q and ℓ_B. However, there is a third length scale:

$$\ell_M = \frac{M^{3/4}}{B^{1/2}}, \qquad (5.3.1)$$

that represents the scale it takes the buoyancy in the buoyant jet to impart a momentum flux M to the fluid. Since ℓ_Q, ℓ_B, and ℓ_M depend on only three quantities, Q, B, and M, the three length scales are not independent and are related through the relationship:

$$\ell_M = \ell_Q^{\frac{3}{2}}\ell_B^{\frac{5}{2}}, \qquad (5.3.2)$$

so that we can use any two of these length scales and write:

$$f\left(\frac{\mu}{B^{1/3}x_3^{5/3}}, \frac{x_3}{\ell_M}, \frac{x_3}{\ell_Q}\right) = 0, \qquad (5.3.3)$$

for large Reynolds number flows.

The quantity x_3/ℓ_Q is usually large since ℓ_Q is a measure of the nozzle size (see (5.1.12)) and this is usually measured in centimeters. Thus for finite x_3 we may assume $x_3/l_Q \gg 1$ so that Q is not an important parameter and we may write:

$$\frac{\mu}{B^{1/3}x_3^{5/3}} = \mathrm{f}\left(\frac{x_3 B^{1/2}}{M^{3/4}}\right). \tag{5.3.4}$$

Now consider the extremes of a pure plume and jet. First, suppose $B \to 0$; the limit of a pure jet. This limit is formally equivalent to taking the limit $x_3 \to 0$ or $M \to \infty$. Under this limit (5.3.4) should become independent of B, see (5.1.17), so that the function in (5.3.4) must be of the form, such that:

$$\frac{\mu}{B^{1/3}x_3^{5/3}} = 0.25\left(\frac{M^{3/4}}{B^{1/2}x_3}\right)^{2/3}, \tag{5.3.5}$$

for $x_3 \ll \frac{M^{3/4}}{B^{1/2}}$ in order that (5.3.4) is the same as (5.1.17).

Now, on the other hand, in the limit: $M \to 0: \; x_3 \gg \frac{M^{3/4}}{B^{1/2}}$, (5.3.4) must reduce to (5.2.11), so that:

$$\frac{\mu}{B^{1/3}x_3^{5/3}} = \mathrm{f}\left(\frac{x_3 B^{1/3}}{M^{3/4}}\right) = 0.15. \tag{5.3.6}$$

Thus the controlling parameter determining whether a buoyant jet is plume or jet like is x_3/ℓ_M. If $x_3/\ell_M \gg 1$ then it behaves as a plume and if $x_3/\ell_M \ll 1$ then it has the features of a jet. It is possible to simplify the formulation a little further by introducing scaled variables for the jet and plume regions. Consider first close to the nozzle, $x_3 \ll \ell_M$ and define the non-dimensional variables:

$$\overline{\mu} = \frac{\ell_Q}{\ell_M R_p}\frac{\mu}{Q} \quad \text{and} \quad \zeta = \frac{0.25x_3}{R_p \ell_m}, \tag{5.3.7}$$

then (5.1.17) becomes:

$$\overline{\mu} = \zeta, \tag{5.3.8}$$

and (5.2.11) becomes, if we note (5.2.21):

$$\mu = \zeta^{5/3}, \tag{5.3.9}$$

valid for $x_3 \gg \ell_M$.

These asymptotic solutions, for the non-dimensional dilution in a vertical round turbulent buoyant jet, are compared in Fischer et al. (1979) with the experimental data of Ricou and Spalding (1961). It is shown that (5.3.8) and (5.3.9) provide reliable estimates for the dilution induced by a buoyant jet; the transition region at $x_3 = \ell_M$ is often quite small and is thus usually neglected in diffuser designs.

5.4. ROUND BUOYANT JET DISCHARGING INTO A STRATIFIED SHEAR FLOW

In this section, we briefly consider three generalizations to the buoyant jet discussed in §5.3; first, a negatively buoyant jet with positive momentum flux into a homogeneous ambient fluid, second a buoyant jet into a stratified ambient receiving water and third, a buoyant jet into a homogeneous receiving water that is moving horizontally with a velocity U.

In §5.3 we showed, from simple dimensional reasoning, that when a vertical buoyant jet issues into a stationary, homogeneous ambient receiving water, the initial momentum dominates the initial entrainment, but after a distance order ℓ_M, given by (5.3.1), the entrainment into the jet is predominantly due to the buoyancy flux, that continues to energize the jet as it rises and the jet behaves as a simple plume. Consider now a negatively buoyant jet, with buoyancy flux $-B$, heavier than ambient fluid, is injected into a stationary homogeneous ambient receiving water with an initial momentum flux M. This is analogous to throwing a ball into the air. It is a common experience that the ball will rise to a certain height, determined by the initial momentum imparted to the ball, and then fall back down due to gravity. In high school, we learned that the height of rise h_T of the ball is given by:

$$h_T = \frac{u^2}{2g}, \tag{5.4.1}$$

where u is the initial upwards velocity and g is the acceleration due to gravity. As we saw in §5.3, the initial momentum induces an entrainment given by (5.1.17), so that the ball in our analogy grows, inflated with ambient fluid as it rises. Given that the initial behavior may again be expected to behave as a growing ball thrust upward, we may use (5.4.1) to determine an expression for the height of rise of a negatively buoyant jet. From (5.1.7) we may write:

$$u \sim \frac{M}{Q}, \tag{5.4.2}$$

and from (5.2.1) it follows that:

$$g' = \frac{\Delta\rho}{\rho_0} g \sim \frac{B}{Q}. \tag{5.4.3}$$

Substituting (5.4.2) and (5.4.3) into (5.4.1) and using (5.1.17) for an estimate of Q yields a scale for the height of rise h_T:

$$h_T \sim \frac{M^{\frac{3}{4}}}{B^{\frac{1}{2}}} = \ell_M. \tag{5.4.4}$$

The above provides a plausible justification, of the height of rise of a negatively buoyant jet, a result that also follows from the observation that the height of rise can only depend on M and B, dimensional reasoning then provides the same scale as (5.4.4). Fischer et al. (1979) provide a detailed discussion of vertical negatively buoyant jets and deduce, from comparison with experimental results, that the constant of proportionality in (5.4.4) is about 2 so that:

$$h_T = 2.0 \frac{M^{\frac{3}{4}}}{B^{\frac{1}{2}}}. \tag{5.4.5}$$

A similar argument appears to hold for negatively buoyant jets that are inclined, to the vertical, provided we use the vertical component M_v of momentum flux in (5.4.5) instead of the total momentum flux M.

For most environmental applications the ambient receiving fluid is stratified; smoke plumes into the atmosphere, wastewater effluent discharges into the coastal ocean and mixing jets in lakes are some typical examples.

Consider now the case of a simple vertical jet injecting neutral water, with a momentum flux, M, into an ambient receiving domain that is linearly stratified:

$$\rho = \rho_0(1 - \varepsilon x_3). \tag{5.4.6}$$

Near the nozzle the jet will be neutrally buoyant, so the entrainment will behave as a simple jet, described in §5.1. The ambient water that is entrained near the level of the nozzle will be carried aloft by the jet momentum and will progressively become heavier relative to the local stratification, simply because the ambient density decreases with height. The upward motion will continue until the fluid in the jet is brought to rest by buoyancy and then the total jet fluid will reverse and tumble back down toward the nozzle level as shown in Fig. 5.4.1. When viewed as described, the jet is similar to an

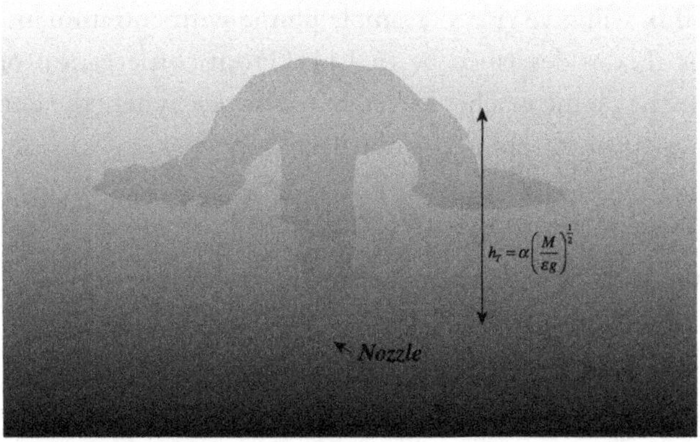

Figure 5.4.1 Schematic of a positively buoyant jet in a stratified ambient fluid.

"upside down" version of the negatively buoyant jet discussed above and we may again expect the height of rise to be given by (5.4.5), except now the buoyancy flux will be given by:

$$B = \frac{\Delta\rho}{\rho_0}gQ \sim \varepsilon g h_T Q \sim \varepsilon g h_T M^{\frac{1}{2}} h_T, \qquad (5.4.7)$$

where we have used (5.1.17) to estimate the volume flux and h_T, is again the height of rise. Substituting this buoyancy flux into (5.4.5) leads to:

$$h_T = \alpha\left(\frac{M}{\varepsilon g}\right)^{\frac{1}{2}}, \qquad (5.4.8)$$

where $\alpha = 3.8$ is a constant determined by fitting (5.4.8) to experimental data (see Fischer et al., 1979).

The more common case is to have a buoyant plume in a stratified ambient environment. For instance fresh water from a wastewater treatment plant enters a temperature or salinity stratified coastal ocean environment. We can use analogous logic as above, but instead of using (5.1.17) to estimate the volume flux we use (5.2.11). Suppose the density of the ambient receiving water again decreases linearly with height, as given by (5.4.6) and we define a buoyancy flux at the origin:

$$B_0 = g\frac{\Delta\rho_0}{\rho_0}Q_0, \qquad (5.4.10)$$

where the zero subscripts stand for the conditions at the nozzle. Initially the buoyancy flux will give rise to a simple plume with entrainment increasing the volume flux as described by (5.2.11). From conservation of mass, we may assume the plume reaches its height of rise, h_T, where the density inside the plume is equal to the local ambient density:

$$B^{\frac{1}{3}} h_T^{\frac{5}{3}} \varepsilon h_T \rho_0 \sim \Delta\rho_0 Q_0. \tag{5.4.11}$$

Rearranging leads to the expression for the height of rise:

$$h_T = \alpha \frac{B_0^{\frac{1}{4}}}{(g\varepsilon)^{\frac{3}{8}}}, \tag{5.4.12}$$

where $\alpha = 3.8$ is a constant determined by fitting (5.4.12) to experimental data (see Fischer et al., 1979).

The last case we briefly examine is when a jet issues into a homogeneous ambient receiving water that is moving horizontally with a constant velocity U. Jets and plumes rarely rise to great heights, so the assumption of a constant cross-stream velocity is usually adequate, but it is also a simple matter to generalize the following result to different ambient horizontal shear profiles. A horizontal cross-current has no impact on the vertical momentum equation, so we may assume (5.1.17) holds and to first order we may further assume, from (5.1.24), the jet area grows as the square of x_3. Hence we may write:

$$\frac{dz}{dt} = \frac{M^{\frac{1}{2}}}{z} \tag{5.4.13}$$

and

$$\frac{dx}{dt} = U, \tag{5.4.14}$$

where z and x are the vertical and horizontal coordinates of the plume trajectory. Integrating (5.4.13) and (5.4.14) and eliminating the time t leads to the trajectory path:

$$\frac{z}{z_M} = \alpha \left(\frac{x}{z_M}\right)^{\frac{1}{2}}, \tag{5.4.15}$$

where $\alpha = 2$ is a constant determined by fitting (5.4.15) to experimental data (see Fischer et al., 1979) and $z_B = \dfrac{M^{\frac{1}{2}}}{U}$.

Similar logic applied to a simple plume leads to:

$$\frac{z}{z_B} = \alpha\left(\frac{x}{z_B}\right)^{\frac{2}{3}}, \tag{5.4.16}$$

where $\alpha = 1$ is a constant determined by fitting (5.4.15) to experimental data (see Fischer et al., 1979) and $z_B = \frac{B}{U^3}$.

5.5. SCALES IN TURBULENT FLOWS

As the fluid velocity increases there comes a point when the flow visually changes from that of a quiescent appearance to one that is clearly "turbulent". Common experience suggests that it is visually relatively easy to distinguish between these two types of flow, but it is rather difficult to provide exact definitions of either type of flow. It is much easier, and probably more instructive, to list the most important characteristics of both types of flow. This is analogous to a patient placing more importance on the symptoms of a disease, rather than the name. The two photographs in Fig. 5.5.1 illustrate the difference between a high Reynolds number turbulent flow of a volcano and the low Reynolds number laminar smoke rising from a cigarette.

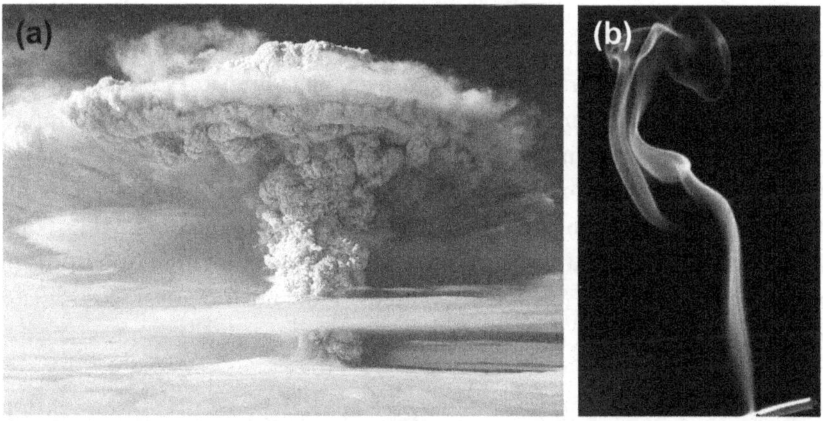

Figure 5.5.1 (a) Puyehue Volcano Chile Eruption, Jun 5, 2011. (b) Smoke rising from a cigarette. (For the color version of this figure, the reader is referred to the online version of this book.)

Laminar Flow

- Smooth in appearance.
- Has well-defined deterministic particle paths.
- Defined by one or two length and time scales that characterize the structure of the velocity field.
- Tracer patches spread slowly with a smooth appearance.
- Dissipates little energy.

Turbulent Flow

- Appearance is irregular, random and containing eddies of many different scales.
- The particle paths are random.
- The flow is characterized by a continuous spectrum of space and time scales from the smallest where dissipation takes place to the large scales that characterize the domain scale motions.
- Tracer patches spread irregularly and rapidly with a dispersive appearance.
- Turbulent flows possess vorticity and the random components are always three-dimensional. Vorticity is stretched and turned forming random intensification in a turbulent flow field. Indeed turbulence relies, for its maintenance, on the intensification of vorticity by what we have called vorticity stretching and tipping §2.12.
- Turbulent flows are strongly dissipative and decay rapidly in the absence of a continual supply of energy.

In order to firm up our ideas about the importance of a small number of length scales in laminar flow and a multitude of scale in turbulent flow, consider, by way of introduction, flow in a long cylindrical straight pipe; we start first with a laminar flow. Since the pipe is assumed to be much longer than its diameter, we may assume the streamlines are parallel and symmetrical about the central axis of the pipe. This means that the fluid acceleration is zero and the only non-zero velocity is the axial component and this is only a function of the radial distance *r*. Under these simplifying conditions the momentum equation (2.11.10), written in cylindrical polar coordinates becomes:

$$0 = -\frac{\partial p}{\partial x} + \mu \frac{1}{r}\frac{\partial}{\partial r}\left(r \frac{\partial v_1}{\partial r} \right). \tag{5.5.1}$$

The only length scale for this problem is the radius R and there is no length scale in the axial direction so the streamlines must be parallel and the pressure gradient must be constant:

$$\frac{1}{\rho_0}\frac{\partial p}{\partial x} = \frac{\Delta P}{L\rho_0} = -\frac{gh_L}{L} = Constant, \tag{5.5.2}$$

where L is the length of the pipe, ΔP is the pressure drop, g is the acceleration due to gravity and h_L is called the head loss. Equation (5.5.1) may be integrated as the first term is constant. Using the no slip boundary conditions at the walls integration leads to:

$$v_1(r) = \frac{gh_L R^2}{4\nu L}\left(1 - \left(\frac{r}{R}\right)^2\right). \tag{5.5.3}$$

The velocity profile is parabolic with a maximum velocity, u_m, at the center of the pipe:

$$u_m = \frac{gh_L R^2}{4\nu L}. \tag{5.5.4}$$

The discharge velocity u_d may be calculated by integrating the velocity (5.5.3) over the pipe cross-sectional area:

$$u_d = \frac{1}{\pi r^2}\int_0^R v_1(r)2\pi r dr = \frac{gh_L R^2}{8\nu L}. \tag{5.5.5}$$

In both pipe flow and open channel hydraulics it is common to formulate a friction factor, f, defined as the coefficient of proportionality between the head loss and the velocity head (see Bernoulli equation (1.12.11):

$$h_L = f\left(\frac{L}{D}\right)\left(\frac{u_d^2}{2g}\right). \tag{5.5.6a}$$

Substituting (5.5.5) into (5.5.6a) for u_d leads to a simple expression for the friction factor for laminar flow in a pipe:

$$f = \frac{64}{Re}, \tag{5.5.6b}$$

where $Re = \frac{UD}{\nu}$ is the pipe flow Reynolds number and D is the pipe diameter. The above results may be interpreted using the bulk momentum equation; the pressure force is resisted by the viscous stress exerted by the walls of the duct on the fluid:

$$2\pi R\tau_w L = (p_1 - p_2)\pi R^2 = -\rho_0 g h_L, \tag{5.5.7}$$

where p_1 and p_2 are the pressure at the entrance and the exit, respectively, and τ_w is the pipe wall stress. Rearranging:

$$\tau_w = \mu \left. \frac{\partial v_1}{\partial r} \right|_{r=R} = -\frac{\rho_0 g R h_L}{2L}, \tag{5.5.8}$$

which may also be obtained directly by taking the derivative of (5.5.3).

Now consider the rate, E_P, of pressure working on the fluid as it enters and leaves the duct:

$$E_P = \int_0^R (p_1 - p_2) v_1 2\pi r dr, \tag{5.5.9}$$

Now since the flow in the pipe is parallel, the pressure will not vary across the pipe area, so:

$$E_P = Q(p_1 - p_2) = \rho_0 g h_L Q \tag{5.5.10}$$

where Q is the discharge through the pipe. Substituting for h_L from (5.5.8) into (5.5.10) and rearranging leads again to (5.5.6); we see that the head loss or pressure drop driving the flow through the pipe is due to the fluid working against the wall stress τ_w at the boundaries of the pipe.

The energy imparted by the pressure working must go somewhere; it is imparted to the fluid as it flows through the duct. This energy is used to overcome the viscous resistance in the fluid and is then converted to heat. To see this, consider an annulus of fluid inside the pipe of thickness dr. The stress acting in the fluid at radius r is given by:

$$\tau_{13} = \mu \frac{\partial v_1}{\partial r}, \tag{5.5.11}$$

and the velocity differential across the annulus is, to first order:

$$\Delta v_1 = \frac{\partial v_1}{\partial r} dr, \tag{5.5.12}$$

hence the rate of working per unit mass, commonly called the dissipation per unit mass, is given by:

$$\varepsilon = \frac{\int_0^R 2\pi r\mu\left(\frac{\partial v_1}{\partial r}\right)^2 \mathrm{d}r}{\pi R^2 \rho_0} = \frac{1}{8}\frac{g^2 h_L^2 R^2}{\nu L^2}, \quad m^2 s^{-3}. \tag{5.5.13}$$

This may be rewritten, using (5.5.5) in terms of the discharge velocity u_d:

$$\varepsilon = 8\nu\left(\frac{u_d}{R}\right)^2 \sim \nu\left(\frac{\partial v_1}{\partial r}\right)^2. \tag{5.5.14}$$

It follows immediately from (5.5.14) that the dissipation of the pressure rate of working occurs at a length scale R and velocity scale u_d; this explains why dissipation is small in laminar flows.

For $\mathrm{Re} > \mathrm{Re_c}$ the flow is turbulent and we must amend the above logic; in a turbulent flow there are many length scales, R is the largest scale and the smallest scale is where viscosity dissipates the rate of working applied to the fluid by the action of the pressure differential pushing the fluid through the pipe. The reader may wish to turn to §2.13 and briefly revise the statistical concepts presented there. Let us visualize a turbulent flow moving through the straight long pipe of length L and radius R. The pressure differential $(p_1 - p_2)$ forcing the fluid through the pipe is assumed to be constant and large enough to maintain a fully turbulent flow. If τ_w is the stress of the fluid on the pipe wall, then this will have a mean component $\overline{\tau}_w$ and a fluctuating component τ'_w and since the mean momentum flux leaving the pipe is the same as entering the pipe, we may write from bulk conservation of momentum:

$$2\pi RL\overline{\tau}_w = (p_1 - p_2)\pi R^2. \tag{5.5.15}$$

By definition, the dissipation per unit mass, ε, must be equal to the rate of pressure working of the unit mass, so that:

$$\overline{\varepsilon} = \frac{2\overline{\tau}_w u_d}{\rho_0 R}. \tag{5.5.16}$$

Given the flow is fully turbulent, the dissipation (5.5.16) must be viewed as the mean local dissipation, obtained by balancing an assumed constant dissipation with the rate of pressure working. Clearly, the actual local dissipation will vary across the pipe radius and also will have a random component as a result of the randomly fluctuating velocity.

As discussed in §2.14, the mean shear stress, $\overline{\tau}_w$, in a turbulent flow is due to the Reynolds stress and from (2.14.4):

$$\overline{\tau}_{ij} = \rho_0\left(\overline{v_i' v_j'}\right). \tag{5.5.17}$$

The velocity field will not just have the single length scale, but rather an infinite continuous range of length scales from the diameter or radius R to a very small-scale where the motion dissipates most of the energy. The overall dynamical processes involved in the turbulent motion were first independently elucidated by Kolmogorov and by Taylor. For the details of these descriptions, the reader is referred to specialty books on Turbulence with a very good introduction being given in Tennekes and Lumley (1972).

In brief, both points of view envisaged that large eddies, in our case of scale R and velocity scale U, that randomly move around in the domain (pipe cross-section) are advected with the mean flow. Given, as we saw from (5.5.14) at these scales little energy is lost to viscous friction, these large scale eddies cannot balance the rate of working imposed by the pressure gradient that forces the flow through the pipe. Taylor and Kolmogorov, independently postulated that such large eddies, the size of the domain scale, lose their energy to smaller scales through an inertial instability cascade, where energy moves down through a continuous spectrum of length scales. This is called the inertial subrange. At the largest scale the turbulent fluid motion is mostly governed by an inertia pressure balance and as we progressively move down scale viscosity becomes more and more important, until we reach the smallest scale of motion, called the Kolmogorov scale, η_0, where the motion becomes a balance between the rate of fluid rate of straining balancing the rate of viscous dissipation. Clearly from (5.5.17) momentum is predominantly transferred by the larger scale eddies and energy is dissipated predominantly at the Kolmogorov scale:

$$\overline{\tau}_{ij} = \rho_0\left(\overline{v_i' v_j'}\right) \sim \rho_0 U^2, \tag{5.5.18}$$

substituting this into (5.5.17) leads to the "outer" or large scale estimate of the dissipation of turbulent kinetic energy:

$$\overline{\varepsilon} \sim \frac{U^3}{R} = \frac{U^2}{R/U} = \frac{U^2}{t_c}, \tag{5.5.19}$$

where t_c is the time scale for an eddy of size R to overturn once. This suggests that large eddies lose a constant, large, fraction of their energy in one revolution.

The mental picture of both Kolmogorov and Taylor, was that the turbulent flows receive their energy at the large scale and the cascade of eddies or scales transfers this energy flux to the Kolmogorov scale where it is dissipated by the smallest shearing scales where viscosity smears the motion, extracting the energy flux and turning it into heat.

Equation (5.5.19) states that the dissipation in the fluid is independent of the fluid viscosity. At first sight this is a contradictory result. However, if we retain the focus on our pipe flow problem, then what (5.5.19) expresses is that, as we increase the pressure differential, more energy goes into the eddy cascade at the pipe radius scale and so this increased energy flux must be dissipated at the smallest scales; this is achieved by adjusting the smallest scales of motion until the viscous friction is sufficient to generate enough heat to dissipate the mechanical energy injected into the fluid by the pressure working. This clearly suggests that the Kolmogorov scale η_0 must only dependent on the viscosity of the fluid and the dissipation, so that we may write:

$$\eta_0 = \eta_0(\nu, \varepsilon). \tag{5.5.20}$$

Simple dimensional analysis then requires that the Kolmogorov scale is given by:

$$\eta_0 = \left(\frac{\nu^3}{\varepsilon}\right)^{\frac{1}{4}}. \tag{5.5.21}$$

In reality this is a simplification, as viscous dissipation acts at all scales, however, partitioning the dynamics as we have done above has been shown to describe experimental observations extremely well.

Similar reasoning leads to the small-scale velocity scale:

$$v_i = (\nu\varepsilon)^{\frac{1}{4}}, \tag{5.5.22}$$

and the small-scale time scale τ is given by:

$$\tau = \left(\frac{\nu}{\varepsilon}\right)^{\frac{1}{2}}. \tag{5.5.23}$$

There are some very interesting consequences of the cascade process that we have described. First, consider the Reynolds number of the smallest scales of motion:

$$Re = \frac{v_i \eta_0}{\nu}. \tag{5.5.24}$$

Upon substitution from (5.5.21) and (5.5.22), we see that this small-scale Reynolds number (5.5.24) has a value of 1, confirming that at these small-scales, the turbulence behaves as the flows described in §4; viscosity dominates the motion at these small-scales.

Second, using (5.5.21) and (5.5.24) the ratio of the smallest length scale to the largest length scale becomes:

$$\frac{\eta_0}{R} = \frac{1}{\mathrm{Re}^{\frac{3}{4}}}, \tag{5.5.25}$$

where, $\mathrm{Re} = \dfrac{UR}{\nu}$, is the large scale Reynolds number. The relationship (5.5.25) clearly shows that the ratio of the smallest scale motion to the largest scale (the domain size) decreases with increasing Reynolds number. In other words, as the large scale Reynolds number is increased, increasing the pressure differential in our pipe flow problems, the smallest scale motions decrease in sizes to provide more frictional dissipation. Relationship (5.5.25) also provides a useful practical guide to the scale of an original scene presented on our television screens as illustrated in Fig. 5.5.1. The picture size marks the scale size of the domain size we are viewing and the smallest scales appear on our screens as a sort of texturing of the image. For instance, when the evening news shows visuals of a volcano eruption as depicted in Fig. 5.5.1a, the plume emanating from the volcano appears with a fine texture. By contrast, the picture in Fig. 5.5.1b shows smooth smoke rising from a cigarette which is a low Reynolds number flow.

Fourth, the Kolmogorov length scale η_0 has another important interpretation as the scale at which the intensification of the vorticity due to stretching just balances the transverse diffusion of vorticity. To see this we compare these two terms from the vorticity equation (see §2):

$$\zeta_i v_{i,j} : \nu \zeta_{i,jj}, \tag{5.5.26}$$

where, from (5.5.21) and (5.5.24), the order of the ith vorticity component is given by:

$$\zeta_i \sim v_{i,j} \sim \frac{(\nu\varepsilon)^{\frac{1}{4}}}{\left(\dfrac{\nu^3}{\varepsilon}\right)^{\frac{1}{4}}} = \left(\frac{\varepsilon}{\nu}\right)^{\frac{1}{2}}, \tag{5.5.27}$$

which is called the strain rate γ due to the turbulent fluctuations. Substituting this into (5.5.26) shows that the two terms are of the same order.

In summary, we have thus close similarity between laminar and turbulent flow at the very largest scales in that the momentum equation reflects a similar force balance (in this case the pressure force being balanced by wall friction), but the two types of flow are totally different in the way the fluid resists the motion and dissipates the energy. In laminar flow, viscous resistance develops by simple shearing motions on scales of the fluid domains or perhaps the scale of the boundary layers (see boundary layers §4.3). By contrast in a turbulent flow, the effect of the cascade mechanism is to create eddies at all scales that are effective in transporting momentum at an enhanced rate toward the wall, leading to an increased frictional resistance.

The above word picture also allows a reasonably accurate understanding of how stratification, common in most environmental flows, influences turbulence. As the stratification increases, vertical motions are inhibited (as discussed in §2.3) and so the eddies discussed above change into more pancake shaped motions. This reduces the vertical Reynolds stresses and the fluid becomes more slippery until, at very strongly stratified conditions, all turbulence is inhibited and the motion is laminar even if the Reynolds numbers are very large; the buoyancy imparted by the stratification prevents vertical motions and eddies become two-dimensional eliminating the possibility of vortex stretching and so preventing an inertial energy cascade as described above.

5.6. TURBULENT FLOW IN A SMOOTH PIPE

We now discuss in more detail the ideas, presented in §5.5, as they apply to flow through a pipe at large Reynolds number, fully turbulent flow. Near the center of the pipe, we may expect the ideas of the energy cascade discussed in §5.5 to apply, but near the pipe wall, the presence of a solid boundary, the pipe wall, may be expected to require the simple cascade to be modified, simply because, near the wall, the scale of the inertia eddies is constrained. Very near the wall, the fluid can only move parallel to the wall and sustain a shear stress through viscous action. This, however, requires a very large rate of strain rate near the wall, so that the viscous stress at the wall can match the large eddy Reynolds stress (5.5.19) that is present in the center of the pipe. Clearly, there must be continuity of stress across the pipe cross-section. It has been observed experimentally, that a thin layer of thickness, δ, exists near the wall, called the "viscous sublayer", which supports the wall shear stress necessary to balance the applied pressure force. Thus very near the wall:

$$\frac{\tau_w}{\rho} = \nu \frac{\partial \overline{v}_1}{\partial x_3} \sim \rho_0 U^2 \tag{5.6.1}$$

where $x_3 = R - r$ is the coordinate perpendicular and away from the wall. Now the quantity τ_w/ρ has units of velocity and we write:

$$u_* = \left(\frac{\tau_w}{\rho}\right)^{\frac{1}{2}}, \tag{5.6.2}$$

and we shall call u_*, the shear velocity.

The thin viscous sublayer near the wall will not feel the large eddies and we may expect, from dimensional analysis, that the thickness will only depend on the viscosity and the shear velocity, so we may write:

$$\delta \sim \frac{\nu}{u_*}. \tag{5.6.3}$$

There are, therefore, not two as in §5.5, but three identifiable limiting length scales in the flow; the radius R of the pipe the scale of the largest eddies, the Kolmogorov scale, η_0, is the smallest scale in the flow in the main part of the pipe, away from the influence of the wall and δ is the scale of the thin high shear layer near the wall. Using (5.6.1) and (5.5.19) we may show:

$$\frac{\delta}{\eta_0} = \frac{1}{Re^{\frac{1}{4}}}. \tag{5.6.4}$$

Hence, for large Reynolds number flows: $\delta < \eta_0 < R$.

We now use these three length scales to construct expressions for the mean velocity profile in the pipe. First, very near the pipe wall, inside the viscous sublayer, it seems reasonable to assume that the flow is parallel and the stress across the layer is constant (see Couette Flow in §2.1):

$$\overline{v}_1 = \frac{u_*^2}{\nu} x_3, \tag{5.6.5}$$

or

$$\frac{\overline{v}_1}{u_*} = \frac{x_3}{\delta}. \tag{5.6.6}$$

Second, immediately outside the viscous sublayer, there will be a region called the "Law of the Wall" where the flow is fully turbulent, but still "feels" the constraints of the pipe wall. In this region, eddies are born from instabilities in the viscous sublayer and grow toward the center of the pipe. However, these eddies have escaped from the influence of the viscous sublayer, so the mean flow is dependent on the fluid viscosity only in the combination of δ, so that:

$$\frac{\overline{v}_1}{u_*} = f\left(\frac{x_3}{\delta}\right). \tag{5.6.7}$$

Third, the flow in the main core of the pipe is fully developed receiving its turbulent energy from the wall region at large scales and maintaining a full energy cascade process to small-scales. It may therefore be expected that the velocity profile will be independent of the viscosity so that:

$$\frac{\overline{v}_1 - \overline{u}_m}{u_*} = g\left(\frac{x_3}{R}\right), \tag{5.6.8}$$

where $\overline{u}_m$ is the maximum velocity at the center of the pipe and R is the radius of the pipe.

Experimental evidence suggests that there is an overlap region where (5.6.7) and (5.6.8) yield an equally valid description of the velocity profile. It is a remarkable fact that this is sufficient to determine the functions f and g.

To see this substitute (5.6.7) into (5.6.8) and then differentiation the resulting expression with respect to ζ, where,

$$\zeta = \frac{zu_*}{\nu} = \frac{x_3}{\delta}. \tag{5.6.9}$$

This yields the result:

$$\zeta\frac{\mathrm{d}f(\zeta)}{\mathrm{d}\zeta} = \xi\frac{\mathrm{d}g(\xi)}{\mathrm{d}\xi}, \tag{5.6.10}$$

where

$$\xi = \frac{x_3}{R} = \frac{x_3 u_*}{\nu}\cdot\frac{\nu}{u_* R}. \tag{5.6.11}$$

Now, as is commonly used in the separation of variables technique, it may be observed that the equality (5.6.10) can only exist if both sides of the equation are equal to (the same) constant. If we call this constant A, then integration leads to:

$$f = A\log\zeta + B, \tag{5.6.12}$$

$$g = A\log\xi + C, \tag{5.6.13}$$

where we have changed from natural logarithm to the base ten logarithmic function for convenience.

Over the years a great deal of data has been collected on the mean velocity distribution in a pipe and these observations are discussed in standard hydraulics texts. A least square fit to the data reveals the value of the constants:

$$\mathrm{A} = 5.65, \tag{5.6.14}$$

$$\mathrm{B} = 5.4, \tag{5.6.15}$$

$$\mathrm{C} = 0. \tag{5.6.16}$$

Substituting (5.6.14)–(5.6.16) into (5.6.7) and (5.6.8) leads to the final expressions for the velocity profiles:

$$\frac{\overline{v}_1}{u_*} = 5.75\log\left(\frac{x_3 u_*}{\nu}\right) + 5.4, \tag{5.6.17}$$

$$\frac{\overline{u}_m - \overline{v}_1}{u_*} = 5.75\log\left(\frac{R}{R-r}\right). \tag{5.6.18}$$

The exact thickness of the viscous sublayer may now be defined as the distance x_3 where (5.6.5) and (5.6.17) cross:

$$\frac{x_3 u_*}{\nu} = 5.75\log\left(\frac{x_3 u_*}{\nu}\right) + 5.4 \tag{5.6.19}$$

This leads to a value for δ:

$$x_3 = 11.6\frac{\nu}{u_*} = 11.68\delta = \delta_\nu \tag{5.6.20}$$

where δ is the scale (ν/u_*).

Combining (5.6.5), (5.6.17) and (5.6.18) provides a composite velocity profile.

To obtain a relationship between the discharge velocity:

$$u_d = [\overline{v}_1], \tag{5.6.21}$$

and the maximum velocity at the center of the pipe requires the integration of the composite profile (5.6.5), (5.6.17) and (5.6.18) such that:

$$u_d = \frac{1}{\pi R^2} \int_0^R \overline{v}_1 2\pi r dr. \tag{5.6.22}$$

An approximate answer may be obtained by substituting only (5.6.18) into (5.6.22) and noting that the singularity at $r = R$ is integrable. This yields:

$$\frac{u_d}{u_m} = \frac{1}{1 + 3.76\dfrac{u_*}{u_d}}. \tag{5.6.23}$$

The relationship between the wall shear velocity and the Reynolds number may be obtained directly by equating (5.6.17) and (5.6.18). If we note that from (5.5.6) and (5.5.8) we may write:

$$u_* = \frac{\sqrt{f}}{2\sqrt{2}} u_d. \tag{5.6.24}$$

Introducing (5.6.24) into (5.6.23) leads to:

$$\frac{u_d}{u_m} = \frac{1}{1 + 1.33\sqrt{f}}. \tag{5.6.26}$$

Now equating (5.6.17)–(5.6.18) and using (4.3.26) leads to the equation for the friction factor of a smooth pipe:

$$\frac{1}{\sqrt{f}} = 2.0 \log \mathrm{Re}\sqrt{f} - 1.0, \tag{5.6.26}$$

where $\mathrm{Re} = \dfrac{u_d D}{\nu}$ pipe flow Reynolds number based on the pipe diameter D.

We thus have two relationships for the friction factor; laminar flow is given by (5.5.6) and turbulent flow is described by (5.6.29). These relationships are shown in Fig. 5.6.1.

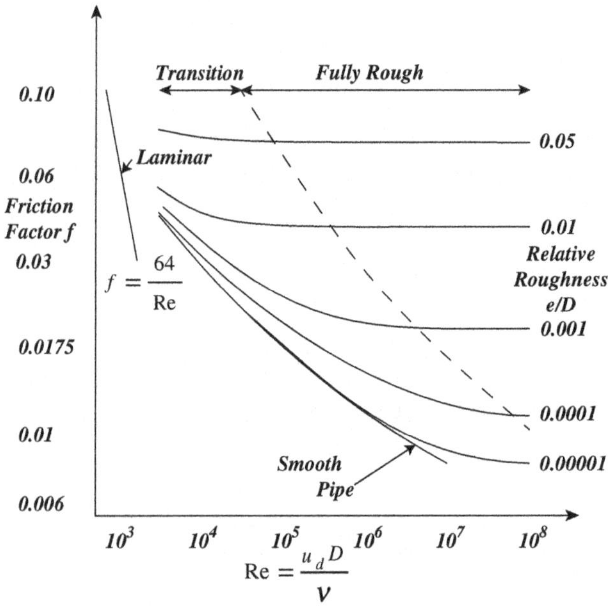

Figure 5.6.1 Friction factor as a function of wall roughness, see laminar flow (5.5.6); smooth walled pipe (5.6.26); rough walled pipe (5.7.11); transition flow (5.7.13).

5.7. TURBULENT FLOW IN A ROUGH PIPE

In §5.6, we saw that the wall of the pipe is covered by a high shear region in which the flow is essentially dominated by viscous behavior. However, it must be remembered that this layer is extremely thin. Suppose we take a simple household supply pipe with a diameter of 0.025 m and a pressure drop of 1 m over 1000 m. From (5.5.8) and (5.6.2) it follows:

$$u_*^2 = \frac{gRh_L}{2L}, \tag{5.7.1}$$

leading to a shear velocity $u_* = 7.8 \times 10^{-3}\ \mathrm{m\,s^{-1}}$ and, from (5.6.20) a viscous sublayer thickness $\delta_\nu = \frac{\nu}{u_*} = 1.5 \times 10^{-3}\mathrm{m}$ (assuming a viscosity $\nu = 10^{-6}\ \mathrm{m^2\,s^{-1}}$).

Experiments show that if the surface roughness $e < 0.2\delta_\nu$, then the surface imperfections (roughness) will be submerged in the viscous sublayer and the surface will appear smooth to the flow, with the discussion in §5.6 remaining valid.

The meaning of surface roughness needs some comment. It is difficult to find an objective measure as the standard deviation of the surface roughness alone is not sufficient since the likelihood of roughness elements triggering an instability in the viscous sublayer depend not only on the height of the roughness, but also on the wavelength of the wall roughness. For this reason engineers have compared actual pipes with smooth pipes that have been artificially roughened with a layer of spherical beads. The absolute roughness of a material is then defined as the diameter of the spheres, e, in our artificially roughened pipe that yield equivalent head losses to the actual pipe.

As the absolute roughness is increased, the viscous sublayer can no longer establish and each roughness element acts to shed eddies in its wake. This occurs when the absolute roughness, $e > 3.3\delta_v$. This marks the boundary when the flow becomes independent of the fluid viscosity.

Once this occurs, the velocity relationship naturally also reflects this independence and the relationships corresponding to (5.6.7) and (5.6.8) become:

$$\frac{\overline{v}_1}{u_*} = f\left(\frac{x_3}{e}\right), \tag{5.7.2}$$

and

$$\frac{\overline{v}_1 - u_m}{u_*} = g\left(\frac{x_3}{R}\right). \tag{5.7.3}$$

Using the same logic as for the smooth pipe, §5.6, it follows directly that,

$$\mathrm{f}\left(\frac{x_3}{e}\right) = A\log\left(\frac{x_3}{e}\right) + B, \tag{5.7.4}$$

and

$$g\left(\frac{x_3}{R}\right) = A\log\left(\frac{x_3}{R}\right) + C. \tag{5.7.5}$$

Once again experimental data may be used to obtain the coefficients such that:

$$\mathrm{A} = 5.65, \tag{5.7.6}$$

$$\mathrm{B} = 8.6 \tag{5.7.7}$$

$$\mathrm{C} = 0 \tag{5.7.8}$$

which leads to the velocity profiles:

$$\frac{\overline{v}_1}{u_*} = 5.75\log_{10}\left(\frac{x_3}{e}\right) + 8.4, \tag{5.7.9}$$

$$\frac{\overline{v}_1 - u_m}{u_*} = 5.75\log_{10}\left(\frac{x_3}{R}\right). \tag{5.7.10}$$

and the friction factor relationship becomes:

$$\frac{1}{\sqrt{f}} = 2.0\log\frac{D}{\varepsilon} + 1.14. \tag{5.7.11}$$

This friction factor is now independent of the Reynolds number; a result that is typical for large Reynolds number flows. Equation (5.7.11) is sketched in Fig. 5.6.1 for different roughness values.

Comparison of (5.7.11) with (5.6.26) reveals that for a smooth walled pipe, the viscous sublayer thickness scale (ν/u^*) acts in an analogous fashion to the roughness length e in a rough walled pipe. This reveals, at least intuitively, that the viscous sublayer acts similarly to the protruding roughness elements in a rough walled pipe suggesting that the viscous sublayer is indeed unstable and gives birth to turbulent motions that scale with of the viscous sublayer thickness; the fluctuations that are born near the wall, either through an instability of the viscous sublayer or through eddy shedding in the lee of wall roughness elements, grow to a scale R before the cascade process robs them of their energy feeding scales of motion back down to a scale of η_0.

In the transition region:

$$38.6 > \frac{u_* e}{\nu} > 2.3, \tag{5.7.12}$$

Colebrook (1939) found, empirically, that the friction factor was given by:

$$\frac{1}{\sqrt{f}} = -2.0\log\left(\frac{e}{7.4R} + \frac{2.51}{\mathrm{Re}\sqrt{f}}\right). \tag{5.7.13}$$

This relation merely patches together the limiting relationships for a smooth pipe (5.6.26) and a fully rough pipe (5.7.11).

5.8. CLASSIFICATION OF OPEN CHANNEL FLOW

Distinct from pipe flow, where the fluid is constrained to move within the walls of the pipe and the pressure gradient pushes the fluid through the pipe, in an open channel or river the free surface adjusts to pressure changes and the depth increases or decreases to accommodate the flow; the pressure at the free surface is always atmospheric. Thus in an open channel flow, the force required to make water move along the channel against friction is supplied by the component of gravity parallel to the bottom slope of the channel, rather than a pressure gradient as is the case in pipe flow. Otherwise the two flows, uniform pipe and channel flow, are very similar; in pipe flow a pressure gradient, uniform across the pipe section, drives the flow through the pipe against friction, whereas a component of gravity, again uniform across the channel cross-section, causes the water to flow downhill. The freedom provided by the free surface, however, allows open channel flows to take on a greater variety of flow configurations and it is convenient, right from the outset, to define particular regimes of flow:

Steady/Unsteady: As in other situations we say an open channel flow is steady when the velocity at every point is constant with time; in turbulent flow, the focus is on the steadiness of the mean velocity and the stationarity of the statistical properties of the turbulent flow field (see also §2.13).

Turbulent/Laminar: Properties of a turbulent flow were discussed in §5.5 and just as in pipe flow, the magnitude of the Reynolds number determines whether the flow in the channel is laminar or turbulent; again there is a critical Reynolds number above which the flow becomes turbulent.

Uniform/Non-Uniform: Uniform flow is when the streamlines are all parallel to the channel bottom. Uniform flow is the exact counterpart to pipe flow and in uniform flows the water velocity in the channel adjusts until the force, due to bottom friction, retarding the flow, just balances the gravity force acting downstream. In a uniform flow, pressure is hydrostatic everywhere and adds little to the flow dynamics. When the streamlines in the channel are not parallel to the bottom we call the flow varied or non-uniform.

Gradually/Rapidly Varied: Non-uniform flows are further divided into gradually varied and rapidly varied flows. Gradually varied flows are varied flows that are dynamically locally uniform and are adequately described by neglecting the vertical acceleration of the water and assuming that the pressure varies hydrostatically over the depth. The flow depth may, however, vary along the channel, adding a hydrostatic pressure force to the gravity force, from where the flow depth is greater to where it is less. In gradually varied flows the equations of motion remain essentially the same as those in uniform flow except that a net pressure force, proportional to the gradient of the water surface, is added to the gravity force. By comparison, rapidly varied flows, are flows where the pressure departs from the hydrostatic variation and some of the pressure force is used to verticaly accelerate the water; in a rapidly varied flow the water free surface responds to the rapidly varying pressure gradient and undergoes a rapid change of height.

Sub/Supercritical: Waves on the surface of a river or channel are common occurrences familiar to all of us. As we shall see in §8, surface waves move over the water surface at a particular speed relative to the speed of the water itself. We call a flow subcritical if the waves can move both upstream and downstream, wave speed is great than the flow velocity and supercritical if the water velocity is larger than the wave speed and all waves generated in the channel are swept down downstream.

The above classification scheme is matrix like, but not all combinations are physically possible. For instance, it is not possible for a flow to be both uniform and unsteady. The great majority of open channel flows are turbulent and either steady and uniform, steady and gradually varied, unsteady and gradually varied, steady and rapidly varied or unsteady and rapidly varied. Rapidly varied flows most often occur as flow adjustments between reaches of gradually varied flows.

5.9. UNIFORM FLOW

The simplest open channel flow configuration is one in which the bottom channel slope and the channel or river cross-section are constant along the length of the channel. Then, if the channel is long, the flow will quickly reach a uniform state where all the streamlines are parallel to the channel bottom. Since the flow is parallel, the pressure will automatically be hydrostatic and the discharge and momentum will be constant along the length of the channel. In other words, the volume, mass and momentum of water flowing into the Eulerian control volume, shown in Fig. 5.9.1, will be exactly equal to the volume, mass and momentum leaving the control volume per unit time.

The conservation laws discussed in §1.8–§1.11 reduce to a simple force balance between gravity acting downhill and bottom friction balancing this

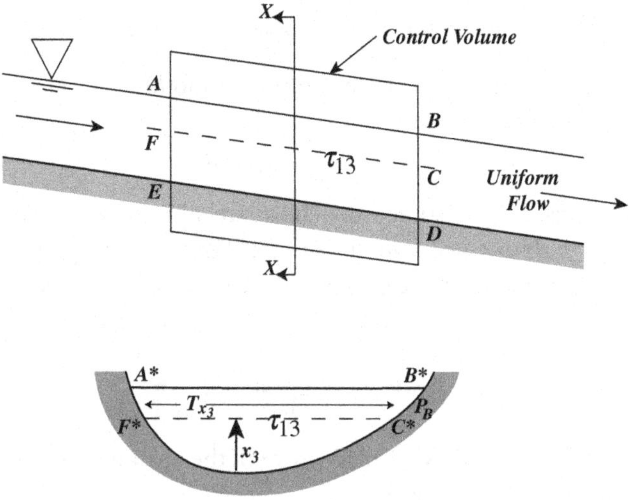

Section X-X

Figure 5.9.1 Schematic of uniform flow.

gravity force; since the friction force increases with the flow velocity, the water velocity will be such as to create a friction force that exactly balances the downstream acting gravity force. This simple force balance may be applied to the control volume ABCF, shown in Fig. 5.9.1, extending from the free surface to a depth $(h - x_3)$, where x_3 is a coordinate pointing upward and perpendicular to the bottom:

$$W\sin\theta = \tau_{13}LT_{x_3} + \tau_B LP_B, \tag{5.9.1}$$

where T_{x_3} is the width of the channel at the height x_3, P_B is the wetted perimeter above x_3, τb is the stress on the channel boundary over the perimeter P_B, τ_{13} is the shear stress in the water at height x_3 in the x_1 direction and L is the length of the control volume in the direction of flow.

The weight of the fluid above the height x_3 is given by:

$$W = \rho g A_{x_3} L, \tag{5.9.2}$$

where A_{x_3} is the cross-sectional area above x_3

Substituting (5.9.2) into (5.9.1) and rearranging terms yields:

$$\tau_{13} = \rho g \sin\theta D_{x_3} + \frac{P_B}{T_{x_3}}\tau_B, \tag{5.9.3}$$

where $D_{x_3} = \frac{A_{x_3}}{T_{x_3}}$. and θ is the bottom angle. For a wide, rectangular channel the last term is very small and may be neglected and $D_{x_3} = (h - x_3)$, so that (5.9.3) reduces to:

$$\tau_{13} = \rho g(h - x_3)\sin\theta. \qquad (5.9.4)$$

In wide rectangular channel, the shear stress in the water column thus varies linearly with distance from the surface. We may also apply (5.9.3) thus over the whole channel by setting $x_3 = 0$. If we further assume that the boundary stress is uniform over the whole wetted perimeter then:

$$\tau_B = \rho g \frac{A_0}{P_0}\sin\theta, \qquad (5.9.5)$$

where A_0 is the flowing cross-sectional area of the channel and P_0 is the total wetted perimeter. The quantity;

$$R = \frac{A_0}{P_0}, \qquad (5.9.6)$$

is called the hydraulic radius and is a measure of the scale of the flow as affected by the side wall friction.

In order to find the velocity of the water in such a channel, we may proceed by analogy with pipe flow where it was assumed, from dimensional reasoning,

$$\tau_b = \frac{f}{8}\rho u_d^2, \qquad (5.9.7)$$

where u_d is the channel discharge velocity $\left(u_d = \frac{Q}{A_0}\right)$, Q is the discharge of water in the channel, f is the friction factor and the factor 8 was added to conform to convention. For flow in a pipe (§4.1) we saw that:

$$f = f\left(\mathrm{Re}, \frac{e}{D}\right), \qquad (5.9.8)$$

where

$$\mathrm{Re} = \frac{u_d D}{\nu}, \qquad (5.9.9)$$

and D is the diameter of the tube. The hydraulic radius for a circular pipe is given by:

$$R = \frac{D}{4}, \qquad (5.9.10)$$

so that the effective Reynolds number for channel flow, if we require compatibility with the friction factor, will be:

$$\mathrm{Re} = \frac{4u_d R}{\nu}, \tag{5.9.11}$$

with the roughness parameter becoming $\frac{e}{4R}$.

Substituting (5.9.5) into (5.9.10) leads to:

$$u_d = \left(\frac{8gR}{f}\right)^{1/2} \sin^{1/2}\theta. \tag{5.9.12}$$

For the simple case of a very wide channel of depth h and discharge per unit width q, this reduced to:

$$h_n = h = \left(\frac{fq^2}{8g\sin\theta}\right)^{\frac{1}{3}}. \tag{5.9.13}$$

This is called the normal flow depth, the depth for the discharge q where downslope gravitational forces just balance the retarding bottom drag forces.

If we note that the head loss in the open channel is $L \sin\theta$, then (5.9.12) may be rewritten in the form:

$$h_L = L\sin\theta = (f)\left(\frac{L}{4R}\right)\left(\frac{u_d^2}{2g}\right), \tag{5.9.14}$$

which is the counterpart to (5.5.6) in pipe flow.

The similarity between (5.5.6) and (5.9.14) is to be expected if we imagine doubling the channel flow by adding the upside down channel flow so that the free surfaces become one. We thus have a solid boundary at the bottom and the top and the channel is now a conduit with twice the area and twice the perimeter, so the hydraulic radius remains the same. There is, however, one important difference between an open channel flow and a pipe flow. In an open channel flow, the free surface allows surface waves to move along the water surface. These waves have an associated velocity field and so may induce velocity fluctuations at the channel bottom, which in turn may influence the bottom stress. In §5.8, we have already discussed the concept of sub and supercritical flows as being determined by the relative speed of the surface waves to that of the velocity u_d of the water surface. From dimensional analysis, the speed of a long surface wave can only depend on the acceleration due to gravity and the depth h of the water column; if the wave is much longer than the water depth, then the water depth is no longer an important parameter, so dimensional reasoning leads to a wave speed $c \sim (gh)^{\frac{1}{2}}$. The ratio of the water velocity to the wave speed is called the surface Froude Number:

$$Fr = \frac{u_d}{(gR)^{\frac{1}{2}}}. \tag{5.9.15}$$

Thus, in general the friction factor will also depend on the Froude Number:

$$f = f\left(\mathrm{Re}, \frac{e}{4R}, Fr\right), \tag{5.9.16}$$

where the Froude Number Fr introduces the influence of surface deformations on the friction factor. However, little information exists on the magnitude of the influence of Fr on the friction factor, but it is clear that whenever the surface undulations become large they will effect the eddy structure, and so the friction factor, in an open channel. In practical applications, the influence of Fr is, however, always neglected.

The above use of the friction factor is the most rational approach for channel flow; it is dimensionally consistent and further has the advantage of unifying pipe and channel flow. However, traditionally hydraulic engineers have used two empirical formulas and these are still used very widely so it is necessary to discuss them here.

The older of these is the Chezy equation, the counterpart to (5.9.12), and reads:

$$U = C(RS)^{1/2}, \tag{5.9.17}$$

where R is the hydraulic radius, S is the slope of the uniform channel and C is the Chezy friction factor. Comparison of (5.9.17) and (5.9.12) reveals the relationship between the friction factor and the Chezy coefficient:

$$C = \left(\frac{8g}{f}\right)^{1/2}. \tag{5.9.18}$$

This clearly shows that the Chezy formulation is similar to the friction factor method, except that C is not dimensionless and so takes on different values for different measurement systems.

The second formula is due to Manning see Chow 1959 who postulated:

$$u_d = \frac{1}{n} R^{2/3} S^{1/2}, \tag{5.9.19}$$

where n is called Manning's n. Many authors (see for example Chow, 1959) give values of n for both artificial channels and natural streams. Once again we may equate (5.9.19)–(5.9.12) in order to obtain a relationship between the friction factor and Manning's n:

$$n = \frac{R^{1/6} f^{1/2}}{(8g)^{1/2}}, \tag{5.9.20}$$

from which we clearly see that the parameter n is not only, not dimensionless, but also depends on the flowing depth; a rather undesirable feature of this formulation.

5.10. VELOCITY PROFILE IN A WIDE CHANNEL WITH UNIFORM FLOW

We now derive an expression for the velocity profile in uniform flow and show how the flow adjusts to develop the necessary friction required to balance the downstream directed gravity force. However, a simple derivation is only possible for the case of a channel that is much wider than it is deep and with a relatively flat bottom. For this limiting case, we may assume (5.9.4) holds and the internal stress varies linearly from zero at the free surface to the bottom stress τ_b at the bottom. This relationship holds for both laminar and turbulent flow. Consider first the simpler case of laminar form where from §2.2 we may write (repeated here for convenience):

$$\tau_{13} = \mu \frac{\partial v_1}{\partial x_3}. \tag{5.10.1}$$

Equating this expression with (5.9.4) and integrating the expression leads to the velocity profile for laminar flow:

$$v_1 = \frac{g \sin\theta h^2}{\nu} \eta \left(1 - \frac{\eta}{2}\right), \tag{5.10.2}$$

where $\eta = \frac{x_3}{h}$. A similar analysis can be carried out for turbulent flow if we consider the mean velocity $\overline{v}_1(x_3)$ depends only on the vertical coordinate. From (5.5.18) and assuming $v_1' \sim u_*$ and $v_3' \sim x_3 \left(\frac{h - x_3}{h}\right) \frac{\partial \overline{v}_1}{\partial x_3}$, then:

$$\tau_{31} = \rho_0 \left(\overline{v_1' v_3'}\right) = v_3' \sim k \rho_0 u_* x_3 \left(\frac{h - x_3}{h}\right) \frac{\partial \overline{v}_1}{\partial x_3}, \tag{5.10.3}$$

where k is the von Karman constant and represents the correlation between the turbulent velocity fluctuation v_1' and v_3'. It assumed here that the vertical velocity fluctuation v_3' must be zero close to the bottom and close to the free surface; the energy in the turbulent fluctuations is insufficient to undulate the free surface. Substituting (5.10.3) into (5.9.4) and using the result that at $x_3 = 0$,

$$\rho_0 u_*^2 = \rho_0 g h \sin\theta, \tag{5.10.4}$$

leads to the differential equation:

$$\frac{u_*}{k x_3} = \frac{\partial \overline{v}_1}{\partial x_3}. \tag{5.10.5}$$

Integrating with respect to x_3 yields

$$\frac{\overline{v}_1}{u_*} = \frac{1}{k} \ell n \frac{x_3}{z_0}, \tag{5.10.6}$$

where z_0 is an integration constant. Equation (5.10.6) is the same form as the formula (5.6.17) for the pipe flow, derived there from purely dimensional analysis and scaling. The same reasoning could have also been applied here, but we have used instead the formulation (5.10.3) to illustrate that the log velocity profile implicitly involves scaling the vertical velocity fluctuations with distance from the bottom and the free surface. The velocity distribution (5.10.6) is called the Prandtl–von Karman universal-velocity distribution law.

When the bottom surface of the channel is smooth then, as in pipe flow,

$$z_0 \sim \frac{\nu}{u_*}, \tag{5.10.7}$$

and therefore:

$$\overline{v_1} = 4.75 u_* \, Log_{10} \frac{9 x_3 u_*}{\nu}, \tag{5.10.8}$$

that is the same as the Law of the Wall for pipes (5.6.17).

For rough beds, by analogy with pipe flow we may assume $z_0 \sim \varepsilon$, so that:

$$\overline{v}_1 = 5.75 u_* \, Log_{10} \frac{30 x_3}{e}. \tag{5.10.9}$$

In both cases, the constants have been derived by comparing the equations with experimental results (see Chow, 1959).

5.11. HYDRAULIC JUMP AND SPECIFIC ENERGY

The hydraulic jump is a very common rapidly varied flow adjustment that we briefly discussed in §1.11 in an application of the bulk momentum equation. It may be observed in rivers, open channels, in street gutters or even on the beach when waves break onto a shoaling beach. The hydraulic jump may be steady or unsteady; the latter is called a bore.

Here we consider, in some detail a steady hydraulic jump as illustrated in Fig. 1.11.3; the channel bottom is horizontal, the width is infinite and the control volume is taken per unit width and large enough so that the flow at §1 and §2 are horizontal and the pressure at §1 and §2 may be assumed to be hydrostatic. In §1.11, we showed that the ratio of the downstream to upstream depth is given by:

$$\beta = \frac{1}{2}\left(\left(1 + 8F_1^2\right)^{1/2} - 1\right). \tag{5.11.1}$$

In Fig. 5.11.1 we schematically illustrate, the flow changes in a hydraulic jump as a function of the Froude Number. The intensity of the turbulence increases as the Froude Number Fr_1 increases and from §5.5, the ratio of the smallest eddies to the scale of the jump itself, the scale of the largest eddies, decreases. The hydraulic jump is one of the few flows where we can explicitly calculate the total energy loss as the water moves through the jump. To do this, we consider energy conservation within the control volume of unit width shown in Fig. 1.11.3.

Consider first the fluxes at §1, Fig. 1.11.3. Rate of working by the pressure force at §1:

$$\dot{W}_P = \int_0^{h_1} \overline{v}_1{}^{(1)} \rho g (h_1 - x_3)\mathrm{d}x_3, \tag{5.11.2}$$

$$\dot{W}_P = \frac{1}{2}\rho g u_d^{(1)} h_1^2, \tag{5.11.3}$$

where it was assumed that the mean velocity is constant with depth and equal to the discharge velocity $u_d^{(1)}$. This assumption may be relaxed, but this adds little to the fundamental understanding of the energetics of the hydraulic jump.

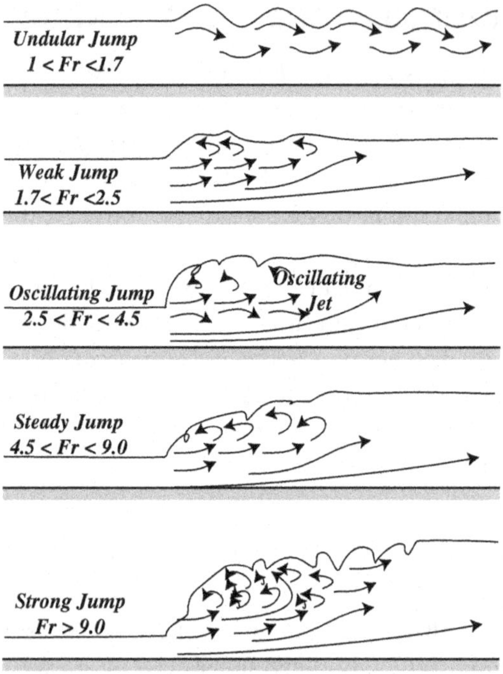

Figure 5.11.1 Forms of hydraulic jump as a function of the entrance Froude Number.

The potential energy flux entering the control volume at §1 (Fig. 1.11.3):

$$\mathrm{PE}_F = \int_0^{h_1} \overline{v}_1 \rho g x_3 \mathrm{d}x_3, \tag{5.11.4}$$

$$\mathrm{PE}_F = \frac{1}{2} u_d^{(1)} \rho g h_1^2, \tag{5.11.5}$$

The kinetic energy flux entering the control volume at §1 (Fig. 1.11.3):

$$\mathrm{KE}_F = \int_0^{h_1} \rho \frac{1}{2} \left(\overline{v}_1^{(1)}\right)^2 \overline{v}_1 \mathrm{d}x_3, \tag{5.11.6}$$

$$\mathrm{KE}_F = \frac{1}{2} \rho \left(u_d^{(1)}\right)^3 h_1. \tag{5.11.7}$$

Similar expressions may be written for §2 (Fig. 1.11.3), simply by interchanging 1 with 2. Thus the net loss of energy ΔE in the control volume, since the flow is steady, is given by the difference of the above energy fluxes between §1 and §2 (Fig. 1.11.3):

$$\Delta E = \frac{1}{2}\rho\left(u_d^{(1)}\right)^3 h_1 - \frac{1}{2}\rho\left(u_d^{(2)}\right)^3 h_2 + \rho g u_d^{(1)} h_1^2 - \rho g u_d^{(2)} h_2^2. \qquad (5.11.8)$$

Defining the discharge per unit width as q then:

$$q = u_d^{(1)} h_1 = u_d^{(2)} h_2, \qquad (5.11.9)$$

allowing (5.11.8) to be recast:

$$\Delta E = q\rho g h_1 \left(\frac{1}{2}F_1^2 - \frac{1}{2}Fr_1^2\left(\frac{h_1}{h_2}\right)^2 + 1 - \frac{h_2}{h_1}\right). \qquad (5.11.10)$$

This expression may be simplified by using the conservation of momentum (5.11.1) to yield:

$$\Delta E = \frac{1}{4}\rho g q \frac{(h_2 - h_1)^3}{h_1 h_2}. \qquad (5.11.11)$$

Thus if $h_2 > h_1$, as is the case for $Fr > 1$, there is a net rate of energy loss in the hydraulic jump. We also immediately see that if $h_2 < h_1$, then there would need to be an energy gain which is not possible; it is for this reason we took the positive root in (1.11.3). A hydraulic jump is thus possible in any supercritical flow, because for such flows there is a conjugate subcritical flow depth for which the flow has the same total momentum flux, but less energy, the energy loss being used to drive the turbulence in the jump.

The above analysis is exact, except for two relatively minor assumptions. First, the shape factors for the kinetic and potential energy and the pressure work are not exactly unity. Second, there are additional fluxes of energy in and out of our control volume. These are the rate at which the turbulent fluctuations do work on the boundaries and the flux of turbulent kinetic energy. Third, in undular hydraulic jumps surface wave form downstream of the jump (see Fig. 5.11.1) and these may, in some instances, propagate away from the jump region taking with them kinetic energy and if large a little momentum.

If we now assume that there is an energy cascade generated by the actions of the hydraulic jump, then we can derive the scale of the Kolmogorov scale as discussed in §5.5, as a function of the Froude Number. To do this, assume that the mass of water actively participating in the energy cascade scales with h_2, then the energy dissipation is given by dividing (5.11.11) by the mass $\rho_0 h_2^2$:

$$\varepsilon = \frac{1}{4} g q \frac{(h_2 - h_1)^3}{h_1 h_2^3}. \tag{5.11.12}$$

From (5.5.22) we have:

$$\frac{\eta_0}{h_2} = \left(\frac{\nu^3}{\varepsilon}\right)^{\frac{1}{4}} = \left(\frac{4 h_1 \nu^3}{g q h_2 (h_2 - h_1)^3}\right)^{\frac{1}{4}} \sim \frac{1}{\mathrm{Re}^{\frac{3}{4}} Fr^{\frac{1}{2}}}, \tag{5.11.13}$$

where $\mathrm{Re} = \frac{\nu}{q}$ and $Fr \gg 1$.

The flow case where the $Fr = 1$, is called critical flow and we now show that associated with critical flow is a minimum in the specific energy. To see this consider the total Bernoulli energy, called the specific energy, along any streamline in a uniform flow:

$$E = \frac{\overline{v}_1^2}{2g} + x_3 \cos\theta + \frac{p}{g \rho_0}, \tag{5.11.14}$$

which is the energy per unit weight relative to the channel bottom, x_3 is the vertical distance normal to the bottom and θ is the angle of the bottom to the horizontal.
Before and after the hydraulic jump the pressure is hydrostatic, so that:

$$p = \rho_0 (h - x_3) g \cos\theta, \tag{5.11.15}$$

where h is the flowing depth, measure perpendicular to the bottom, and so the specific energy becomes:

$$E = \frac{\overline{v}_1^2}{2g} + h\cos\theta, \tag{5.11.16}$$

independent of the height x_3 of the fluid perpendicular to the bottom, within the water column. If we now make the simplifying assumption that the velocity of the channel is uniform with depth, then:

$$\overline{v}_1 = u_d = \frac{Q}{A(h)}, \tag{5.11.17}$$

where Q is the flow in the channel, constant along the channel in steady flow, and $A(h)$ is the vertical flowing area, a function of the flowing depth. Substituting (5.11.17) into (5.11.6) leads to the expression:

$$E = \frac{Q^2}{2gA^2} + h\cos\theta. \tag{5.11.18}$$

The condition for the energy minimum may be found by equating the derivative with respect to the depth to zero:

$$\frac{dE}{dh} = \frac{Q^2}{2g}\left(- 2A(h)^3 \frac{dA(h)}{dh} \right) + \cos\theta = 0. \tag{5.11.19}$$

If we note:

$$\frac{1}{A}\frac{dA}{dh} = \frac{T}{A} = \frac{1}{D} \tag{5.11.20}$$

where T is the top width of the flow and D is called the hydraulic depth, then (5.11.19) implies:

$$Fr^2 = \frac{u_d^2}{gD\cos\theta} = 1, \tag{5.11.21}$$

where Fr is the Froude Number generalized to a flow in a channel of arbitrary cross-section and with a bottom slope. For a horizontal, rectangular channel (5.11.21) reduces to the two-dimensional Froude Number (5.9.14).

As seen from Fig. 5.11.2, super critical flow has a specific energy given by the lower branch of the energy curve, from A to M. Each energy point

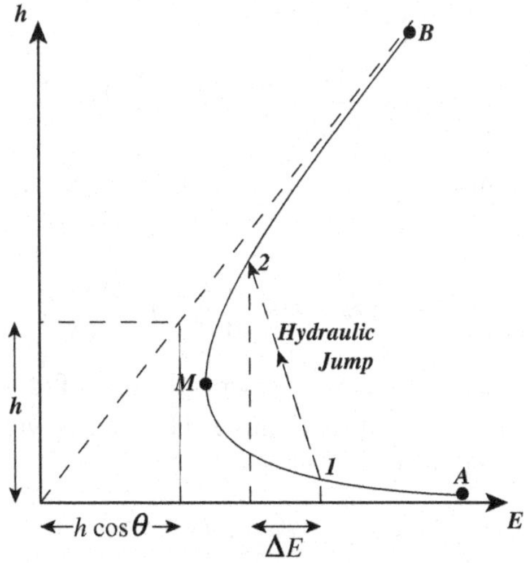

Figure 5.11.2 Specific energy as a function of depth.

on this lower branch has a conjugate depth, with same specific energy, a jump could occur to this conjugate depth if no dissipation were present. This is not possible, so jumps occur to conjugate depths, as shown in Fig. 5.11.2 where just the right amount of energy is lost to feed the turbulent dissipation induced by the energy cascade.

5.12. GRADUALLY VARIED FLOW

Consider a channel, of arbitrary section, with a flow that is adjusting its depth gradually in order to match the friction force at the bottom with the downstream acting gravity force; we wish to find the change of depth with distance downstream. The total energy head, defined as the energy per unit weight, as in (6.5.6), becomes:

$$H = h_b + h\cos\theta + \beta\left(\frac{u_d}{2g}\right)^2 = h_b + E, \tag{5.12.1}$$

where the pressure is assumed to be hydrostatic, β is a velocity profile shape factor relating the kinetic energy derived from the discharge velocity to the actual integrated cross-sectional variable kinetic energy, h is the flowing depth normal to the bed slope, h_b is the height of the bottom and E is the specific energy given by (5.11.16) (see also Fig. 5.11.2). The rate of change of the total head H with respect to downstream distance x_1, may be obtained by introducing the discharge Q to replace the velocity and then differentiating (5.12.1) with respect to x_1:

$$\frac{dH}{dx_1} = \frac{dh_b}{dx_1} + \cos\theta\frac{dh}{dx_1} + \frac{\beta}{2g}Q^2\left(-2A^{-3}\right)\frac{dA}{dx_1}. \tag{5.12.2}$$

Substituting from the definition of the Froude Number (5.11.21) into (5.12.2) we get,

$$\frac{dH}{dx_1} = -S_b + \cos\theta\frac{dh}{dx_1} - \beta Fr^2\frac{dh}{dx_1}. \tag{5.12.3}$$

The left hand side of (5.12.3) represents the loss of total energy per unit weight of water per unit length of channel. This may be viewed as an energy slope S which we define by:

$$S = -\frac{dH}{dx_1}. \tag{5.12.4}$$

Substituting for S in (5.12.3) and rearranging yields an expression for the rate of change of the flow depth with respect to downstream distance:

$$\frac{\mathrm{d}h}{\mathrm{d}x} = \frac{S_b - S}{\cos\theta - \beta Fr^2}. \qquad (5.12.4)$$

Gradually varied flow is locally uniform so the energy slope S may be estimated from the uniform flow frictional resistance laws (5.9.7). This means that the local discharge velocity determines the bottom friction losses that are, however, not necessarily equal to the bed slope.

In §5.9, we discussed three different friction formula of the uniform flow and each will lead to a slightly different formulation for the depth of flow variation. Consider first the friction factor formulation (5.9.7):

$$S = \frac{fU^2}{8gR}. \qquad (5.12.5)$$

Substituting (5.12.5) into (5.12.4) leads to:

$$\frac{\mathrm{d}h}{\mathrm{d}x_1} = \frac{S_b - \dfrac{fU^2}{8Rg}}{\cos\theta - \beta Fr^2}. \qquad (5.12.6)$$

We may rewrite this by defining the normal flow depth for the same discharge such that:

$$Q = A_n\left(\frac{8gR_n}{f}\right)^{1/2} S_b^{1/2}, \qquad (5.12.7)$$

where the subscript n refers to normal flow conditions that hold for uniform flow.

Substituting (5.12.7) into (5.12.6) yields:

$$\frac{\mathrm{d}h}{\mathrm{d}x_1} = \frac{S_b\left(1 - \dfrac{R_n A_n^2}{RA^2}\right)}{\cos\theta - \beta Fr^2}. \qquad (5.12.8)$$

Similarly we may introduce the critical flow depth as the depth for which the Froude number is unity for the same discharge:

$$\frac{Q^2}{gD_c A_c^2} = 1, \qquad (5.12.9)$$

where the subscript c refers to critical flow conditions at the same discharge Q. If we further assume that the slope is small so that:

$$\cos\theta = 1, \tag{5.12.10}$$

then substituting for Q from (5.12.9) into (5.12.8) yields:

$$\frac{dh}{dx} = S_b \frac{\left(1 - \dfrac{R_n A_n^2}{R A^2}\right)}{\left(1 - \dfrac{D_c A_c^2}{D A^2}\right)}, \tag{5.12.11}$$

a relationship that relates the rate of change of depth to the ratios of the actual depth to normal and critical depths at the given discharge.

Alternatively, we may introduce the normal and critical discharges Q_n and Q_c, at the given flow depth, then

$$Fr^2 = \frac{Q^2}{gDA^2} = \frac{Q_c^2}{gDA^2}\frac{Q^2}{Q_c^2} = \frac{Q^2}{Q_c^2}, \tag{5.12.12}$$

since by definition,

$$\frac{Q_c^2}{gDA^2} = 1, \tag{5.12.13}$$

and similarly:

$$S = \frac{fQ^2}{8RgA^2} = \frac{fQ_n^2}{8RgA^2}\cdot\frac{Q^2}{Q_n^2} = S_B\frac{Q^2}{Q_n^2}, \tag{5.12.14}$$

since

$$S = \frac{fQ_n^2}{8RgA^2}. \tag{5.12.15}$$

Now substituting these expressions into (5.12.4) leads to:

$$\frac{dh}{dx} = S_B \frac{\left(1 - \dfrac{Q}{Q_n^2}\right)}{\left(1 - \dfrac{Q}{Q_c^2}\right)}, \tag{5.12.16}$$

where Q_n and Q_c are the normal and critical discharges at the actual flow depths.

For the simple case of a rectangular channel:

$$D = h, \tag{5.12.16}$$

and

$$A = Th, \tag{5.12.17}$$

so that (5.12.11) and (5.12.16) reduce to:

$$\frac{dh}{dx_1} = S_b \left(\frac{1 - \left(\frac{h_n}{h}\right)^3}{1 - \left(\frac{h_c}{h}\right)^3} \right) \tag{5.12.18}$$

$$\frac{dh}{dx_1} = S_b \left(\frac{1 - \left(\frac{q}{q_n}\right)^2}{1 - \left(\frac{q}{q_c}\right)^2} \right) \tag{5.12.19}$$

where q is the discharge per unit width.

Before discussing these relationships consider what happens if instead of using (5.12.5) we use the Manning's equation (5.9.18):

$$Q = \frac{1}{n} R^{\frac{2}{3}} A S^{\frac{1}{2}}, \tag{5.12.20}$$

so that the energy slope is given by:

$$S = \frac{n^2 Q^2}{R^{\frac{4}{3}} A^2}, \tag{5.12.21}$$

then (5.12.11) becomes:

$$\frac{dh}{dx_1} = S_b \frac{\left(1 - \frac{R_n^{\frac{4}{3}} A_n^2}{R^{\frac{4}{3}} A^2}\right)}{\left(1 - \frac{D_c A_c^2}{D A^2}\right)}, \tag{5.12.22}$$

but (5.12.16) remains unchanged.

For the simple rectangular channel (5.12.22) reduces to:

$$\frac{dh}{dx_1} = S_b \frac{\left(1 - \dfrac{h_n^{\frac{10}{3}}}{h^{\frac{10}{3}}}\right)}{1 - \dfrac{h_c^3}{h^3}}, \tag{5.12.23}$$

which is slightly different to (5.12.18), but qualitatively yields similar behavior.

The depth variation with distance along the flow channel requires the integration of these simple differential equations. The reader is referred to advanced open channel books such as Chow (1959), here, in keeping with the environmental theme, we shall focus on discussing the form of the solutions, as a precursor to the Environmental Hydraulics discussion in the next chapter. For this discussion we set $\beta = 1$ and assume the channel is rectangular in shape, so that (5.12.18) applies.

Case A: Horizontal channel bottom: $S_b = 0$. For this case the normal depth $h_n = \infty$ as there is no gravity force in the flow direction and the water moves either because it has momentum imparted upstream or because the free surface has a slope. It is most convenient to go back to (5.12.11) and set $S_b = 0$ then:

$$\frac{dh}{dx} = \frac{\dfrac{-fQ^2}{8RgA^2}}{1 - \dfrac{Q^2}{gDA^2}} = \frac{-f\dfrac{Q^2}{8h^3 g}}{1 - \left(\dfrac{h_c}{h}\right)^3}. \tag{5.12.24}$$

H2 Profile: $h > h_c$; $\dfrac{dh}{dx} < 0$: The depth decreases with x_1 until $h \to h_c$ where $\dfrac{dh}{dx_1} \to -\infty$ leading to a vertical free surface where the interface crosses the critical depth. Such a depth variation is referred to as a *H2* profile and may be observed when water leaves a reservoir over a spillway as shown in Fig. 5.12.1a.

H3 Profile: $h < h_c$; $\dfrac{dh}{dx} > 0$: The depth increases with x_1 and when $h \to h_c$, $\dfrac{dh}{dx_1} \to +\infty$, indicating that the water surface must enter a hydraulic jump. This profile is referred to as an *H3* profile and shown in Fig. 5.12.1b. An example of a supercritical flow coming from under a sluice gate on

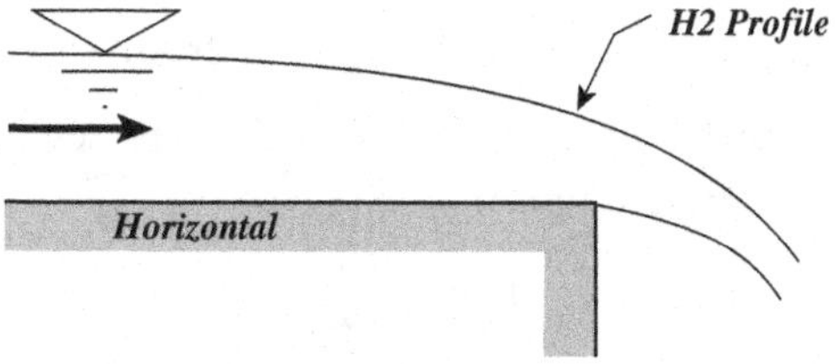

Figure 5.12.1a Schematic of *H2* profile.

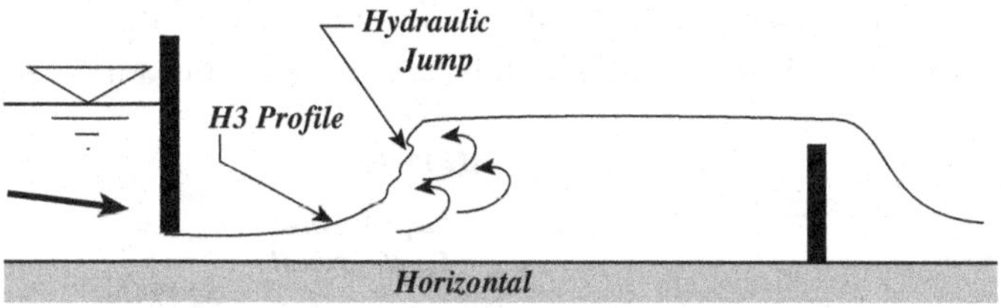

Figure 5.12.1b Schematic of *H2* profile.

a horizontal bed, deepening before entering a hydraulic jump is shown in Fig 5.12.1b; the deepening is necessary as the depth entering the jump must have the exit depth equal to the conjugate depth as given by (5.11.1) as there is no H profile to adjust the depth after the hydraulic jump.

Case B: Mild downstream slope: $h_n > h_c$: The term mild slope is used for bottom slopes that lead to a uniform flow, at the particular discharge, where the normal depth is larger than the critical depth at the same discharge.

M1 Profile: $h > h_n > h_c$; $\dfrac{dh}{dx} > 0$: With increasing x_1, $\dfrac{dh}{dx_1} \rightarrow S_B$ so the flow downstream must approach a horizontal water surface. An example of such a flow is shown in Fig. 5.12.1c; the *M1* profile is usually referred to a backwater curve.

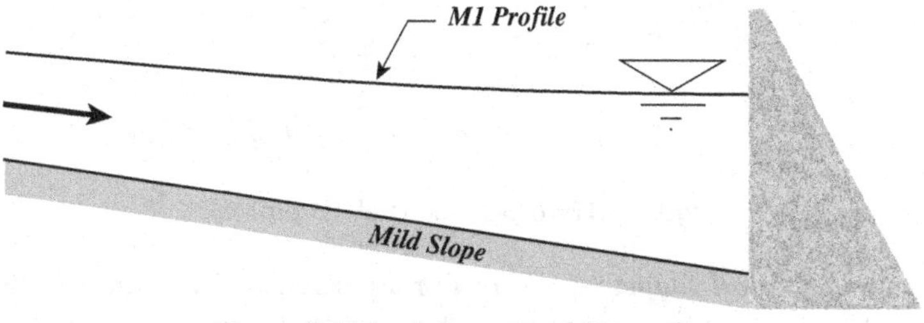

Figure 5.12.1c Schematic of *M1* profile.

M2 Profile: $h_n > h > h_c$; $\dfrac{dh}{dx} < 0$: As seen from (5.12.18) the depth, relative to the channel bottom, progressively decreases with downstream distance until $h \rightarrow h_c$ where upon:

$$\frac{dh}{dx_1} \rightarrow -\infty, \tag{5.12.27}$$

and so as the water surface approaches the critical depth, the surface slope tends to $-\infty$, as schematically shown in Fig. 5.12.1d. Such profiles occur when a uniform subcritical flow adjust to a supercritical flow at a sudden increase in bottom slope as could be the case at the start of rapids in a river.

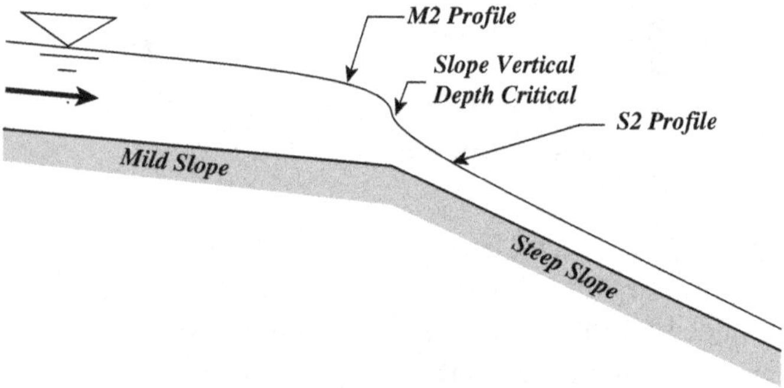

Figure 5.12.1d Schematic of *M2* and *S2* profiles.

M3 Profile: $h_n > h_c > h$; $\dfrac{dh}{dx} > 0$: The water surface steepens upwards as the depth approaches the critical depth and the water surface slope becomes vertical, indicating profile must merge into a hydraulic jump; an example of such a flow is the approach to a hydraulic jump as shown in Fig. 5.12.1e.

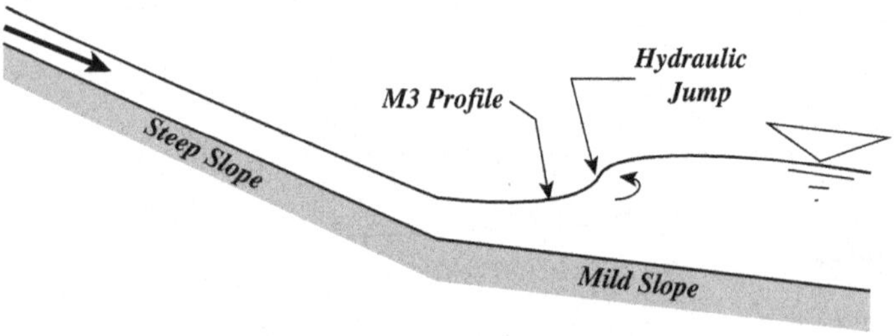

Figure 5.12.1e Schematic of *M3* profile.

Case C: Critical Slope: $h_n = h_c$: For a particular discharge Q, the bed bottom is critical when the uniform flow depth h_n is equal to the critical depth h_c. For this case we have two flow profiles; when $h = h_n = h_c$ the water surface remains unchanged. In both cases, the water surface becomes horizontal, $\frac{dh}{dx_1} \rightarrow S_B$ as we approach downstream.

C1 Profile: $h > h_n = h_c$ (Fig. 5.12.1f).

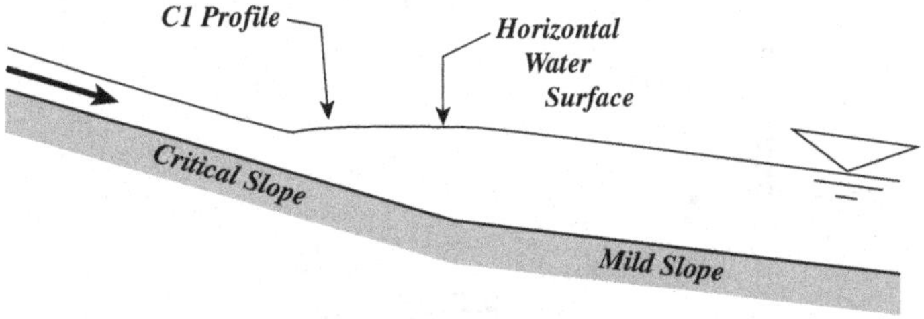

Figure 5.12.1f Schematic of *C1* profile.

C3 Profile: $h_n = h_c > h$ (Fig. 5.12.1g).

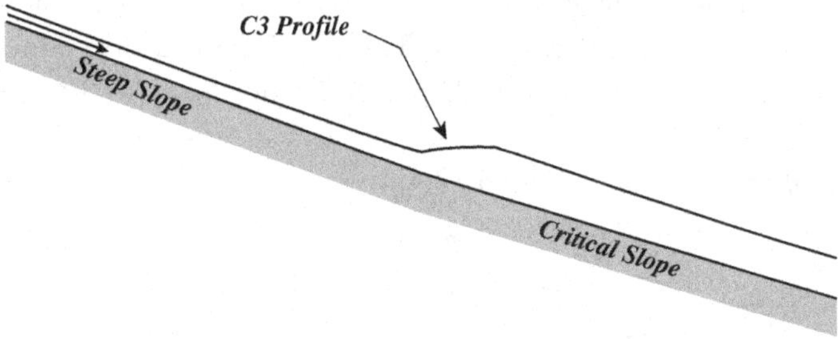

Figure. 5.12.1g Schematic of *C3* profile.

Case C: Steep Slope: $h_c > h_n$: For steep slopes we have $h_n < h_c$ so and three profiles are possible. In all cases $\frac{dh}{dx_1} \rightarrow S_B$ and the flow moves downstream and the water surface becomes horizontal.

S1 Profile: $h > h_n > h_n$ (Fig. 5.12.1h).

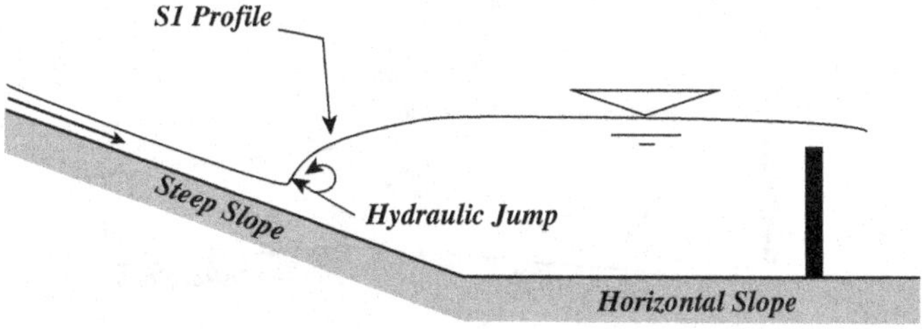

Figure 5.12.1h Schematic of *S1* profile.

S2 Profile: $h_c > h > h_n$: As seen in Fig. 5.12.1b, the flow must start vertical and approach $\frac{dh}{dx_1} \to 0$ and the flow approaches a constant depth downstream.

S3 Profile: $h_c > h_n > h$: An illustration of this S3 profile is shown in Fig. 5.12.1i.

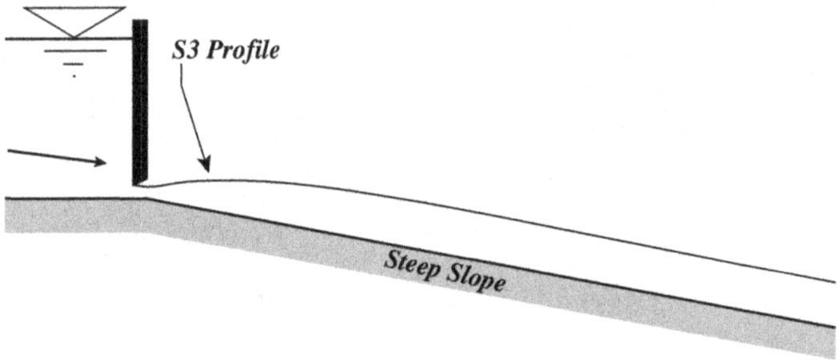

Figure 5.12.1i Schematic of *S3* profile.

Case D: Adverse Slope: $h_n = \infty$: The last case is where the bottom slope is uphill in the direction of flow. This is called an adverse slope. In this case we again cannot have a case where $h > h_n$ since $h_n = \infty$. Two cases are, however, possible:

A2 Profile: $h_n > h > h_c$ (Fig. 5.12.1j).

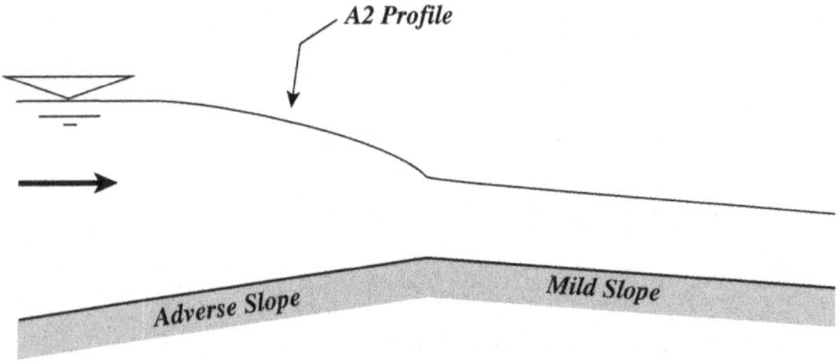

Figure 5.12.1j Schematic of *A2* profile.

S3 Profile: $h_n > h_c > h$ (Fig. 5.12.1k).

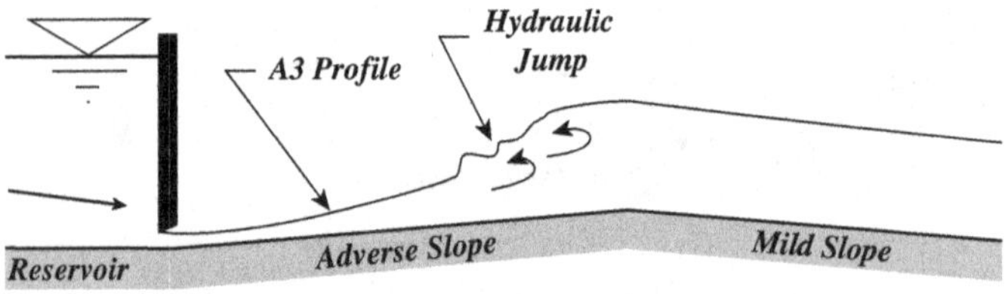

Figure 5.12.1k Schematic of *A3* profile.

REFERENCES

Chow, V.T., 1959. Open_Channel Hydraulics. McGraw-Hill, p. 680.

Coiebrook, C.F., 1939. Turbulent flow in pipes, with particular reference to the transition region between the smooth and rough pipe laws. Journal of Institution of Civil Engineers (London) 11, 133–156.

Fischer, et al., 1979. Mixing in Inland and Coastal Waters. Academic Press, p. 483.

Ricou, F.P., Spalding, D.B., 1961. Measurements of entrainment by axisymmetric turbulent jets. Journal of Fluid Mechanics 11, 21–32.

Tennekes, H., Lumley, J.L., 1972. A First Course in Turbulence. MIT Press, Cambridge, Massachusetts, p. 300.

CHAPTER 6

Environmental Hydraulics

Contents

By contrast to traditional hydraulics, introduced in chapter 5, the realization that stratification of the water column may have a dominant influence on the motion in water bodies came much later. Oceanographers had long realized that the earth's rotation played a determining role in ocean circulation, but it was not until the early 1960s that engineers realized that, temperature and or salinity variations in lakes, estuaries and coastal seas inhibited vertical motions at all scales, from basin scales to turbulent straining at the smallest scales. In simple terms, a density stratification adds a buoyancy force that the motion must overcome each time the fluid moves vertically. Many of the concepts in this chapter have similarities with the hydraulic concepts discussed in chapter 5, if we liken the free surface to a density interface.

6.1. GRADUALLY VARIED TWO LAYER FLOW

Consider the flow of two fluids in a rectangular channel and suppose the density of the upper fluid is ρ_1 and the density of the lower fluid is ρ_2 as shown in Fig. 6.1.1.

For simplicity, we shall assume that the two fluids are immiscible and the velocities $v_1^{(1)}$ and $v_1^{(2)}$ in fluid layer one and two are uniform with depth. To analyze the flow, we proceed in a similar fashion to the discussion in §5.12.

Environmental Fluid Dynamics
ISBN 978-0-12-088571-8, DOI: 10.1016/B978-0-12-088571-8.00006-1

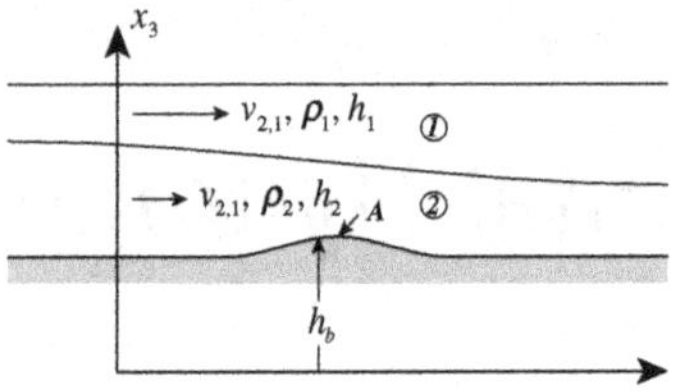

Figure 6.1.1 Two layer flow over a sill.

Let H_1 be the total head of a fluid particle at position (1) (Fig. 6.1.1) then

$$H_1 = \frac{p_1}{\rho_1 g} + x_3^{(1)} + \frac{v_{1,1}^2}{2g}, \tag{6.1.1}$$

where $x_3^{(1)}$ is the elevation of the fluid particle at the point (1). If we assume the flow varies gradually then we assume that the pressure is hydrostatic so that:

$$p_1 = \rho_1 g(h_b + h_2 + h_1 - x_3^{(1)}), \tag{6.1.2}$$

where atmospheric pressure has been removed, as it acts on all surfaces equally, h_1 and h_2 are the flowing layer depths and h_b is the height to the bottom. Substituting (6.1.2) into (6.1.1) and introducing the discharge per unit width q:

$$q_1 = v_{1,1} h_1, \tag{6.1.3}$$

equation (6.1.1) becomes:

$$H_1 = h_b + h_1 + h_2 + \frac{q_1^2}{2gh_1^2}. \tag{6.1.4}$$

Similarly we may write an expression for the total head in layer 2:

$$H_2 = h_b + h_1 + h_2 - \frac{\Delta\rho_{12} h_1}{\rho_2} + \frac{q_2^2}{2gh_2^2}, \tag{6.1.5}$$

where the density difference,

$$\Delta\rho_{12} = \rho_2 - \rho_1. \tag{6.1.6}$$

In order to get the variation of the layer thickness as a function of the downstream distance, x_1, it is convenient to differentiate (6.1.4) and (6.1.5) with respect to x_1, remembering that the discharge in each layer is constant with respect to x_1:

$$\frac{dH_1}{dx_1} = \frac{dh_b}{dx_1} + \frac{dh_2}{dx_1} + \frac{dh_1}{dx_1} - \frac{q_1^2}{gh_1^3}\frac{dh_1}{dx}, \tag{6.1.7}$$

and

$$\frac{dH_2}{dx_1} = \frac{dh_b}{dx_1} + \frac{dh_2}{dx_1} + \frac{dh_1}{dx_1} - \frac{\Delta\rho_{12}}{\rho_2}\frac{dh_1}{dx_1} - \frac{q_2^2}{gh_2^3}\frac{dh_2}{dx_1}. \tag{6.1.8}$$

Assuming that there are no energy losses:

$$\frac{dH_1}{dx_1} = \frac{dH_2}{dx_1} = 0, \tag{6.1.9}$$

and defining the Froude numbers:

$$Fi_1^2 = \frac{q_1^2}{g'_{12}h_1^3}, \tag{6.1.10}$$

$$Fi_2^2 = \frac{q_2^2}{g'_{12}h_2^3}, \tag{6.1.11}$$

where

$$g'_{12} = \frac{\Delta\rho_{12}}{\rho_2}g, \tag{6.1.12}$$

and then subtracting (6.1.8) from (6.1.7) leads to:

$$0 = -Fi_1^2\frac{dh_1}{dx_1} + Fi_2^2\frac{dh_2}{dx_1} + \frac{dh_1}{dx_1}. \tag{6.1.13}$$

Once again if the flow is gradually varied and $\Delta\rho_{12}/\rho_1 \ll 1$ then the free surface is approximately horizontal so that we may write

$$h_1 + h_2 + h_b = \text{constant}, \tag{6.1.14}$$

so that:

$$\frac{dh_2}{dx_1} = -\left(\frac{dh_1}{dx_1} + \frac{dh_b}{dx_1}\right). \tag{6.1.15}$$

Substituting (6.1.15) into (6.1.13) and rearranging the terms leads to the simple equation:

$$\left(1 - G^2\right)\frac{dh_1}{dx_1} = F_{i2}^2 \frac{dh_b}{dx_1}, \qquad (6.1.16)$$

where,

$$G^2 = F_{i1}^2 + F_{i2}^2 . \qquad (6.1.17)$$

Equation (6.1.16) may be compared to (5.12.3) and clearly shows that if $\frac{dh_b}{dx_1} \equiv 0$ then $\frac{dh_1}{dx_1} = 0$ and the flow is uniform. On the other hand, if the bottom is in the form of a sill, as shown in Fig. 6.1.1, where at the apex (point A in Fig. 6.1.1) of the bottom mound, locally $\frac{dh_b}{dx_1} = 0$ then (6.1.16) becomes:

$$G^2 = \frac{q_1^2}{g'_{12} h_1^3} + \frac{q_2^2}{g'_{12}(H - h_b - h_1)} = 1, \qquad (6.1.18)$$

that relates the two discharges to the depth of flow h_1 and the water surface elevation H.

6.2. RAPIDLY VARIED TWO LAYER FLOW: INTERNAL HYDRAULIC JUMP

As in the case of open channel hydraulics, the interface between two fluid layers may change rapidly, similarly to the hydraulic jump discussed in §5.11. Consider the internal hydraulic jump shown in Fig. 6.2.1.

For simplicity, we shall assume that the upper layer is stationary and the layer properties are as shown in Fig. 6.2.1. The reader may wish to show that

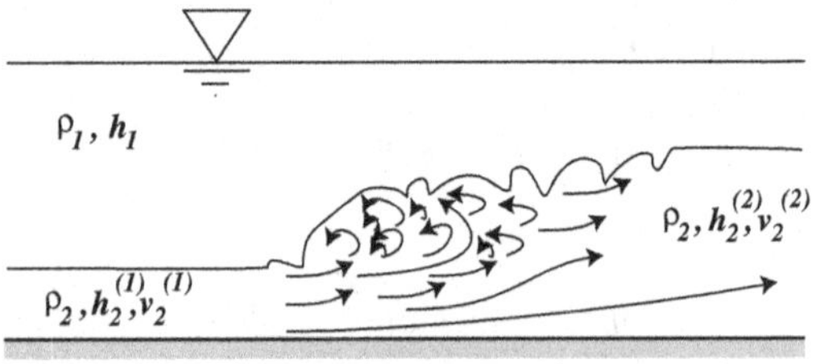

Figure 6.2.1 Schematic of an internal hydraulic jump.

a similar analysis follows if the upper layer is allowed to have a non-zero velocity.

Conservation of volume states:

$$h_2^{(1)} v_{2,1}^{(1)} = h_2^{(2)} v_{2,1}^{(2)}, \tag{6.2.1}$$

where it is assumed that the turbulence at the jump interface does not lead to appreciable entrainment of the upper fluid into the lower fluid and $h_1^{(i)}$ is the depth of layer 1 at station i, $h_2^{(i)}$ is the depth of layer 2 at station i, and $v_{2,1}^{(1)}$; $v_{2,1}^{(2)}$ are the horizontal velocity in layer 2, at stations (1) and (2).

Conservation of Momentum states:

$$\rho_2 h_2^{(2)} \left(v_{2,1}^{(2)}\right)^2 - \rho_2 h_2^{(1)} \left(v_{2,1}^{(1)}\right)^2 = \frac{1}{2}\rho_{12} g \left(h_2^{(1)}\right)^2 - \frac{1}{2}\rho_{12} g \left(h_2^{(2)}\right)^2. \tag{6.2.2}$$

Equations (6.2.1) and (6.2.2) may be combined to form an equation for the depth ratio:

$$\Delta = \frac{h_2^{(2)}}{h_2^{(1)}}, \tag{6.2.3}$$

$$\Delta^2 + \Delta - Fi_1^2 = 0, \tag{6.2.4}$$

where

$$Fi_1^2 = \frac{q_2^2}{g'_{12}\left(h_2^{(1)}\right)^3}, \tag{6.2.5}$$

and g'_{12} is defined as in (6.1.12).

The quadratic (6.2.4) may be solved for Δ:

$$\Delta = \frac{\left(1 - 8Fi_1^2\right)^{1/2} - 1}{2}, \tag{6.2.6}$$

which is the counterpart to (5.11.1).

6.3. SURFACE LAYER HYDRAULICS

We now consider the case discussed above, but where the bottom layer is stationary, homogeneous and very deep and the upper layer is in contact with the atmosphere imparting heat fluxes to the water and a surface wind stress. We consider the flow to be horizontally uniform and one dimensional with the objective of ascertaining the rate of mixing at the interface induced by the surface heat, momentum and energy fluxes. This is the turbulent counterpart to the Couette flow already discussed in §3.2, but with a slightly denser layer acting as the duct bottom. Surface layers are fundamental to many geophysical situations and the analysis below is common to all cases, except that the momentum equation will differ depending on the situation being studied. The surface layer may, for convenience, be divided into the surface shear layer, the surface layer core, the mixing layer at the base of the core region and the deep underlying water (see Fig. 6.3.1).

We shall now apply the equations developed in §2.15 to the water column surface layer and then vertically integrate the resulting equations across the surface layer using the assumed density and velocity profiles

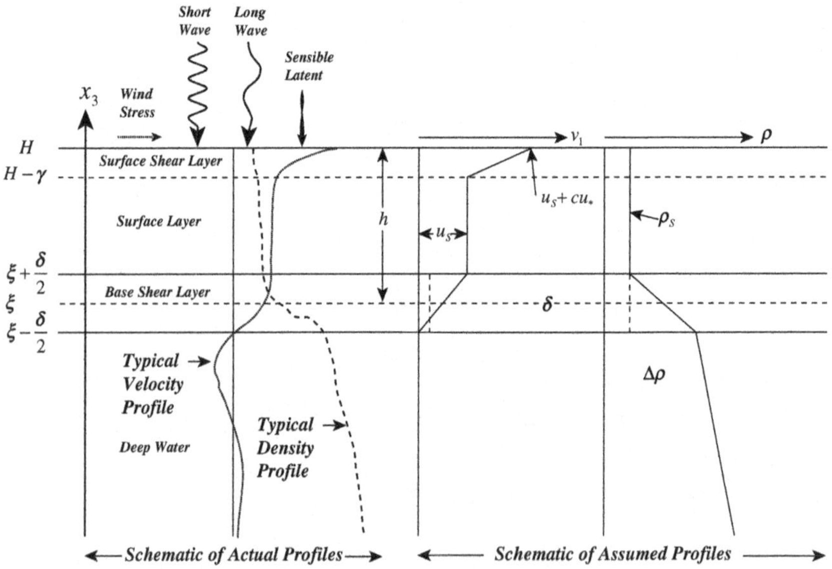

Figure 6.3.1 Surface layer definition sketch.

shown in Fig. 6.3.1; the water column is assumed homogeneous in the horizontal direction:

Deep Water: $x_3 < H - h - \frac{\delta}{2}$

$$\overline{v}_1 = 0 \tag{6.3.1}$$

$$\overline{\rho} = \rho_b, \tag{6.3.2}$$

$$\overline{\theta} = \theta_b. \tag{6.3.3}$$

Base Shear Layer: $H - h - \frac{\delta}{2} < x_3 < H - h + \frac{\delta}{2}$

$$v_1 = u_s(t)\left(\frac{x_3}{\delta} - \frac{1}{\delta}\left(H - h - \frac{\delta}{2}\right)\right), \tag{6.3.4}$$

$$\overline{\rho} = \rho_s + \Delta\rho(t)\,\frac{1}{\delta}\left(H - h + \frac{\delta}{2} - x_3\right), \tag{6.3.5}$$

$$\overline{\theta} = \theta_s + \Delta\theta(t)\,\frac{1}{\delta}\left(H - h + \frac{\delta}{2} - x_3\right). \tag{6.3.6}$$

Surface Layer: $H - h + \frac{\delta}{2} < x_3 < H - \gamma$

$$v_1 = u_S(t), \tag{6.3.7}$$

$$\overline{\rho} = \rho_s(t), \tag{6.3.8}$$

$$\overline{\theta} = \theta_s(t). \tag{6.3.9}$$

Surface Shear Layer: $H - \gamma < x_3 < H$

$$v_1 = u_s + \frac{Cu_*}{\gamma}(x_3 - H + \gamma), \tag{6.3.10}$$

$$\overline{\rho} = \rho_s(t), \tag{6.3.11}$$

$$\overline{\theta} = \theta_s(t), \tag{6.3.12}$$

where $u_s(t)$ and $\rho_s(t)$ are the velocity and density in the core of the surface layer, $\Delta\rho = \rho_b - \rho_s$ is the density difference, δ is the base layer thickness and γ is the surface shear layer thickness ($\gamma \ll h$), h is the surface layer thickness and C is a constant yet to be determined.

The flow is assumed to be parallel, the turbulent field homogeneous and all the surface transfers independent of x_1. With these assumptions all the divergence terms in the conservation equations are zero.

Consider first the mean temperature equation (§2.9) where we have retained the radiation fluxes:

$$\frac{\partial \bar{\theta}}{\partial t} = -\frac{\partial(\overline{\theta' v_3'})}{\partial x_3} - \frac{1}{\rho_0 C_p}\frac{\partial q}{\partial x_3}, \tag{6.3.13}$$

where q is the radiation heat flux at depth x_3 assumed positive upwards, ρ_0 is a background density and C_p is the specific heat of water.

The equation for the mean density field follows directly from (6.3.13) and the simplified equation of state:

$$\frac{\rho'}{\rho_0} = -\alpha\theta', \tag{6.3.14}$$

so that:

$$\frac{\partial \bar{\rho}}{\partial t} = -\frac{\partial(\overline{\rho' v_3'})}{\partial x_3} - \frac{\alpha}{C_p}\frac{\partial q}{\partial x_3}. \tag{6.3.15}$$

The parallel flow assumption and neglecting Coriolis forces simplifies the mean momentum equation (§2.11) so that:

$$\frac{\partial \bar{v}_1}{\partial t} = -\frac{\partial(\overline{v_1' v_3'})}{\partial x_3}. \tag{6.3.16}$$

Similarly, the turbulent kinetic energy equation (§2.15) becomes:

$$\frac{\partial}{\partial t}\left(\frac{\overline{E'}}{2}\right) = -\left(\overline{v_1' v_3'}\right)\frac{\partial \bar{v}_1}{\partial x_3} - \frac{\partial}{\partial x_3}\left(\overline{v_3'\left(\frac{p'}{\rho_0} + \frac{E'}{2}\right)}\right) - \frac{g}{\rho_0}\left(\overline{\rho' v_3'}\right) - \varepsilon, \tag{6.3.17}$$

where

$$\varepsilon = \mu\overline{v_{i,j}' v_{i,j}'}, \tag{6.3.18}$$

$$E' = (v_i' v_i') = v_1'^2 + v_2'^2 + v_3'^2, \tag{6.3.19}$$

$$\overline{\rho' v_3'} = -\alpha \rho_0 (\overline{\theta' v_3'}), \tag{6.3.20}$$

$$\alpha = \text{coefficient of thermal expansion.} \tag{6.3.21}$$

The boundary conditions at the water surface may be written:

$$-\overline{v_1' v_3'} = u_*^2 = \frac{\tau_S}{\rho_0}, \tag{6.3.22}$$

$$\rho_0 C_p (\overline{\theta' v_3'}) = Q_L + H_L + H_S, \tag{6.3.23}$$

where Q_L = net long-wave radiation, H_S = sensible heat transfer, and H_L = latent heat transfer.

The heat fluxes are assumed to be positive when the heat flux is upwards, out of the water surface.

The above conservation equation (6.3.13), (6.3.15), (6.3.16) and (6.3.17) can now be integrated with respect to x_3, using the assumed profiles (6.3.1)–(6.3.12) and integrating each equation from $x_3 = H - h - \delta/2$ to $x_3 = H$ and then letting $\delta \to 0$. The integration requires the use of the Leibnitz' rule (Appendix) which states:

$$\int_{a(t)}^{b(t)} \frac{\partial g(x_3)}{\partial t} \mathrm{d}x_3 = \frac{\mathrm{d}}{\mathrm{d}t} \int_{a(t)}^{b(t)} g(x_3)\mathrm{d}x_3 - g(b,t)\frac{\mathrm{d}b}{\mathrm{d}t} + g(a,t)\frac{\mathrm{d}a}{\mathrm{d}t}. \tag{6.3.24}$$

Integrating the heat equation (6.3.13) leads to:

$$\underset{1}{h\frac{\mathrm{d}\theta_S}{\mathrm{d}t}} = \underset{2}{-\Delta\theta\frac{\mathrm{d}h}{\mathrm{d}t}} - \underset{3}{\frac{(Q_L + H_L + H_S + q_s - q_b)}{C_p \rho_0}}, \tag{6.3.25}$$

where

$$q_s = q(H), \tag{6.3.26}$$

and

$$q_b = q(H - h - \delta/2). \tag{6.3.27}$$

Term 1, in (6.3.25) is the rate of change of heat content (divided by $C_p\rho_0$) of the surface layer, term 2 is the heat entrained at the base of the surface layer (divided by $C_p\rho_0$) and term 3 is the net heat flux at the surface minus that lost at the base.

The same procedure, applied to the momentum equation (6.3.16), yields an equation for u_s, the velocity of the surface layer:

$$\frac{\mathrm{d}}{\mathrm{d}t}(u_s h) = u_*^2, \tag{6.3.28}$$

that simply states that the momentum of the surface layer increases in proportion to the applied wind stress; the flow is assumed parallel so there are no other forces.

The integration of the turbulent kinetic energy equation (6.3.17) is again identical in principle and leads to:

$$\underset{1}{\frac{\mathrm{d}}{\mathrm{d}t}\left(\frac{E_s h}{2}\right)} = \underset{2}{\frac{g_s' h}{2}\frac{\mathrm{d}h}{\mathrm{d}t}} + \underset{3}{\frac{u_s^2}{2}\frac{\mathrm{d}h}{\mathrm{d}t}} + \underset{4}{Cu_*^3} - \left(\underset{5}{\frac{\overline{v_3' p'}}{\rho_0}} + \underset{6}{\frac{\overline{v_3' E'}}{2}}\right)\Bigg|_{x_3=H} + \underset{7}{\frac{w_*^3}{2}} - \underset{8}{\varepsilon_s h}, \tag{6.3.29}$$

The interpretation of each term is as follows:

Term 1: Rate of change of turbulent kinetic energy in the surface layer, where $E_S = \left[\overline{E'}\right]_{over\ core}$

Term 2: Rate of working needed to lift the entrained volume of water $\frac{\mathrm{d}h}{\mathrm{d}t}$ a distance $\frac{h}{2}$ through a density differential $\Delta\rho$, where $g_s' = \frac{\Delta\rho}{\rho_0}$.

Term 3: Rate of shear production of turbulent kinetic energy at the base of the surface layer.

Term 4: Rate of introduction of turbulent kinetic energy at the surface by shear production due to the wind working in the thin layer: $H - \gamma < x_3 < H$.

Term 5: The rate of working by the wind pressure fluctuations at the water surface.

Term 6: The net flux of turbulent kinetic energy due to surface wave activity.

It is common to group terms 4, 5 and 6, together into a single term $C_N^3 u_*^3/2$, where C_N is a new coefficient, yet to be determined.

Term 7: The rate of introduction of potential energy (positive or negative) due to surface heat transfers. Long-wave and sensible and latent heat transfers are assumed to act at the water surface, whereas short-wave radiation penetrates the water column, and adds a buoyancy flux dependent on the rate of extinction of solar radiation; the greater the turbidity of the water the greater the stability near the surface imparted by the short-wave radiation:

$$w_*^3 = \frac{g\alpha\left(h+\frac{\delta}{2}\right)}{\rho_0 C_P}\left\{q_S(H) - q_S\left(\xi - \frac{\delta}{2}\right) + Q_L + H_S + H_L - \frac{2}{\left(h+\frac{\delta}{2}\right)}\int_{\xi-\frac{\delta}{2}}^{H} q(x_3)\mathrm{d}x_3\right\}. \qquad (6.3.30)$$

Term 8: Rate of dissipation of turbulent kinetic energy.

Before we can proceed to solve for the unknowns E_S, h, θ_s and u_s, we need a further equation as we also do not yet know ε_S. This may be derived from the scaling analysis discussed in §5.5, where it was shown for a homogeneous fluid:

$$\varepsilon_S \sim \frac{C_E}{2}\frac{E_S^{\frac{3}{2}}}{h}. \qquad (6.3.31)$$

where C_E is a coefficient and the factor 2 has been added to conform to the remaining terms.

Conceptually the energy budget expressed by (6.3.29) is as follows. Energy introduced by the wind (4, 5 and 6) and surface heating and cooling (7) is used to spin up the turbulent kinetic energy in the layer (1), but much of this energy is lost to dissipation (8). A small fraction, the difference between these two groups, is exported to the base of the surface layer were it is augmented by the shear production (3) to entrain and lift heavier fluid from just below the surface layer to the centroid of the surface layer (2). It is convenient to split (6.3.29) into two equations, one for energy balance in the core of surface layer and one for the entrainment action at the base of the surface layer. This is done by splitting the dissipation in (6.3.29) into a part, $C_E E_S^{\frac{3}{2}}$, that represents dissipation in the core and another part, $C_F E_S^{\frac{3}{2}}$, that goes to augment working in the base layer

$$\text{Core}: \; h\frac{\mathrm{d}E_S}{\mathrm{d}t} = -(C_F + C_E)E_S^{\frac{3}{2}} + C_N^3 u_*^3 + w_*^3. \qquad (6.3.32)$$

$$\text{Base}: \quad E_S\frac{\mathrm{d}h}{\mathrm{d}t} + g'_S h\frac{\mathrm{d}h}{\mathrm{d}t} = C_F E_S^{\frac{3}{2}} + C_S u_S^2 \frac{\mathrm{d}h}{\mathrm{d}t}, \tag{6.3.33}$$

where two coefficients have been introduced. The flow of energy, as conceptualized by this model, is shown in Fig. 6.3.2.

If we assume $\frac{\mathrm{d}E_s}{\mathrm{d}t.}$ is small and can be neglected, then (6.3.32) and (6.3.33) may be rewritten in the form:

$$\frac{\mathrm{d}h}{\mathrm{d}t} = \frac{C_F}{(C_F + C_E)}\frac{(C_N^3 u_*^3 + w_*^3)}{(g'_S h + E_S - C_S u_S^2)}, \tag{6.3.34}$$

which forms a simple entrainment law.

Equations (6.3.25), (6.3.28), (6.3.29) and (6.3.34) form a complete set of equations for u_s, θ_s, h and E_S. The values of the coefficients C_N, C_F, C_E and C_S have been determined by many researchers and a good summary is given by Spigel et al. (1986) who showed that laboratory and field data suggest:

$$\begin{aligned} C_N &= 1.33 \\ C_F &= 0.25 \\ C_E &= 1.15 \\ C_s &= 0.20 \end{aligned} \tag{6.3.35}$$

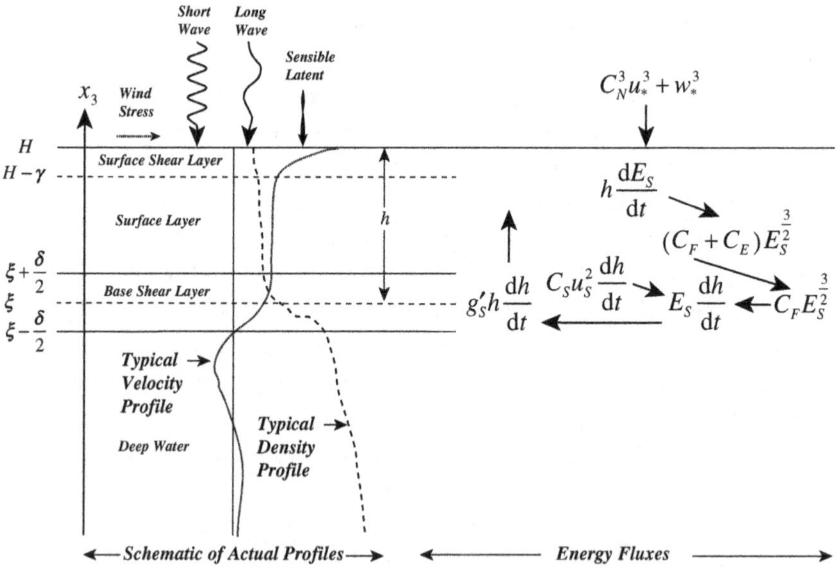

Figure 6.3.2 Schematic of energy flux path.

To illustrate the role of the various energy sources we now consider a number of limiting cases:

Wind Stirring Deepening: Deepening by wind stirring only where $w_* = 0$ and the surface layer deepens into a weak linear temperature gradient. Suppose initially the stratification reaches all the way to the surface. Since there is no heat exchange we may write

$$g_S' = \frac{\Delta\rho}{\rho_0} g = \frac{N^2 h}{2}, \tag{6.3.36}$$

where $N^2 = -\frac{g}{\rho_0}\frac{d\rho}{dz}$ is the square of the buoyancy frequency of the background stratification.

This result follows directly from geometry. Now suppose further that the wind stress shear velocity u_* commenced instantaneously and so u_s is very small. If u_* is constant then we may expect E_s to be constant so that (6.3.32) reduces to:

$$E_S^{\frac{3}{2}} = \frac{C_N^3 u_*^3}{(C_F + C_E)}, \tag{6.3.37}$$

and (6.3.34) may be simplified to a dominant balance between stirring energy eroding the interface and the buoyancy flux needed to raise the entrained fluid to the center of gravity of the surface layer:

$$C_F E_S^{\frac{3}{2}} = \frac{N^2 h^2}{2}\frac{dh}{dt}. \tag{6.3.38}$$

Combining these two equations and integrating the resulting equation with respect to time, leads to:

$$h = \left\{ \frac{C_F C_N^3 u_*^3}{(C_F + C_E) N^2} \right\}^{\frac{1}{3}} t^{\frac{1}{3}}. \tag{6.3.39}$$

Substituting for the coefficients from (6.3.35) yields the general non-dimensional deepening law:

$$\frac{hN}{u_*} = 1.4(Nt)^{\frac{1}{3}}. \tag{6.3.40}$$

The speed of the surface layer follows directly from (6.3.28):

$$\frac{u_s}{u_*} = 0.73(Nt)^{\frac{2}{3}}. \tag{6.3.41}$$

Deepening by shear production: Once again let us assume $w_* = 0$, the water column is linearly stratified and the wind is applied suddenly and then held constant. Determine the deepening rate due solely to the turbulent kinetic energy generated at the base of the surface layer, by shear production across the base shear layer. As in the example above, let us assume that E_S, the turbulent kinetic energy store is small, then (6.3.32) becomes

$$C_s u_s^2 \frac{dh}{dt} = g' h \frac{dh}{dt}, \tag{6.3.42}$$

that together with the momentum equation (6.3.28), leads to:

$$\frac{hN}{u_*} = 0.57(Nt)^{\frac{1}{2}}, \tag{6.3.43}$$

and

$$\frac{u_S}{u_*} = 1.80(Nt)^{\frac{1}{2}}. \tag{6.3.44}$$

By comparing (6.3.40) with (6.3.43) we see that deepening by shear production begins to dominate deepening by stirring after a non-dimensional time of about $(Nt) = 224$.

Deepening by natural convection. In this example assume $u_* = 0$ and w_* is made up of a heat loss at the surface. We will make again the same simplifications as in the above two examples so that (6.3.40) simplifies to:

$$E_S^{\frac{3}{2}} = \frac{w_*^3}{(C_F + C_E)}, \tag{6.3.45}$$

and (6.3.41) becomes:

$$\frac{h^2 dh}{dt} = \frac{2C_F}{(C_F + C_E)} \frac{w_*^3}{N^2}, \tag{6.3.46}$$

but from (6.3.30):

$$w_*^3 = \frac{g\alpha Q}{\rho_0 C_P} h, \tag{6.3.47}$$

where Q is the net heat loss at the surface. Substituting (6.3.47) into (6.3.46) and integrating yields:

$$\frac{h}{h_0} = 0.74(Nt)^{\frac{1}{2}}, \tag{6.3.48}$$

where

$$h_0 = \left(\frac{\alpha g Q}{\rho_0 C_P N^3} \right)^{\frac{1}{2}}. \quad (6.3.49)$$

6.4. METEOROLOGICAL BOUNDARY LAYER

In the above discussion of the surface layer §6.3, we assumed that the surface meteorological fluxes were given. In principle, we could now carry out an analogous analysis to the meteorological boundary layer as given in 6.3 and derive the fluxes by coupling the two layers via the air–water fluxes. This is, however, a complicated procedure and the fluxes of momentum, energy and mass across the air–water interface are more simply derived by following the procedure described in §5.6 and §5.7 and use the "Law of the Wall" derived by meteorologists (see Stull, 1988) that allows the local heat and momentum fluxes to be estimated from data from a meteorological station mounted above the water surface.

The short-wave radiation Q_S (300–1000 nm) is usually measured directly and the long-wave radiation Q_L (greater than 1000 nm) emitted from clouds and atmospheric water vapor can also be measured directly or calculated from cloud cover, air temperature, and humidity. A good overview of how such fluxes may be estimated when no measurements are available is given by Tennessee Valley Authority (1972). The reflection coefficient, or albedo, for the short-wave and long-wave radiation varies from one water body to the next, and depends on the angle of the sun, the color of the water and the surface wave state. Back radiation Q_B from the warm water surface may be calculated from the black body radiation law, or may be measured by pointing a radiation instrument toward the water surface.

To calculate the momentum, the sensible heat, and the latent heat fluxes aerodynamic bulk formulas have been developed that take the form (Stull, 1988):

$$u_*^2 = \frac{\tau_s}{\rho_a} = -\overline{v_1' v_3'} = C_D U_z^2, \quad (6.4.1)$$

$$\frac{H_S}{\rho_0 C_P} = \overline{\theta' v_3'} = -u_* \theta_* = -C_H U_z (\theta_z - \theta_s), \quad (6.4.2)$$

$$\frac{H_L}{\rho_0 L_v} = \overline{q' v_3'} = -u_* q_* = -C_W U_z (q_z - q_s), \tag{6.4.3}$$

where τ_s is the surface stress, ρ_a is the density of the air, v_1' and v_3' are the horizontal and vertical fluctuation of velocity and the over bar is a time average long enough to gain statistical confidence, but short enough to allow for diurnal variability, usually 10 min is a good interval. The shear velocity in the air is u_*, C_D is the momentum or drag coefficient, U_z is the velocity of the air at a certain height z above the water surface, H_S is the sensible heat transfer, C_p is the specific heat of water, θ' is the temperature fluctuation, θ_* is the temperature scale, C_H is the sensible heat transfer coefficient, θ_s is the water surface temperature, θ_z is the temperature of the air at the height z above the water, H_L is the latent heat flux, L_v is the latent heat of vaporization, q' is the specific humidity fluctuation, q_* is the specific humidity scale, C_W is the latent heat transfer coefficient, q_z is the specific humidity at a height of z above the water surface, and q_s is the specific humidity at saturation pressure at the water surface temperature. Since the water moves, U_z should be replaced by U_z-U_s, although for most wind speeds, this is a small correction.

The value of the transfer coefficients C_H and C_W for such bulk modeling has received a great deal of attention. If the meteorological sensors are placed within the law-of-the-wall region of the internal boundary layer, then it is possible to make use of the equilibrium profiles for the law-of-the-wall boundary layer to compute the surface fluxes from point sensor measurements. It is convenient to first introduce the characteristic length scale of the law-of-the-wall region in the meteorological boundary layer, called the Monin–Obukhor length scale (Monin and Obukhov, 1954). This is the length scale that characterizes the relative importance of the convective and stress-working energy sources and takes the place of the pipe radius in §5.6. To see this, consider air flowing over a water surface and assume that the surface stress is characterized by the air shear velocity $u*$ and the net buoyancy flux due to a sensible heat transfer H_S (W m^{-2}) and a latent heat flux H_L (W m^{-2}) is given by B (m^2 s^{-3}). The momentum and buoyancy flux enter the air of the meteorological boundary layer at the air–water interface and produce density and velocity fluctuations in the law-of-the-wall region of the boundary layer. In general, the intensity of the velocity fluctuations is determined by the air shear velocity $u*$ and the induced density anomaly in the atmospheric boundary layer is ρ_*. First, consider the case where u_* is zero and B is heating the bottom of the atmospheric boundary layer. Under such circumstances, air in contact with the water, will be heated and convectively rise in the form of plumes say with a characteristic velocity $w*$. This is similar to the simple plume discussed in §5.2. For such convection the heat must be convected away from the air–water interface so that,

$$w_* \rho_* = \frac{\rho_a B}{g}, \tag{6.4.4}$$

and since the air rises under gravity:

$$w_*^2 = \frac{g}{\rho} \rho_* z, \tag{6.4.5}$$

where z is the height above the air–water interface.

Combining (6.4.4) and (6.4.5) we get an expression for w_*:

$$w_* = (Bz)^{\frac{1}{3}}. \tag{6.4.6}$$

Thus when a cool wind moves cool air over a warm lake, we have two velocity scales $u*$ and $w*$. These two are equal at a height z = L, where,

$$L = \frac{u_*^3}{B} \tag{6.4.7}$$

This is called the Monin–Obukhov length scale and represents the height below which the velocity fluctuations are dominated by the shear velocity and above which the fluctuations are dominated by the buoyancy flux *B*. The reader may wish to compare this discussion with the behavior of a buoyant jet described in §5.3.

The air moving over water is often moist and so the temperature alone does not determine the density, and we need to include the effect of the specific humidity q. The computation of the density is facilitated by introducing the virtual potential temperature that is the temperature of dry air at atmospheric pressure that has the same density as the moist air at height z. As shown in §1.3 we saw

$$\theta_v = \theta_a(1 + 0.61q). \tag{6.4.8}$$

So provided we use θ_v instead of θ we may treat the moist air as dry air, so that (6.4.7) becomes:

$$L = \frac{u_*^3}{g\alpha(\overline{\theta'_v v'_3})}. \qquad (6.4.9)$$

Now if we use the logic of §2.13 and decompose our variables into a fluctuating and mean component we may derive an expression for θ'_v in terms of θ'_a and q′ and then substitute the result into (6.4.9):

$$L = \frac{u_*^3 \overline{\theta}_v}{g\alpha\left((1+0.61q)\dfrac{H_S}{\rho_0 C_P} + 0.61\left(\dfrac{\overline{\theta} H_L}{\rho_0 L_v}\right)\right)}. \qquad (6.4.10)$$

Equation (6.4.10) conveniently allows L to be evaluated from a knowledge of the properties of the air over the lake and the surface fluxes.

If $L > 0$ then the air is being heated at the air–water interface which leads to active convective mixing in the meteorological boundary layer for $z > L$. If $L = 0$ then air–water fluxes are zero and the meteorological boundary layer is termed neutral; buoyancy plays no role in the structure of the boundary layer and we may define "neutral exchange coefficients". On the other hand if $L < 0$ then there is a net cooling of the air at the air–water interface and the air becomes stably stratified with the associated attenuation of the turbulence; the air will slip more easily over the water surface. Clearly, for $L > 0$, the exchange of heat and momentum will be augmented by convection, fueling the turbulent exchange so that the exchange coefficients will be larger than those at neutral conditions and for $L < 0$, we may expect the exchange coefficient to be somewhat smaller than the neutral values.

Monin and Obukhov (1954), introduced similarity relationships, similarly to the discussion in §5.6, assumed to hold in the law-of-the-wall region:

$$\frac{kz}{u_*}\frac{dU}{dz} = \phi_M\left(\frac{z}{L}\right), \qquad (6.4.11)$$

$$\frac{kz}{\theta_*}\frac{d\overline{\theta}}{dz} = \phi_H\left(\frac{z}{L}\right), \qquad (6.4.12)$$

$$\frac{kz}{q_*}\frac{d\overline{q}}{dz} = \phi_W\left(\frac{z}{L}\right). \qquad (6.4.13)$$

The functions ϕ_M, ϕ_H, and ϕ_W have been derived in form from similarity arguments and then the coefficients identified by matching the found equations with experimental data. We refer the reader to specialist books for the details, it suffices here to present the final, generally accepted forms, of the final solutions, obtained by integrating (6.4.11), (6.4.12) and (6.4.13) once the functions have been identified:

$$U = \frac{u_*}{k}\left(\ln\left(\frac{z}{z_M}\right) - \psi_M\left(\frac{z}{L}\right)\right), \tag{6.4.14}$$

$$\theta - \theta_s = \frac{\theta_*}{k}\left(\ln\left(\frac{z}{z_H}\right) - \psi_H\left(\frac{z}{L}\right)\right), \tag{6.4.15}$$

$$q - q_s = \frac{q_*}{k}\left(\ln\left(\frac{z}{z_W}\right) - \psi_W\left(\frac{z}{L}\right)\right), \tag{6.4.16}$$

where z_M, z_H and z_W are roughness lengths equivalent to e in the pipe flow example from §5.7.

Unstable case:

$$\psi_M(x) = 2\ln\left(\frac{1+x}{2}\right) + \ln\left(\frac{1+x^2}{2}\right) - 2\tan^{-1}(x) + \frac{\pi}{2}, \tag{6.4.17}$$

$$\psi_H(x) = \psi_W(x) = 2\ln\left(\frac{1+x}{2}\right), \tag{6.4.18}$$

where

$$x = \left(1 - 16\left(\frac{z}{L}\right)\right)^{\frac{1}{4}}. \tag{6.4.19}$$

Stable case:

$$\psi_M\left(\frac{z}{L}\right) = \psi_H\left(\frac{z}{L}\right) = \psi_W\left(\frac{z}{L}\right) = -5\left(\frac{z}{L}\right); \quad 0 < \left(\frac{z}{L}\right) < 0.5, \tag{6.4.20}$$

$$\psi_{M,H,W}\left(\frac{z}{L}\right) = -0.5\left(\frac{z}{L}\right)^{-2} - 4.25\left(\frac{z}{L}\right)^{-1} - 7\ln\left(\frac{z}{L}\right) - 0.852; \quad 0.5 < \left(\frac{z}{L}\right) < 10.0, \tag{6.4.21}$$

$$\psi_{M,H,W}\left(\frac{z}{L}\right) = \ln\left(\frac{z}{L}\right) - 0.76\left(\frac{z}{L}\right) - 12.093; \quad \left(\frac{z}{L}\right) > 10.0. \tag{6.4.22}$$

These relationships have been tested, by various researchers, over the range from $-15 < z/L < 15$. Equations (6.4.17)–(6.4.22) can now be used to relate the neutral to the actual transfer coefficients, leaving only the surface roughness to be estimated. The roughness lengths are related directly to the drag coefficients, as can be seen by substituting (6.4.17)–(6.4.22) into (6.4.1)–(6.4.3). For non-neutral conditions, after considerable algebra, this procedure yields:

$$\frac{C_{D,H,W}}{C_{D,H,W;N}} = \left[1 + \frac{C_{D,H,W;N}}{k^2}\left(\psi_M\psi_{M,H,W} - \frac{k\psi_{M,H,W}}{C_{D;N}^{\frac{1}{2}}} - \frac{k\psi_M C_{D;N}^{\frac{1}{2}}}{C_{D,H,W;N}}\right)\right], \tag{6.4.26}$$

The approach suggested by Hicks (1975) to obtain values for the non-neutral exchange coefficients, C_D, C_H and C_W involves an iterative procedure, where neutral values of C_{DN}, C_{HN} and C_{WN} are chosen and neutral fluxes are calculated by (6.4.1)–(6.4.3). These values are then substituted into (6.4.10)–calculate the Monin–Obukov length L and thus z/L. New values of the transfer coefficients, now partially corrected for stability, may be obtained from (6.4.26). This procedure may be repeated and the scheme as converges. For moderate wind speeds, less than 15 ms^{-1}, a good value $C_{D;N}$, $C_{H;N}$ and $C_{W;N}$ is 1.3×10^{-3}. The reader is referred to Imberger and Patterson (1990), for an detailed discussion of the above procedure.

6.5. GRAVITY CURRENT HYDRAULICS

A flow that is common in coastal oceanography, estuarine flows and river inflows to lakes is that of a heavier fluid underflowing a standing water. The inflow may be heavier because it is sediment laden, saltier or colder than the receiving water. In such cases, the inflow plunges and then underflows as an entraining, gradually varied flow as shown in Fig. 6.5.1. The analysis of such flows parallels that discussed in §6.1, except here we allow for entrainment between the two fluid layers.

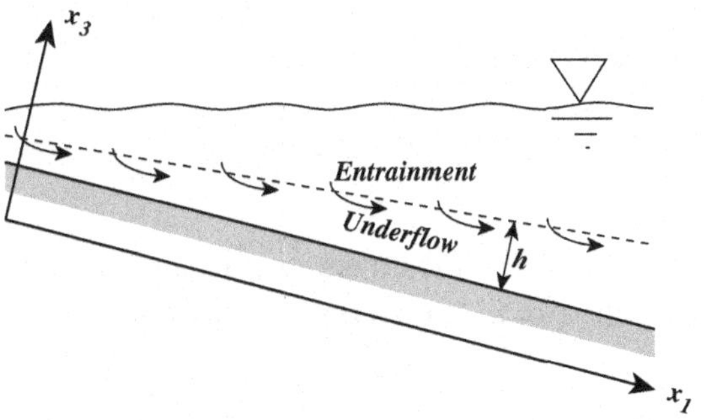

Figure 6.5.1 2D underflow schematic.

The simplest model of such an underflow is to assume that the along slope velocity is uniform within the underflow:

$$\overline{v}_1(x_1, x_3) = u_s(x_1), \tag{6.5.1}$$

and the fluid in the upper layer is stationary. Conservation of volume (§2.1) then requires:

$$\overline{v}_3(x_1, x_3) = -x_3 \frac{\mathrm{d}u_s(x_1)}{\mathrm{d}x_1}. \tag{6.5.2}$$

Similarly, we shall assume that the mean density in the underflow:

$$\overline{\rho} = \rho_s(x_1), \tag{6.5.3}$$

is only a function of the downslope coordinate. Now conservation mass requires:

$$\overline{v}_1 \frac{\partial \overline{\rho}}{\partial x_1} + \overline{v}_3 \frac{\partial \overline{\rho}}{\partial x_3} = -\frac{\partial(\overline{v_3'\rho'})}{\partial x_3} \tag{6.5.4}$$

substituting from (6.5.1), (6.5.2) and (6.5.3) leads to:

$$(\overline{v_3'\rho'}) = u_s x_3 \frac{\mathrm{d}\rho_s}{\mathrm{d}x_1}. \tag{6.5.5}$$

However, since no new mass enters the underflow through the entraining surface

$$\frac{\mathrm{d}(\rho_s u_s h)}{\mathrm{d}x_1} = 0, \tag{6.5.6}$$

where the volume flux $(u_s h)$ increases with downstream distance due to the entrainment of reservoir water into the down flow:

$$\frac{\mathrm{d}(u_s h)}{\mathrm{d}x_1} = E(Ri)u_s, \tag{6.5.7}$$

where $E(Ri)$ is called the entrainment coefficient, that from dimensional reasoning, is a function of the down flow Richardson number Ri, where:

$$Ri = \frac{g_s' h}{u_s}, \tag{6.5.8}$$

and where:

$$g_s' = \left(\frac{\rho_s - \rho_0}{\rho_0}\right) g. \tag{6.5.9}$$

Conservation of mean momentum states:

$$\underset{1}{\frac{\mathrm{d}}{\mathrm{d}x_1}\left(hu_s^2\right)} = \underset{2}{-u_{*b}^2} + \underset{3}{\cos\theta \frac{\mathrm{d}}{\mathrm{d}x_1}\left(\frac{g_s' h^2}{2}\right)} + \underset{4}{hg_s' \sin\theta}, \tag{6.5.10}$$

where θ is the bottom slope angle and u_{*b}^2 is the shear velocity squared, at the bottom:

$$u_{*b}^2 = C_D u_s^2, \tag{6.5.11}$$

where C_D is the bottom drag coefficient. Term 1, in (6.5.10) is the change of momentum flux, term 2 represents the retarding bottom stress, term 3 is the net hydrostatic pressure force due to the increasing flowing depth arising from the entrainment and term 4 is the gravitational force acting downslope; only the density difference over and above the upper fluid leads to a gravitational force.

By combining (6.5.6) and (6.5.7) and introducing the Richardson number, (6.5.10) may be rearranged:

$$-\frac{h}{u_s}\frac{\mathrm{d}u_s}{\mathrm{d}x_1} = \frac{h}{3Ri}\frac{\mathrm{d}Ri}{\mathrm{d}x_1} = \frac{C_D + E(Ri)\left(1 - \frac{Ri}{2}\right) - Ri\sin\theta}{(1 + Ri\cos\theta)}. \tag{6.5.12}$$

In the discussion of a simple plume §5.2 we concluded that the plume Richardson number was constant. From (6.5.13) it is also apparent that if the numerator becomes zero, the Richardson number of the down flow is a constant:

$$Ri = \frac{2(C_D + E(Ri))}{E(Ri) + 2\sin\theta}. \qquad (6.5.13)$$

The solution to (6.5.13) is called the normal Richardson number Ri_n. When the flow Richardson number is equal to the normal Richardson number, then the interface has adjusted so that the down slope gravitational force is balanced by the retarding bottom frictional drag, the pressure difference due to a sloping interface and the "pseudo" stress at the interface resulting from entraining zero momentum reservoir water into the underflow. The plume Richardson number (5.2.21) and the normal flow depth in uniform flow in a channel (5.9.13) are counterparts to (6.5.13).

To solve (6.5.14) and determine the normal Richardson number, we need first to derive an expression for the entrainment coefficient as a function of the Richardson number. This may be done by integrating the turbulent kinetic energy equation, across the underflow in an analogous way that was done for the surface layer §6.3, equation (6.3.49). The turbulent kinetic energy divided by the density, E_S, may be assumed to be constant over the depth of the underflow:

$$E_s = [\overline{v_i' v_i'}], \qquad (6.5.14)$$

where the square brackets denote an average over the depth. The conservation of turbulent kinetic energy for the flow in the underflow may then be written:

$$\underset{1}{\frac{d}{dx_1}(u_s E_s h)} = \underset{2}{-\frac{1}{2} h g_s' u_s E \cos\theta} + \underset{3}{u_s u_*^2} + \underset{4}{u_s^3 E} - \underset{5}{\varepsilon_s h}. \qquad (6.5.15)$$

The interpretation of each term in (6.5.15) is as follows:

Term 1: The increase in turbulent kinetic energy as the flow moves downstream. Given we are examining a gradually varied flow the dominant contribution to this term is the spin up of the entrained water from the stationary reservoir. To a good approximation E_S may be assumed approximately constant, so that:

$$\frac{d}{dx_1}(u_s E_s h) = E_s \frac{d}{dx_1}(u_s h) = E_s E u_s. \qquad (6.5.16)$$

Term 2: Rate of energy required to mix lighter reservoir water entrained into the underflow to the center of gravity of the underflow.

Term 3: Rate of production of turbulent kinetic energy by bottom drag.

Term 4: Rate of shear production of turbulent kinetic energy from the mean flow.

Term 5: Rate of dissipation of turbulent kinetic energy, ε_s is the mean dissipation in the underflow.

We may thus follow the logic presented in §6.3 and assume that the rate of turbulent kinetic energy dissipation is proportional to the turbulent kinetic energy production with a proportionality constant of C_E so that:

$$\varepsilon_s h = C_E(Eu_s^3 + u_*^2 u_s), \tag{6.5.17}$$

and parameterize E_S:

$$E_s = C_S(Eu_s^3 + u_*^2 u_s). \tag{6.5.18}$$

Substituting (6.5.16)–(6.5.18) into (6.5.15) and rearranging leads to the expression of the entrainment coefficient E in terms of flow variables:

$$E(Ri) = \frac{(1 - C_E - C_S)C_D}{\left(1 - \frac{1}{2}Ri\cos\theta - C_E - C_S\right)}, \tag{6.5.19}$$

which is the required equation that, together with (6.5.13) determines the underflow entrainment. Underflows that have reached a normal state, the Richardson number Ri is constant:

$$\frac{dRi}{dx_1} = 0, \tag{6.5.20}$$

and together with (6.5.6) and (6.5.7) it follows that u_S is a constant reducing (6.5.7) to:

$$\frac{dh}{dx_1} = E, \tag{6.5.21}$$

the underflow depth increases linearly with distance downslope.

6.6. SELECTIVE WITHDRAWAL

In order to illustrate the concept of selective withdrawal from a stratified fluid, we now analyze the simple configuration of the sudden initiation of a 2D sink, with discharge q m^2 s^{-1} in a linearly stratified fluid of infinite extent. We may anticipate the motion stages. As the fluid is assumed to be incompressible, on initiation of the sink a radial pressure field will instantaneously set up to infinity that then causes the fluid to move radially toward the sink. As the vertical motion proceeds, buoyancy will constrain the vertical motion and the fluid will progressively move horizontally until all vertical motion has disappeared and the flow consist solely of the thin layer of fluid moving horizontally at the level of the sink. After all unsteady motions have disappeared we may expect the final solution to be given by:

$$v_1(x_3) = -\frac{q}{2}\delta(x_3) \quad x_1 > 0, \tag{6.6.1}$$

where $\delta(x_3)$ is the Dirac delta function (see appendix) and the flow is symmetric about the x_1 axis. From this word picture, we see that is unlikely that the non-linear advective terms, the viscous terms or the temperature diffusion terms are of any importance before the buoyancy has collapsed the flow to a thin layer. We thus use the conservation of volume equation (2.11.1) and the linearized versions of the conservation of momentum (2.11.10):

$$\frac{\partial v_1}{\partial x_1} + \frac{\partial v_3}{\partial x_3} = 0, \tag{6.6.2}$$

$$\rho_0 \frac{\partial v_1}{\partial t} = -\frac{\partial p}{\partial x_1}, \tag{6.6.3}$$

$$\rho_0 \frac{\partial v_3}{\partial t} = -\frac{\partial p}{\partial x_3} - \rho g, \tag{6.6.4}$$

$$\frac{\partial \rho}{\partial t} + v_3 \frac{\partial \rho_e}{\partial x_3} = 0, \tag{6.6.5}$$

where the total density ρ_T has been partitioned such that:

$$\rho_T = \rho_0 + \rho_e(x_3) + \rho_0(x_1, x_3, t) \tag{6.6.6}$$

and the total pressure ρ_T is the sum of the hydrostatic pressure and the pressure due to the initiation of the sink at time $t = 0$:

$$p_T = p_e(x_3) + p(x_1, x_3, t), \tag{6.6.7}$$

$$\frac{dp_e}{dx_3} = -(\rho_0 + \rho_e)g, \tag{6.6.8}$$

Eliminating the pressure between (6.6.2) and (6.6.3) and introducing the stream function, $\psi(x_1, x_3, t)$ such that:

$$v_1(x_1, x_3, t) = \frac{\partial \psi(x_1, x_3, t)}{\partial x_3}, \tag{6.6.9}$$

$$v_3(x_1, x_3, t) = -\frac{\partial \psi(x_1, x_3, t)}{\partial x_1}, \tag{6.6.10}$$

leads to the single equation for $\psi(x_1, x_3, t)$:

$$\frac{\partial^2}{\partial t^2}\left(\frac{\partial^2 \psi}{\partial x_1^2} + \frac{\partial^2 \psi}{\partial x_3^2}\right) + N^2 \frac{\partial^2 \psi}{\partial x_1^2} = 0, \tag{6.6.11}$$

with the initial/boundary condition

$$\psi(0, x_3, t) = -\frac{q}{2} \operatorname{sgn}(x_3) H(t), \tag{6.6.12}$$

and where,

$$N^2 = -\frac{g}{\rho_0} \frac{d\rho_e}{dx_3}, \tag{6.6.13}$$

is the buoyancy frequency, N, squared.

Let ℓ be a length scale of the ensuing motion and introduce the non-dimensional variables:

$$\xi_1 = \frac{x_1}{\ell}, \tag{6.6.14}$$

$$\xi_3 = \frac{x_3}{\ell}, \tag{6.6.15}$$

$$\tau = tN, \tag{6.6.16}$$

$$\varphi = \frac{\psi}{q}, \tag{6.6.17}$$

Introducing (6.6.14)–(6.6.15) into (6.6.16) and (6.1.17) into (6.6.11) yields the non-dimensional equations:

$$\frac{\partial^2}{\partial \tau^2}\left(\frac{\partial^2 \varphi}{\partial \xi_1^2} + \frac{\partial^2 \varphi}{\partial \xi_3^2}\right) + \frac{\partial^2 \varphi}{\partial \xi_1^2} = 0, \tag{6.6.18}$$

$$\psi(0, \xi_3, t) = -\frac{1}{2}\text{sgn}(\xi_3)H(\tau). \tag{6.6.19}$$

Equation (6.6.18) is most easily solved using the Laplace transform technique where we introduce the new variable:

$$\overline{\varphi}(x_1, x_3, s) = \int_0^\infty \varphi e^{-st} d\tau \tag{6.6.20}$$

and then multiply (6.6.18) and (6.6.19) by $e^{-s\tau}$ and integrate the resulting equations with respect to τ from 0 to ∞. This leads to the equations:

$$\frac{\partial^2 \overline{\varphi}}{\partial \xi_3^2} + \left(\frac{s^2+1}{s^2}\right) \frac{\partial^2 \overline{\varphi}}{\partial \xi_1^2} = 0, \tag{6.6.21}$$

$$\overline{\varphi}(0, \xi_3, s) = -\frac{1}{2}\frac{1}{s}\text{sgn}(\xi_3). \tag{6.6.22}$$

Equation (6.6.21) may be simplified by introducing the stretched variable

$$\eta_3 = \left(\frac{s^2+1}{s^2}\right)^{\frac{1}{2}} \xi_3, \tag{6.6.23}$$

so that (6.6.21) becomes:

$$\frac{\partial^2 \overline{\varphi}}{\partial \eta_3^2} + \frac{\partial^2 \overline{\varphi}}{\partial \xi_1^2} = 0. \tag{6.6.24}$$

Equation (6.6.24) is the Laplace equation subject to a sink at the origin in terms of the (ξ_1, η_3) coordinates. The solution is simply radial flow toward the origin so that:

$$\overline{\varphi} = \frac{1}{\pi s}\tan^{-1}\left(\left(\frac{1+s^2}{s^2}\right)^{\frac{1}{2}} \frac{\xi_3}{\xi_1}\right) = \frac{1}{\pi s}\tan^{-1}(\alpha\xi_3), \tag{6.6.25}$$

where

$$\alpha = \left(\frac{1+s^2}{s^2}\right)^{\frac{1}{2}} \frac{1}{\xi_1}. \tag{6.6.26}$$

The complete solution is given by the Laplace inverse of (6.6.25). For the purpose of illustrating the evolution of the velocity field and showing that simple configurations in a stratified flow collapse to a horizontal motion it is convenient to derive the inverse of (6.6.25) in terms of the horizontal velocity $v_1(x_1, x_3, t)$. Given (6.6.9) it follows that:

$$\overline{v}_1(\xi_1, \xi_3, s) = \frac{\partial \overline{\varphi}}{\partial \xi_3} = \frac{1}{\pi s} \frac{1}{(1 + \alpha^2 \xi_3^2)} \frac{\alpha}{\xi_1} = \frac{\xi_1}{\pi(\xi_1^2 + \xi_3^2)} \frac{(1 + s^2)^{\frac{1}{2}}}{(\sin^2\theta + s^2)}, \tag{6.6.27}$$

where

$$\sin\theta = \frac{\xi_3}{(\xi_1^2 + \xi_3^2)^{\frac{1}{2}}} = \beta. \tag{6.6.28}$$

The general inverse of (6.2.6) is not straight forward and we content ourselves here with evaluating the behavior of the velocity field at small and large times in order to verify the premise we started this section with; radial flow at small time and horizontal motion at large times.

For small time the inverse of (6.6.27) is given by the inverse of:

$$\overline{v}_1 \sim \frac{\xi_1}{\pi(\xi_1^2 + \xi_3^2)} \frac{1}{s}, \tag{6.6.29}$$

and the inverse of $\frac{1}{s}$ is simply $H(t)$, so that

$$v_1 \sim \frac{1}{\pi(x_1^2 + x_3^2)^{\frac{1}{2}}} \cos\theta, \tag{6.6.30}$$

which represents radial flow toward the sink.

In the theory of Laplace transforms, it is shown that the inverse of (6.6.27), for large time τ, is given by the inverse of any singularities of (6.6.27). In this case (6.6.27) has a simple pole at:

$$s = i \sin\theta, \tag{6.6.31}$$

where the function behaves as:

$$\overline{v}_1 \sim \frac{\cos^2\theta}{\pi(\xi_1^2 + \xi_3^2)^{\frac{1}{2}}} \frac{1}{(\beta^2 + s^2)}. \tag{6.6.32}$$

Now the inverse of $\dfrac{\beta}{\beta^2 + s^2}$ is given by $H(\tau)\sin\beta\tau$ (Campbell and Foster, 1948) so for large times:

$$v_1(\xi_1, \xi_3, \tau) \sim \frac{\cos^2\theta}{\pi(\xi_1^2 + \xi_3^2)^{\frac{1}{2}}} H(\tau)\,\frac{\sin\beta\tau}{\beta}. \qquad (6.6.33)$$

Now it is shown in books on generalized functions (Lighthill 1962) that:

$$\lim_{\tau\to\infty} \frac{\sin\beta\tau}{\beta} = \delta(\beta) = \delta\left(\frac{\xi_3}{(\xi_1^2 + \xi_3^2)^{\frac{1}{2}}}\right). \qquad (6.6.34)$$

Substituting the expression (6.6.34) into (6.6.33) yields:

$$\underset{t\to\infty}{v_1(x_1, x_3, t)} \sim -\frac{q}{2}\delta(x_3); \quad x_1 > 0, \qquad (6.6.35)$$

showing that, for long time, the velocity field collapses into a line flow at the elevation of the sink; it is for this reason such flows are called "selective withdrawal". The sink flow discussed here clearly shows that buoyancy acts only after there has been some motion and then ultimately the flow collapses to horizontal flow unless prevented to do so by viscosity or convective inertia.

6.7. UNSTEADY SINK FLOW IN A LINEARLY STRATIFIED FLUID CONTAINED IN A HORIZONTAL DUCT: SHEAR WAVES

In the above section, we examined the collapse of a flow field into a simple horizontal line flow into the sink. Another fundamental concept of stratified flow is what happened when the above sink configuration is embedded into a horizontal duct as shown in Fig. 6.7.1 By analogy with the above, on initiation of the sink discharge, the first flow configuration will be potential flow into the sink, radial close to the sink and duct flow at infinity. The curved streamlines of such a potential flow will bend the horizontal constant density lines. This will continue until the buoyancy forces, so invoked, become comparable to the inertia forces. There will be some internal overshoot and buoyancy will cause the displacement to bounce back, this suggests some type of wave motion. We now investigate these waves a little more closely as they are the precursor to most steady motions of stratified fluids in vertically bounded domains. The case we will now examine is that of flow into a sink, at the center of a semi-infinite duct as shown in Fig. 6.7.1.

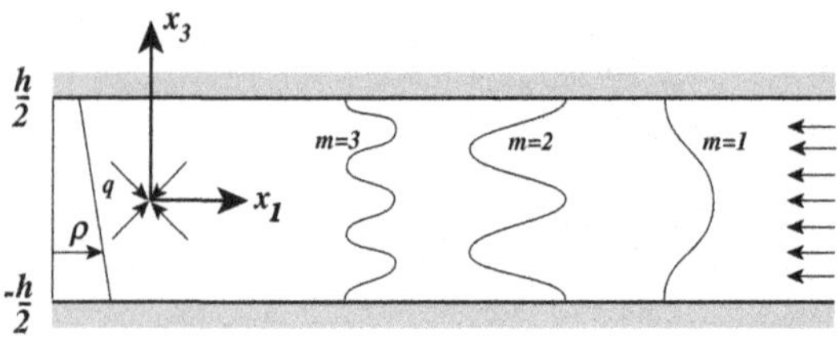

Figure 6.7.1 Schematic of the point sink in a semi-infinite duct.

Once again, as in §6.6, at time $t=0$, a sink flow is initiated at a rate $q\,\mathrm{m}^2\,\mathrm{s}^{-1}$ and the duct, of height h, is filled with a linearly stratified fluid. The fluid is assumed to be inviscid, non-diffusive and advective acceleration forces are neglected. Under these circumstances, the equation for the stream function is the same as (6.6.11) repeated here for convenience

$$\frac{\partial^2}{\partial t^2}\left(\frac{\partial^2\psi}{\partial x_1^2}+\frac{\partial^2\psi}{\partial x_3^2}\right)+N^2\frac{\partial^2\psi}{\partial x_1^2}=0 \tag{6.7.1}$$

with the initial/boundary condition:

$$\psi(0,x_3,t) = -\frac{q}{2}\operatorname{sgn}(x_3)H(t), \tag{6.7.2}$$

$$\psi\left(x_1,\frac{h}{2},t\right) = \frac{q}{2}, \tag{6.7.3}$$

$$\psi\left(x_1,\frac{-h}{2},t\right) = \frac{-q}{2}, \tag{6.7.4}$$

where N is the buoyancy frequency (6.6.13).

The solution to (6.7.1) subject to (6.7.2)–(6.7.4) may be achieved by first taking a Laplace transform in time and then seek a Fourier series solution in x_3. However, the resulting equations are rather difficult to invert. The mechanism for the establishment of the final flow configuration can be effectively investigated by noting that the scale of the flow in the horizontal direction is much larger than that in the vertical direction that is at most h. With this assumption, equation (6.7.1) becomes:

$$\frac{\partial^2}{\partial t^2}\left(\frac{\partial^2\psi}{\partial x_3^2}\right)+N^2\frac{\partial^2\psi}{\partial x_1^2}=0. \tag{6.7.5}$$

This is most easily solved by taking the Laplace transform:

$$s^2\frac{\partial^2\overline{\psi}}{\partial x_3^2}+N^2\frac{\partial^2\overline{\psi}}{\partial x_1^2} = 0, \tag{6.7.6}$$

where

$$\overline{\psi}(x_1,x_3,s) = \int_0^\infty \psi e^{-st}\mathrm{d}\tau. \tag{6.7.7}$$

Now scaling the x_1 coordinate so that

$$\xi_1 = \frac{s}{N}x_1 \tag{6.7.8}$$

Equation (6.7.6) becomes:

$$\frac{\partial^2\overline{\psi}}{\partial x_3^2}+\frac{\partial^2\overline{\psi}}{\partial x_1^2} = 0. \tag{6.7.9}$$

The boundary conditions transform to:

$$\psi(0,x_3,t) = -\frac{q}{2s}\mathrm{sgn}(x_3), \tag{6.7.10}$$

Together (6.7.9) and (6.7.10) describe a simple impulsively initiated sink flow in a duct filled with a homogeneous fluid in the (ξ_1,x_3,τ) coordinate system. The solution may be obtained through a separation of variables:

$$\overline{\psi}(x_1,x_3,s) = -\frac{qx_3}{hs}-\sum_{m=1}^{m=\infty}\frac{q}{m\pi s}e^{-\frac{2\pi m}{N}\frac{x_1}{h}s}\sin\left(\frac{2m\pi x_3}{h}\right). \tag{6.7.11}$$

Noting that the Laplace inverse of the $\frac{1}{s}$ is H (t) and the exponential in (6.7.11) simply adds a shift to the inverse, the result for the stream function becomes:

$$\psi(x_1,x_3,t) = -\frac{qx_3}{h}H(t)-\sum_{m=1}^{m=\infty}\frac{q}{m\pi}\sin\left(\frac{2m\pi x_3}{h}\right)H\left(t-\frac{2m\pi x_1}{Nh}\right). \tag{6.7.12}$$

This solution has a simple interpretation. The first term is a uniform flow toward the sink, set up instantaneously after the initiation of the discharge, a consequence of the incompressibility of the fluid. The terms in the series represent shear waves as shown in Fig. 6.7.1 traveling with a speed

$$c_m = \frac{Nh}{2m\pi}, \tag{6.7.13}$$

each leaving, in its wake, a flow concentration of thickness:

$$\delta_m = \frac{h}{4m}. \tag{6.7.14}$$

In the limit as $t \to \infty$, the solution again approaches a delta function in x_3, given by (6.6.35).

REFERENCES

Campbell, G.A., Foster, R.M., 1948. Fourier integrals for practical applications D. Van Nostrand Co., p 177.

Hicks, B.B., 1975. A procedure for the formulation of bulk transfer coefficients over water. Boundary Layer Met., 315–324.

Imberger, J., Patterson, J.C., 1990. Physical limnology. In: Wu, T. (Ed.), Advances in Applied Lighthill, M.J., 1962. Introduction to Fourier Analysis and Generalised Functions Cambridge University Press, p. 79.

Mechanics, vol. 27. Academic Press, Boston, pp. 303–475.

Monin, A.S., Obukhov, A.M., 1954. Basic laws of turbulent mixing in the atmosphere near the ground. Jr. Akad. Nauk SSSR Geofiz. Inst. 24 (151), 163–187.

Spigel, R.H., Imberger, J., Rayner, K.N., 1986. Modeling the diurnal mixed layer. Limnol. Oceanogr. 31 (3), 533–556.

Stull, R.B., 1988. An Introduction to Boundary Layer Meteorology. Kluwer Academic Publisher, p. 666.

Tennessee Valley Authority, 1972. Heat and Mass Transfer Between a Water Surface and the Atmosphere, Lab. Report No. 14. Norris Tennessee.

CHAPTER 7

Mixing in Environmental Flows

Contents

Mixing is probably the most important fluid attribute for the environmental engineer. Nature uses the fluid as a holding medium, but most importantly aquatic living organisms have evolved in water bodies relying on a certain amount of dilution of waste products from other organisms, so that only in extreme cases are chemicals very concentrated, nearly always the dispersion and mixing dilutes chemicals in the water column to a point where they can be assimilated as food by organisms of the food chain. In this chapter, we focus on mixing and dispersion in a water column where temperature and or salinity stratification leads to a density stratification that inhibits vertical mixing.

7.1. TURBULENCE IN A STRATIFIED SHEAR FLOW

In §2.14, §2.15 and §5.5, we discussed the main features of turbulent motions in a homogeneous fluid and showed how energy flows from large scale eddies to ever smaller scales until viscous action finally converts the kinetic energy into heat. The water in most natural systems is stratified with warm or fresh surface water over-lying cool or salty bottom water. This means we must modify the ideas developed for homogeneous fluids because in a stratified water column mixing involves lifting the heavier, cooler water residing underneath lighter layers of water to the center of gravity of the mixed patch. This takes energy. In order to understand what effect this change of potential energy has on the mixing process, consider a very simple

Environmental Fluid Dynamics
ISBN 978-0-12-088571-8, DOI: 10.1016/B978-0-12-088571-8.00007-3

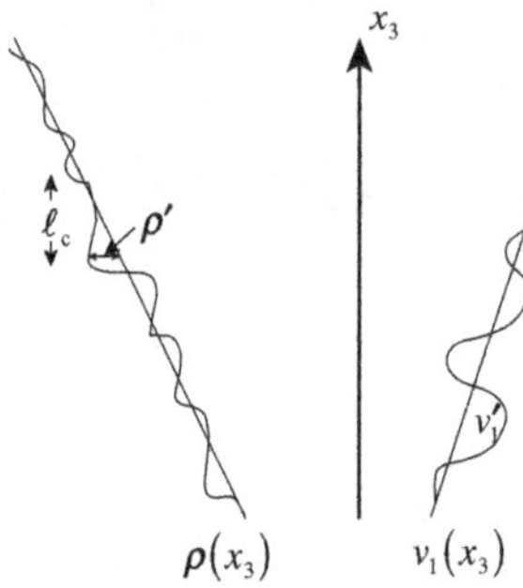

Figure 7.1.1 Schematic defining the density ρ' and velocity v_1' perturbations and the associated length scale ℓ_C in a turbulent stratified shear flow.

configuration, that of a uniform mean shear flow in a linearly stratified fluid. The problem (Fig. 7.1.1) is completely specified by the set of parameters:

$$\underset{s^{-1}}{S}, \quad \underset{s^{-2}}{N^2}, \quad \underset{m^2 s^{-1}}{\nu}, \quad \underset{m^2 s^{-1}}{\kappa}, \quad \underset{s}{t}, \quad \underset{m s^{-1}}{q'(0)}, \quad \underset{m,}{\ell_c(0)}$$

where S is the shear, N is the buoyancy frequency, ν is the viscosity, κ is the molecular diffusion coefficient, t is the time, $q'(0)$ is the initial RMS of the velocity fluctuations and $\ell_c(0)$ is the initial RMS scale of the motion.

All characteristics of the mean flow and the turbulence must be functions of the above variables and the initial conditions that trigger the turbulence. The velocity shear $S = \dfrac{\partial \bar{v}_1}{\partial x_3}$ is independent of x_3 since we have assumed that $\bar{v}_1$ is linear in x_3. The buoyancy frequency N is defined through the relationship:

$$N^2 = \frac{-g}{\rho_0}\frac{\partial \bar{\rho}}{\partial x_3}, \tag{7.1.1}$$

where g is the acceleration due to gravity, ρ_0 is the mean density and the minus sign has been incorporated to make N^2 positive for a stable gradient. The acceleration due to gravity g enters our problem only in combination with $\bar{\rho}$ since it is the weight of a fluid parcel that determines the dynamical behavior of the particle.

The physical meaning of N may be best understood by analyzing the motion of a parcel of fluid that has been moved out of its stable equilibrium as shown in Fig. 7.1.1. If ℓ_C is the distance the particle has been moved from the original position then the density anomaly ρ' is given by:

$$\rho' \sim \frac{\partial \bar{\rho}}{\partial x_3}\ell_c, \tag{7.1.2}$$

so that the effective acceleration due to gravity,

$$g' = \frac{\rho' g}{\rho_0} = N^2 \ell_c. \qquad (7.1.3)$$

Now the time taken for this "heavier" particle to fall back to its original position under the action of gravity is given by:

$$T_f \sim \left(\frac{\ell_c}{g'}\right)^{1/2} = N^{-1}, \qquad (7.1.4)$$

where we have assumed that the particle simply falls in a gravitational field.

The parameter N^{-1} is thus the time scale for particles to return to their stable equilibrium position by falling under gravity. Internal wave motions are not simply particles falling, but also involve horizontal accelerations of the fluid that must make way for the falling particles. This horizontal acceleration retards the falling particles and so internal wave periods are always slower than N^{-1}.

There are seven independent variables on which the characteristics of the flow may depend, so that there are five dimensionless groups that determine the flow. Following convention, we choose the first three as:

$$Ri = \frac{N^2}{S^2}, \qquad (7.1.5)$$

$$Pr = \frac{\nu}{\kappa}. \qquad (7.6.6)$$

$$T^* = St \qquad (7.1.7)$$

The Richardson number Ri has a simple physical interpretation if one examines the energetics of the mixing. Suppose we have a layer of fluid of thickness ℓ_c that is partially mixed as shown in Fig. 7.1.2 and suppose further that after the mixing event has finished we may idealize the new buoyancy frequency in the patch by N_2 and the new shear by S_2. The change from state 1 to state 2 in potential energy, relative to the bottom of the mixing zone, is given by:

$$\Delta PE = \int_0^{\ell_c} \Delta\rho g x_3 \mathrm{d}x_3 = \frac{\gamma}{12}\rho_0 \ell_c^3 N^2, \qquad (7.1.8)$$

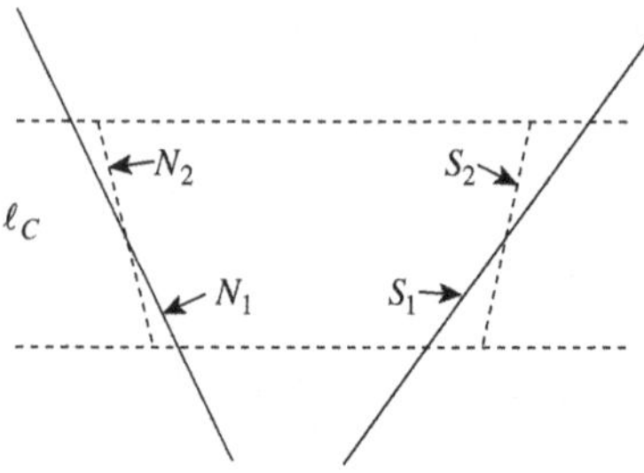

Figure 7.1.2 Changes in the density and velocity profiles due to partial mixing over a length scale ℓ_c.

where $\Delta\rho$ is the change in density between state 1 and 2 where $N_2^2 = (1-\gamma)N^2$, so that for a complete mixing $\gamma = 1$ and for no mixing $\gamma = 0$.

If we assume that as the mixing takes place the velocity profile is also partially mixed so that the shear changes from S_1 to S_2 and we write:

$$S_2 = (1-\beta)\left(\frac{\partial \bar{v}_1}{\partial x_3}\right)_1 = (1-\beta)S, \tag{7.1.9}$$

then the release of kinetic energy implied by this change in the velocity profile is given by:

$$\Delta KE = \int_0^{\ell_c} \frac{1}{2}\rho\left(\left(\bar{v}_1^{(2)}\right)^2 - \left(\bar{v}_1^{(1)}\right)^2\right)\mathrm{d}x_3 = \beta\left(\frac{2-\beta}{24}\right)\left(\frac{\partial \bar{v}_1}{\partial x_3}\right)^2 \rho_0 \ell_c^3. \tag{7.1.10}$$

Thus, the ratio of the potential energy required to mix the water column to that available from the mean kinetic energy released by the mixing is given by:

$$\frac{\Delta PE}{\Delta KE} = \frac{2\gamma N^2}{\beta(2-\beta)S^2} = \frac{2\gamma}{\beta(2-\beta)}Ri. \tag{7.1.11}$$

Thus if $Ri < \dfrac{(2-\beta)\beta}{3\gamma}$ there is a sufficient energy available from the mean kinetic energy field to overcome the necessary potential energy required to achieve the mixing. However, this does not mean that mixing will necessarily occur, as we have so far not accounted for the energy necessary to overcome the dissipation of turbulent kinetic energy due to viscous action at the smaller scales. Complete homogenization ($\beta = \gamma = 1$) requires

$Ri < 1/2$, but partial mixing is possible ($\gamma = \beta \to 0$) whenever $Ri < 1$. However, in both of these limiting examples we must still account for dissipation that may be estimated by noting that the dissipation per unit mass must scale, from §5.5, as $\frac{\rho_0 u^3}{\ell_C}$, where u is a velocity scale. The time scale for releasing the kinetic energy is given as ℓ_c/u (§5.5) so the energy lost to dissipation over the time it takes to mix the fluid is given by:

$$\rho_0 \varepsilon \ell_c \frac{\ell_c}{u} \sim \rho_0 \frac{u^3}{\ell_c} \cdot \ell_c \cdot \frac{\ell_c}{u} \sim \rho_0 u^2 \ell_c \sim \rho_0 \left(\frac{\partial u}{\partial x_3}\right)^2 \ell_c^3 \sim \Delta KE, \tag{7.1.12}$$

The above argument shows that a certain fraction of the kinetic energy is always lost to dissipation. A well-known result from stability analysis of an inviscid shear flow (Yih, 1980) tells us that the flow is stable for,

$$Ri > 1/4. \tag{7.1.13}$$

Suppose the value 1/4 forms the stability boundary then according to the above model, whenever there is a complete mixing of the fluid column, 50% of the kinetic energy is released to dissipation and 50% is used to lift the mixed fluid parcel resulting in a change of potential energy. On the other hand, when mixing is only partial ($\gamma = \beta \to 0$), then 75% of the released kinetic energy goes to dissipation and only 25% of the released energy is utilized to lift the fluid parcel. This is an important result as most experimental evidence on mixing efficiency point to a conversion ratio close to the latter value (Imberger and Ivey, 1993).

The other two dimensionless groups have straightforward interpretations. The group St is the time measured in units of the shear time or the number of times an eddy has turned over. The value of the Prandtl number, Pr, determines the relative importance of the loss by molecular diffusion of momentum of a fluid particle to the loss of the density anomaly of the buoyancy driving forces.

The evolution of the flow must also depend on the initial conditions. Thus, unless the turbulence quickly "forgets" the history of its origin, the initial conditions will be important in determining the flow characteristics. The initial conditions should involve a specification of the complete properties of the initial turbulent field, but this would require a complete specification of all the joint probability functions of the initial velocity and density fields. It is unlikely that the structural details of the initial conditions will play an important role in the evolution of the flow and the RMS magnitudes of the velocity fields $u'(0)$ and the length scale $\ell_C(0)$ will be sufficient to capture the initial turbulence properties. The initial acceleration perturbation $g'(0)$ may be estimated from:

$$g'(0) = \frac{\rho'(0)}{\rho_0} g = N^2 \ell_c(0). \tag{7.1.14}$$

If ℓ is the ensemble averaged length scale of the largest eddies, then dimensional analysis suggests

$$\ell = \ell_p f_1 \left(Ri,\ St,\ \mathrm{Pr};\ Fr_t(0),\ \mathrm{Re}_t(0)\right), \tag{7.1.15}$$

where f_1 is a universal function and the effect of the initial disturbance has been captured by the inclusion of the initial value of the Froude and Reynolds numbers:

$$Fr_t(0) = \frac{u(0)}{N^{1/2}\,\ell(0)}, \tag{7.1.16a}$$

$$\mathrm{Re}_t(0) = \frac{u(0)\,\ell(0)}{\nu}. \tag{7.1.16b}$$

Similarly, we may write

$$u = u_p f_2(Ri, St, \mathrm{Pr}; Fr_t(0), \mathrm{Re}_t(0)), \tag{7.1.17}$$

and

$$\varepsilon = \varepsilon_p f_3\,(Ri, St, \mathrm{Pr}; Fr_t(0), \mathrm{Re}_t(0)), \tag{7.1.18}$$

where

$$\ell_p = \left(\frac{\nu}{N}\right)^{1/2}, \tag{7.1.19}$$

$$u_p = (\nu N)^{1/2}, \tag{7.1.20}$$

and

$$\varepsilon_p = \nu N^2, \tag{7.1.21}$$

are called the primitive length, velocity and dissipation scales.

The vertical eddy diffusion coefficient may be derived from the relationship:

$$\overline{\rho' v'_3} = \kappa_\rho \left(\frac{\partial \bar{\rho}}{\partial x_3}\right) = \frac{\rho_0 \kappa_\rho}{g} \frac{g}{\rho_0} \left(\frac{\partial \bar{\rho}}{\partial x_3}\right) = -\frac{\rho_0 \kappa_\rho}{g} N^2. \qquad (7.1.22)$$

Now we may write:

$$\frac{g(\overline{w'\rho'})}{\rho_0 \kappa N^2} = C(\mathrm{Ri_g}) \frac{g u' \rho'}{\rho_0 \kappa N^2} = C(\mathrm{Ri_g}) \frac{u\ell}{\kappa} = C(\mathrm{Ri_g}) \mathrm{Pr}\, \mathrm{Re_t}, \qquad (7.1.23)$$

where $C(\mathrm{Ri_g})$ is the correlation function between the vertical velocity and density fluctuations and where we have used (7.1.13) to estimate the density fluctuation. Direct measurements of the buoyancy flux using a laser Doppler forward velocimeter and a temperature microstructure sensor have enabled (7.1.23) to be evaluated (Yeates et al., 2012). These authors showed that for particular mixing events the buoyancy flux is sometimes down gradient as implied by (7.1.22) and sometimes when the profiler pierced a collapsing mixing event, the flux was up gradient. The details of this go beyond this introductory text, but the reader may picture, once again the analogy with the negatively buoyant simple plume as discussed in §5.4. When the turbulence is actively developing and mixing the stratified fluid, we would expect the buoyancy flux to be down gradient, however, when either the background shear relaxes or the mixing event has over mixed, buoyancy forces will collapse the mixed patch, layering the fluid back toward its original stratification; this appears as a re-stratification and so the buoyancy

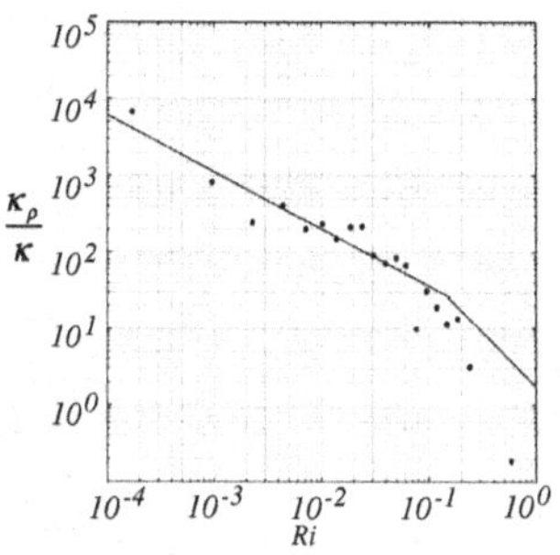

Figure 7.1.3 Normalized effective eddy diffusivity as a function of *Ri*. The shaded area is net down gradient eddy diffusion coefficient from Yeates et al. (2012) obtained from direct buoyancy flux measurements in stratified lakes and estuaries. The dashed line is $\frac{\kappa_\rho}{\kappa} = 1 + \frac{80}{Ri^{\frac{2}{3}}}$; $Ri < 0.1$ and $\frac{\kappa_\rho}{\kappa} = 1 + \frac{1}{20\, Ri^4}$; $Ri \geq 0.1$

flux will be up gradient. Direct measurements of the net buoyancy flux averaged over many events in bins of *Ri* numbers are shown in Fig. 7.1.3.

7.2. THE KINEMATICS OF DISPERSION

In §5.5 we stated that one of the main characteristics that distinguishes turbulent from laminar motion is the very much faster rate of dispersion in a turbulent flow compared to that in laminar flow. In this section we show, with a purely kinematical argument, why this is so and what is the main feature of a turbulent flow that leads to an enhanced rate of dispersion.

When a tracer is introduced into a turbulent fluid the motion distorts the dye volume and changes, say, an initially spherical volume into a sheet like volume that is continuously stretched, twisted and folded (Fig. 7.2.1). The volume of fluid that contains the extremities of the dye sheets increases rapidly with time, but the volume of dyed fluid will only increase by molecular diffusion. The density of dye sheets per unit volume will vary greatly, but when viewed from a distance the volume containing dye sheets will appear like a spreading diffuse cloud. In the absence of molecular diffusion there are thus two distinct entities. First, the actual volume of dyed fluid in the form of stretching and folding surfaces; this volume remains constant with time. Second, the volume of fluid containing the extremities of the dye sheets will continue to grow. The magnitude of this volume or cloud is determined by the maximum separation of the dye sheets. The size of this cloud can, conveniently, be measured by computing the moments of inertia about the centroid of the cloud and in a turbulent flow this will increase rapidly with time; we will call the growth of this cloud dispersion.

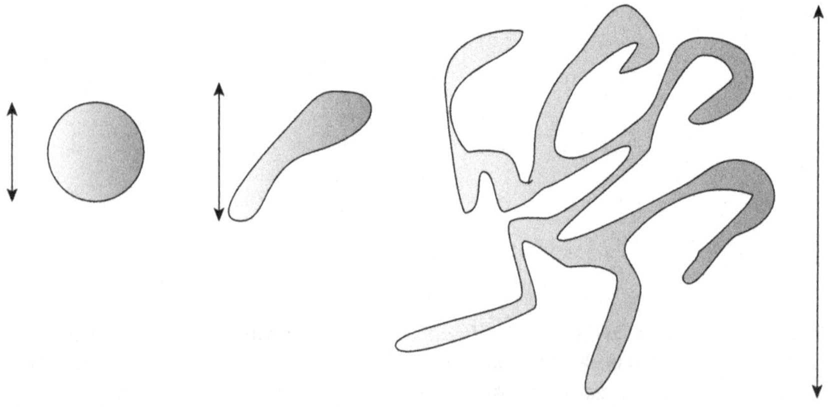

Figure 7.2.1 Schematic of the distortion of an initially small sphere of tracer in a turbulent flow.

The stretching of the dye surface by the motion decreases the thickness of the surface and increases the dye gradient normal to the surface. On the other hand, folding increases the surface density in a particular volume and increases the proximity of surfaces to each other. Both processes allow molecular diffusion to smear the surfaces into a continuous mass. We shall call the combination of folding, stretching and diffusion, mixing as these three combine to homogenize the tracer; dispersion will be reserved to describe the spreading of the cloud by folding and stretching.

The scale at which diffusion takes place may again conveniently be estimated by postulating a simple balance, at the smallest scale, between the action of the dye sheet being stretched and the transverse molecular diffusion:

$$v_j \frac{\partial C}{\partial x_j} \sim \kappa \frac{\partial^2 C}{\partial x_i^2}, \tag{7.2.1}$$

where C is the concentration of the dye so that:

$$x_1 \sim \eta_\beta = \left(\frac{\kappa^2 \nu}{\varepsilon}\right)^{1/4}, \tag{7.2.2}$$

where κ is the molecular diffusivity of the tracer, ν is the fluid kinematic viscosity, ε is the rate of dissipation of turbulent kinetic energy and where we have related the velocity scale to the dissipation via (5.5.17). The scale η_β is called the Batchelor scale for the dye tracer and represents the scale at which the rate of thinning of dye sheets is just balanced by the rate of thickening by transverse molecular diffusion. Note for most tracers or temperature the Batchelor scale is considerably smaller than the Kolmogorov's scale (5.5.19).

From this discussion it is clear that the rapid dispersion of tracers in a turbulent flow is predominantly the result of the rapid spreading of the tracer sheets, in other words the dispersion; mixing is important for the homogenization of the tracer and this is important for any chemical reactions that may possibly take place with the tracer.

The rate of spreading, the result of fluid particles carrying the tracer to increasing separation distances may conveniently be estimated with an elegant analysis due to Taylor (1921). For simplicity, we present the results here only for a one-dimensional spreading of a cloud in a homogeneous turbulent field. Let $U(t)$ be a realization of the x_1 random velocity component of a fluid particle (following the motion) and assume that the field is stationary with a zero mean (see §2.13).

The x_1 displacement X in time t of a particular particle is then given by:

$$X = \int_0^t U(t')\mathrm{d}t'. \tag{7.2.3}$$

Now since integration is a linear operation it follows that the mean displacement,

$$\overline{X} = \int_0^t \overline{U}(t')\mathrm{d}t' = 0, \tag{7.2.4}$$

where the over bar is the mean or expectation of the quantity.

Similarly, the variance of the displacement is given by,

$$\begin{aligned}\overline{X}^2 &= \overline{\int_0^t U(t')\mathrm{d}t' \int_0^t U(t'')\mathrm{d}t''} \\ &= \int_0^t \int_0^t \overline{U(t')U(t'')}\mathrm{d}t'\mathrm{d}t'',\end{aligned} \tag{7.2.5}$$

where the integrand is the autocovariance of the Lagrangian velocity $U(t)$. If the velocity field is stationary then this can only be a function of the time difference

$$\tau = t'' - t'. \tag{7.2.6}$$

By introducing the auto-correlation function:

$$R(\tau) = \frac{\overline{U(t')U(t'')}}{\overline{U^2(0)}}, \tag{7.2.7}$$

we may rewrite (7.2.5) in the form:

$$X^2 = \overline{U^2(0)} \int_0^t \int_0^t R(\tau)\mathrm{d}t'\mathrm{d}t''. \tag{7.2.8}$$

Changing the integration variables to,

$$\tau = (t'' - t'), \tag{7.2.9}$$

$$\zeta = (t'' + t'), \tag{7.2.10}$$

leads to:

$$\overline{X^2} = 2\overline{U^2(0)} \int_0^t (t - \tau) R\left(\frac{\tau}{\sqrt{2}}\right) d\tau. \tag{7.2.11}$$

Now suppose that the velocity field is such that the velocity at t'' becomes uncorrelated to that at t' for large separations, then equation (7.2.11) becomes:

$$\overline{X^2} = 2\overline{U^2(0)} \left\{ T_c t - 4 \int_0^\infty \tau R(0) d\tau \right\}, \tag{7.2.12}$$

where T_c is given by the expression,

$$T_c = \sqrt{2} \int_0^\infty R(\tau) d\tau. \tag{7.2.13}$$

The integral in (7.2.12) is a constant so that:

$$\overline{X^2} = 2\overline{U^2(0)} T_c t + \text{constant}, \tag{7.2.14}$$

and from this it follows that,

$$\frac{d < X^2 >}{dt} \sim 2\overline{U^2(0)} T_C, \tag{7.2.15}$$

which may be rewritten in the form:

$$\frac{d < X^2 >}{dt} = 2(\overline{U^2(0)})^{1/2} L_c, \tag{7.2.16}$$

where the length scale L_C is given by:

$$L_c = (\overline{U^2(0)})^{1/2} T_c. \tag{7.2.17}$$

This length scale represents the average distance a particle moves in the flow field in one correlation time. In simple terms, this represents the average size of the largest eddy in the flow. Hence, provided we are able to determine the correlation time scale and the standard deviation of the Lagrangian velocity, we are able to estimate the rate of dispersion of particles.

If we now envisage a series of closely spaced particles making up a tracer cloud, then linear superposition of these particles implies that the variance of the cloud concentration is the same as the variance of a particular particle; equation (7.2.14) thus implies that the cloud will disperse so that the variance grows linearly with time and the constant of proportionality is given by the autocovariance of the Lagrangian velocity field.

This analysis, due to Taylor (1921), clearly shows the role of the Lagrangian kinematics of the motion in the dispersion of fluid particles and hence mass, momentum and other fluid properties. However, note that the analysis makes no mention of how the particles mix with surrounding fluid, it merely provides an estimate of how fast a set of particles separate or disperse; dispersion, however, determines the global rate of increase of a pollutant cloud. Mixing which is the action of locally homogenizing the particles with the surrounding fluid depends on the smallest scales of motion.

Unfortunately, in practice it is not often possible that we can evaluate the correlation coefficient, as it is not possible to extract the necessary information about the Lagrangian velocity field $U(t)$. However, the importance of (7.2.14) is that it clearly shows the role of the velocity field with its action of stretching and folding in determining the rate of spread of a tracer; the rate of spread of the variance is determined by the velocity scale times the size of the largest eddy.

One very important case where we can apply (7.2.14) as the Lagrangian properties may be evaluated is in pipe flow §5.6 and open channel flow §5.9, irrespective of whether the flow is laminar or turbulent. In laminar flow through a pipe the velocity profile, (§5.6), is parabolic and mixing is via molecular diffusion. In laminar flow, we may think about tracer molecules migrating throughout the pipe cross-section as the fluid in the pipe moves through the pipe. A tracer particle will thus "sample" velocities ranging from zero when near the pipe wall and u_m, the center line maximum velocity when near the center of the pipe. If we assume that the molecules move randomly through the pipe cross-section then it is clear that the Lagrangian velocity $U(t)$ is a stationary random function and we may apply the above theoretic development with:

$$\overline{U^2(0)} \sim u_m^2 \tag{7.2.18}$$

and since the mixing of the tracer across the pipe cross-section is given by molecular diffusion,

$$T_c \sim \frac{R^2}{\kappa}. \tag{7.2.19}$$

Substituting (7.2.18) and (7.2.19) into (7.2.15) leads to a rate of dispersion in a pipe given by:

$$\frac{\mathrm{d}< X^2 >}{\mathrm{d}t} \sim \frac{R^2 u_m^2}{\kappa}. \tag{7.2.20}$$

In §7.5, we derive these results by solving the full diffusion equation and from this we show that:

$$\frac{\mathrm{d} < X^2 >}{\mathrm{d}t} = \frac{R^2 u_m^2}{192\kappa}. \tag{7.2.21}$$

Now consider the same problem, but instead of laminar flow in the pipe suppose the flow is turbulent as discussed in §5.6 and §5.7. The molecule of tracer, discussed above, must now be replaced by a fluid particle and the cross-sectional diffusion or mixing is now by turbulent diffusion as discussed in §5.5. Further, the characteristic scale for the velocity fluctuations is now no longer u_m, but rather u_*, the turbulent wall shear velocity, given by:

$$u_* = \left(\frac{\tau_w}{\rho}\right)^{\frac{1}{2}}, \tag{7.2.22}$$

where τ_w, is the shear stress on the pipe wall. The cross-sectional effective diffusion coefficient that takes the place of κ in (7.2.21) now becomes:

$$\kappa_t \sim u_* R, \tag{7.2.23}$$

so that:

$$T_C \sim \frac{R}{u_*}. \tag{7.2.24}$$

Substituting (7.2.24) and using the shear velocity in (7.2.15) leads to:

$$\frac{\mathrm{d} < X^2 >}{\mathrm{d}t} \sim u_* R. \tag{7.2.25}$$

Comparison with experimental results provides the coefficient of proportionality (Fischer et al., 1979):

$$\frac{\mathrm{d} < X^2 >}{\mathrm{d}t} = 10.1 u_* R. \tag{7.2.26}$$

7.3. FUNDAMENTAL SOLUTION OF THE DIFFUSION EQUATION

We have shown that the temperature and salinity fields in a fluid are governed by the diffusion equations (see §2.9). The vorticity equation is more complicated, but it again has a diffusion term that is responsible for spreading the vorticity. Above in §7.2, we showed that the variance of a tracer cloud spreading in a turbulent fluid increases linearly with time once the cloud becomes larger than the largest eddy of the motion. We shall now show that a tracer distribution that satisfies the diffusion equation posses a variance that grows linearly with time.

To show this, consider once again a one-dimensional flow advecting with a constant velocity U. Suppose a tracer in the flow is governed by the one-dimensional diffusion equation:

$$\frac{\partial C}{\partial t} + v_1 \frac{\partial C}{\partial x_1} = \kappa_c \frac{\partial^2 C}{\partial x^2}, \tag{7.3.1}$$

where t is the elapsed time and x_1 is the direction of flow. Now the variance of any tracer distribution $C(x, t)$ is defined as:

$$V_C = \frac{\int\limits_{-\infty}^{\infty} (x_1 - x_c)^2 C(x_1, t) \mathrm{d}x_1}{\int\limits_{-\infty}^{\infty} C(x_1, t) \mathrm{d}x_1}, \tag{7.3.2}$$

where x_c is the distance to the centroid of the distribution defined by:

$$x_c = \frac{\int\limits_{-\infty}^{\infty} x_1 C(x_1, t) \mathrm{d}x_1}{\int\limits_{-\infty}^{\infty} C(x_1, t) \mathrm{d}x_1}. \tag{7.3.3}$$

In order to show that the variance grows linearly with time we multiply (7.3.1) by $(x - x_c)^2$ and integrate with respect to x from $-\infty$ to ∞:

$$\left\{\frac{\partial}{\partial t}(V_c)\right\} \int\limits_{-\infty}^{\infty} C(x, t)\mathrm{d}x = \kappa_c \int\limits_{-\infty}^{\infty} (x - x_c)^2 \frac{\partial^2 C}{\partial x^2} \mathrm{d}x, \tag{7.3.4}$$

where the advective term vanishes as a direct consequence of (7.3.3). The RHS of (7.3.4) may be simplified by integration by parts twice and noting that C and $\frac{\partial C}{\partial x}$ are zero of $\pm\infty$. This leads to the result:

$$\frac{\partial V_c}{\partial t} = 2\kappa_c, \tag{7.3.5}$$

showing that for any tracer distribution $C(x,t)$ satisfying (7.3.1) the variance grows linearly with time and the rate of growth is twice the diffusion coefficient.

This is the same result that was derived in §7.2 for a tracer cloud spreading in a stationary homogeneous turbulent fluid after the cloud has grown sufficiently to encompass many of the largest eddies; in other words, eddies can only stretch and fold portions of the cloud and not transport the cloud as a whole. This is reminiscent of the actions of molecules in molecular diffusion.

The converse is most likely also true. If a concentration satisfies (7.3.5) then as distribution is governed by (7.3.1), we shall only present a plausible argument, not a proof. Remembering that a particle is a turbulent flow has a position given by:

$$x_1(t) = \int_0^t v_1(t)\mathrm{d}t, \tag{7.3.6}$$

then if $v_1(t)$ is a random variable, $x_1(t)$ is the sum or integral of a random variable and so in itself random. Now by the central limit theorem, provided $v_1(t)$ satisfies some weak independence conditions $x_1(t)$ becomes, as $t \to \infty$, a normal random variable and thus its distribution is governed by normal distribution. It is likely, although not proven, that the Lagrangian velocity field for stationary homogeneous turbulence satisfies the necessary statistical constraints.

Thus, by comparison with (7.2.14), we may say that tracers spreading in such turbulent fields obey the diffusion equation with an effective diffusion coefficient:

$$\kappa_c = \overline{U^2(0)}T_c. \tag{7.3.7}$$

The result holds for times larger than T_c.

The above discussion shows the importance of the diffusion equation for both laminar and turbulent flows as well as shear dispersion. For this reason, we now derive the fundamental solution to (7.3.1); the tracer concentration arising from the instantaneous injection of a mass P of tracer per unit area at time zero to in an infinite domain of fluid moving with a constant velocity $v_1 = U$.

Since U is a constant we can remove its influence by moving with the water. This is achieved by introducing a new set of coordinates τ and ζ:

$$\zeta = x_1 - Ut, \tag{7.3.8}$$

$$\tau = t, \tag{7.3.9}$$

so that (7.3.1) become:

$$\frac{\partial C}{\partial \tau} = \kappa_c \frac{\partial^2 C}{\partial \zeta^2}. \tag{7.3.10}$$

The solution to (7.3.10) in an infinite domain may in general be written:

$$C = C(\kappa_c, P, \zeta, \tau), \tag{7.3.11}$$

where P is the mass of tracer injected at time zero.

Now equation (7.3.10) is linear so that C must depend linearly on P. However, in order to compare P with the concentration C it is necessary to write both these quantities in the same units; that is we must change P into a concentration by dividing by a length scale. Scanning through the variables in (7.3.11) reveals that $(\kappa_c\tau)^{1/2}$ has units of length, and invoking the above mentioned linearity we may seek a solution to (7.3.10) in the form:

$$C = \frac{P}{(\kappa_c\tau)^{1/2}} f(\eta), \tag{7.3.12}$$

where

$$\eta = \frac{\zeta}{(\kappa_c\tau)^{1/2}}. \tag{7.3.13}$$

It is important to point out that the choice of the above length scale was not unique and from dimensional considerations alone we could have chosen ζ just as well. As we shall see below, the above choice was motivated by the desire to simplify the differential equation resulting from substituting

(7.3.12) into (7.3.10); often the final choice involves trial and error and it is not possible to make the correct choices a priori.

By the chain rule for differentiation we get from (7.3.12):

$$\frac{\partial C}{\partial \tau} = -\frac{P}{2(\kappa_c \tau)^{1/2}} \frac{1}{\tau} f(\eta) - \frac{1}{2} \frac{P}{2(\kappa_c \tau)^{1/2}} \frac{\eta}{\tau} \frac{df(\eta)}{d\eta}, \tag{7.3.14}$$

and

$$\frac{\partial^2 C}{\partial \tau^2} = -\frac{P}{(\kappa_c \tau)^{1/2}} \frac{1}{\kappa_c \tau} \frac{d^2 f(\eta)}{d\eta^2}. \tag{7.3.15}$$

Substituting (7.3.14) and (7.3.15) into (7.3.10) leads to the ordinary differential equation:

$$2\frac{d^2 f}{d\eta^2} + \eta \frac{df}{d\eta} + f = 0. \tag{7.3.16}$$

The simplicity of this ordinary differential equation is the result of the correct choice of variables in (7.3.12). Integrating (7.3.16) once, with respect to η, yields

$$2\frac{df}{d\eta} + \eta f = Const. \tag{7.3.17}$$

From symmetry, the tracer will spread equally in both directions relative to the origin so that a boundary condition at the origin becomes

$$\eta = 0; \quad \frac{df}{d\eta} = 0, \tag{7.3.18}$$

and thus the constant in (7.3.47) is zero.

Integrating once more yields

$$f = f_0 e^{-\frac{\eta^2}{4}}, \tag{7.3.19}$$

where f_0 is again an integration constant that may be determined by noting that the total mass of pollutant in the water must be conserved so that:

$$\int_{-\infty}^{\infty} C(\zeta\tau)d\zeta = P. \tag{7.3.20}$$

If we note that (Appendix),

$$\int_{-\infty}^{\infty} e^{\frac{-\eta^2}{2}} d\eta = \sqrt{2\pi}, \tag{7.3.21}$$

then

$$C(\zeta, \tau) = \frac{P}{(4\pi\kappa_c\tau)^{1/2}} e^{-\frac{\zeta^2}{4\kappa_c\tau}}, \tag{7.3.22}$$

which is called the fundamental solution to the diffusion equation (7.3.10).

The important feature of this solution is that it describes the way diffusion leads to a redistribution of mass and a growth of the tracer cloud. The size of the cloud is usually defined by distance in which the concentration is larger than a certain fraction of the peak concentration. If we choose say 2% as the cut off then from (7.3.22) the size is given by:

$$\zeta_2 = \pm 2\sqrt{4\kappa_c t}, \tag{7.3.23}$$

and the cloud is contained in the space:

$$-2\sqrt{4\kappa_c t} \leq x - Ut = 2\sqrt{4\kappa_c t}, \tag{7.3.24}$$

that shows that the scale for a cloud growing by pure diffusion is given by $(\kappa_c\tau)^{\frac{1}{2}}$.

In terms of the original variables, (7.3.22) becomes:

$$C(x_1, t) = \frac{P}{(4\pi\kappa_c t)^{\frac{1}{2}}} e^{\frac{(x_1 - Ut)^2}{4\kappa_c t}}. \tag{7.3.25}$$

This shows that the cloud is simply advected with the flow, diffusing as it goes.

7.4. SHEAR DISPERSION

The mixing of a tracer in a turbulent flow is due to a combination of the spreading action by the correlated velocity fluctuations and the mixing induced by the gradient enhancement due to stretching, blending due to the folding and homogenization by molecular diffusion. Together these mechanisms impart the characteristic of greatly enhanced dispersion to a turbulent flow. A single scale version of the same type of combination of mechanisms, called shear dispersion, consists of a combination of a single scale stretching, that intensifies the tracer gradients, intensifying the transverse diffusion until it balances the flux due to the enhanced gradients. This process is a single building block of the more general turbulent diffusion. Clearly, if the scale of the motion causing the stretching and folding is relatively large then the effect of dispersion is rapid, but the time before the process becomes fully effective will also be long. Often, since the stretching occur at very large scales, the transverse diffusion is not a molecular process, but is itself turbulent dispersion supported by the smaller scale turbulent motions.

To understand shear dispersion, consider a simple parallel turbulent (or laminar) flow in the x_1 direction as shown in Fig. 7.4.1. For convenience, we bound the domain at the top and bottom by solid boundaries, separated by a distance h to form a duct. Suppose at some early time a tracer was introduced into the flow between the boundaries and has, by simple laminar or turbulent diffusion, spread to fill the space between the boundaries. Diffusion will continue to spread the cloud of tracer longitudinally. The concentration C (the mean concentration in the case of turbulent flow) will thus be governed by the equation.

$$\frac{\partial C}{\partial t} + v_1 \frac{\partial C}{\partial x_1} = \frac{\partial}{\partial x_1}\left(\kappa_1 \frac{\partial C}{\partial x_1}\right) + \frac{\partial}{\partial x_3}\left(\kappa_3 \frac{\partial C}{\partial c_3}\right), \tag{7.4.1}$$

where it is stressed again that for turbulent flow C is the mean concentration and κ_1 and κ_3 are the turbulent diffusion coefficient and v_1 is the mean velocity. Over bars have been omitted for convenience of writing only as we wish to emphasis that the mechanism is not dependent on whether the flow is laminar or turbulent.

We may split the longitudinal velocity into two components:

$$v_1 = u_d + v_d, \tag{7.4.2}$$

where u_d is the cross-sectional average, or the discharge velocity, and v_d is the deviation from the cross-sectional average. Now, in order to follow the

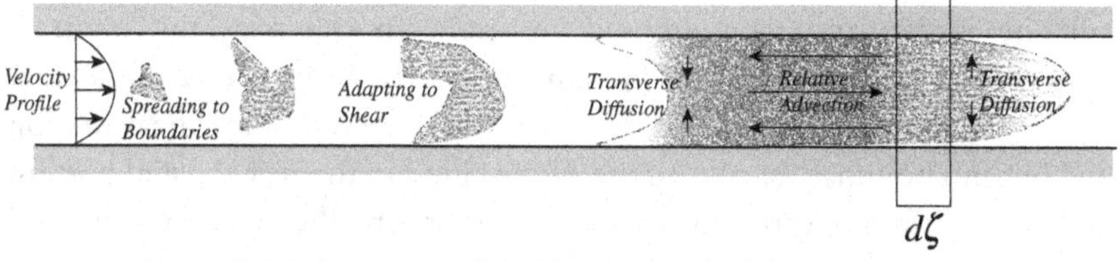

Figure 7.4.1 Schematic of the development of shear dispersion.

cloud of tracer down the two-dimensional duct we introduce new coordinates:

$$\zeta = x_1 - u_d t, \tag{7.4.3}$$

and

$$\tau = t. \tag{7.4.4}$$

In terms of this new coordinate systems (7.4.1) reduces to:

$$\frac{\partial C}{\partial \tau} + v_d \frac{\partial C}{\partial \zeta} = \frac{\partial}{\partial \zeta}\left(\kappa_1 \frac{\partial C}{\partial \zeta}\right) + \frac{\partial}{\partial x_3}\left(\kappa_3 \frac{\partial C}{\partial x_3}\right). \tag{7.4.5}$$

The solution of this equation is difficult, but we can make progress toward a solution by considering the mechanism of tracer redistribution. The second term in (7.4.5) describes the advection, by the mean flow, of the tracer relative to an observer moving with the discharge velocity. As seen in Fig. 7.4.1, in the center of the duct the fluid advects the tracer downstream while at the top and bottom the advection, relative to the discharge velocity, is upstream. Now, if on average, the tracer concentration has a negative gradient (decreasing with increasing ζ), then this term will increase the tracer concentration in the center of the duct and decrease it near the top and bottom boundaries; this is the case for the downstream side of the cloud. The reverse is true for the upstream side. The changes brought about by the advection term are shown in Fig. 7.4.1.

The advection thus causes, on the downstream and upstream sides, a vertical concentration gradient to develop that enhances vertical diffusion; the build up continues until the term $\frac{\partial}{\partial x_3}\left(\kappa_3 \frac{\partial C}{\partial x_3}\right)$ becomes large enough to balance the term $v_d \frac{\partial C}{\partial \zeta}$. Such a model means that, as the cloud moves downstream, the tracer is continuously stretched into the shape of the velocity profile and then diffused vertically. In this way, the tracer is constantly cycled through the cloud, being ejected at the center of the duct on the downstream side, mixed vertically, advected to the rear near the top and bottom boundaries, ejected out of the cloud in the rear top and bottom and then diffusing vertically to the center where the tracer is advected forward relative to the cloud. This recycling is the primary process.

Now for vertical diffusion to act strongly enough, advection must build up a sufficiently large vertical concentration gradient. The tracer distribution conforming to this simple two-process model is thus given, to first order, by the solution of:

$$v_d \frac{\partial C}{\partial \zeta} = \frac{\partial}{\partial x_3}\left(\kappa_3(x_3)\frac{\partial C}{\partial x_3}\right). \tag{7.4.6}$$

Let us seek a solution of the form,

$$C = g(\zeta, \tau) + h_d(x_3), \tag{7.4.7}$$

where $g(\zeta, \tau)$ is the average over a vertical plane cross-section and $h_d(x_3)$ is the deviation concentration.

Substituting (7.4.7) into (7.4.6) yields:

$$v_d(x_3)\frac{\partial g(\zeta, \tau)}{\partial \zeta} = \frac{\partial}{\partial x_3}\left(\kappa_{c3}(x_3)\frac{\partial h_d(x_3)}{\partial x_3}\right). \tag{7.4.8}$$

This equation may be integrated once to yield:

$$\kappa_3(x_3)\frac{\partial h_d(x_3)}{\partial x_3} = \frac{\partial g(\zeta)}{\partial \zeta}\int_0^{x_3} v_d(z')\mathrm{d}z', \tag{7.4.9}$$

where we have chosen the constant of integration to be zero since at $x_3 = 0$

$$\frac{\partial h_\mathrm{d}}{\partial x_3}(0) = 0, \tag{7.4.10}$$

as there is no flux across the boundary.

This equation may be integrated once again so that:

$$h_d(x_3) = \frac{\partial g(\zeta, \tau)}{\partial \zeta}\int_0^{x_3}\frac{1}{\kappa_3(z')}\int_0^{z'} v_d(z'')\mathrm{d}z'\,\mathrm{d}z'' + A \tag{7.4.11}$$

where A is an integration constant; thus once $v_d(x_3)$ and $\kappa_3(x_3)$ are given, it is possible to determine the diffusion tracer field $h_d(x_3)$.

It is now possible to calculate the mass flux M of the tracer at any cross-section ζ. This is given by:

$$M = \int_0^d v_d h_d \mathrm{d}x_3 = \frac{\partial g(\zeta, \tau)}{\partial \zeta}\int_0^d v_d(z')\int_0^{z'}\frac{1}{\kappa_3(z'')}\int_0^{z''} v_d(z''')\mathrm{d}z''\,\mathrm{d}z'. \tag{7.4.12}$$

Suppose we introduce the scales U and κ_c so that $u_d \sim U$ and $\kappa_3 \sim \kappa_c$ and we use these to define the non-dimensional functions:

$$f(\eta) = \frac{u_d(\eta)}{U},$$

$$\kappa = \frac{\kappa_3}{\kappa_c}, \tag{7.4.13}$$

and

$$\eta = \frac{x_3}{h}, \tag{7.4.14}$$

then the mass flux due to the primary circulation is given by:

$$M = -\frac{U^2}{\kappa} h^3 \beta \frac{\partial g(\zeta, \tau)}{\partial \zeta}, \tag{7.4.15}$$

where

$$\beta = -\int_0^1 f(\eta') \int_0^{\eta'} \frac{1}{\kappa(\eta'')} \int_0^{\eta''} f(\eta''')\mathrm{d}\eta'''\, \mathrm{d}\eta''\, \mathrm{d}\eta', \tag{7.4.16}$$

is a shape function.

Given (7.4.15) we can now write a conservation of mass equation for the channel as a whole. Consider the control volume of width $\mathrm{d}\zeta$ shown in Fig. 7.4.1. Equating the net mass flux into the control volume to the rate of change of tracer mass yields:

$$\left(\frac{\partial g}{\partial \tau}\right) \mathrm{d}t\, \mathrm{d}\zeta = -\mathrm{d}M = -\frac{\partial M}{\partial \zeta}\mathrm{d}\zeta. \tag{7.4.17}$$

Substituting from (7.4.15) leads to an equation for $g(\zeta, \tau)$:

$$\frac{\partial g}{\partial \tau} = \frac{U^2 h^2 \beta}{\kappa} \frac{\partial^2 g}{\partial \zeta^2}. \tag{7.4.18}$$

This is the diffusion equation, with a longitudinal dispersion coefficient K_L:

$$K_L = \frac{U^2 h^2 \beta}{\kappa}, \tag{7.4.19}$$

that now depends on the velocity scale U, the depth of the duct h and inversely on the turbulent or laminar diffusion coefficient κ_c. At first sight this may seem like a contradiction, but it must be remembered (see Fig. 7.4.1) that (7.4.18) is a model equation describing the behavior of the cross-sectional averaged tracer concentration that is the net result of the combined effects of an internal circulation and vertical diffusion. As the diffusion coefficient decreases, the cloud becomes more elongated or stretched before vertical diffusion can compete with the horizontal advection. This enhanced stretching appears as net dispersion in (7.4.18).

Throughout this analysis, we have assumed that the tracer cloud is advected through itself at a rate that is much faster than the rate that the cloud diffuses as given by (7.4.18). We can now investigate what constraints this assumption places on the analysis. Consider the ratio:

$$\frac{\dfrac{\partial g}{\partial \tau}}{\dfrac{\partial}{\partial x_3}\left(\kappa_3 \dfrac{\partial h_d}{\partial x_3}\right)} \sim \frac{Uh^2}{\kappa L}, \tag{7.4.20}$$

where L is the scale of the cloud length and where we have used (7.4.18) to estimate the scale for τ and (7.4.11) to obtain the scale for h_d. For the ratio to be small, we must therefore have

$$L > U\frac{H^2}{\kappa_c}, \tag{7.4.21}$$

meaning that the cloud must be longer than the distance the fluid travels in the time it takes the tracer to diffuse across the channel. Comparing this statement with the physics displayed in Fig. 7.4.1, we see that the constraints (7.4.21) simply requires the cloud to be long enough to allow the tracer to be dispersed across the whole channel.

In practice it is found that (7.4.18) becomes a good description of the cross-sectional average of the tracer for times that are larger than

$$T_c = 0.4\frac{H^2}{\kappa_c}, \tag{7.4.22}$$

from the time of the tracer release (Fischer et al., 1979).

As a simple example of the evaluation of (7.4.16), consider a linear velocity distribution in the channel and a constant diffusion coefficient, the problem discussed in §3.2.

$$u = \frac{u_0 x_3}{d}. \tag{7.4.23}$$

From (7.4.2),

$$u_d = \frac{u_0}{2}, \tag{7.4.24}$$

$$v_d = u_0\left(\frac{x_3}{d} - \frac{1}{2}\right), \tag{7.4.25}$$

and from (7.4.12):

$$f(\eta) = \left(\eta - \frac{1}{2}\right). \tag{7.4.26}$$

Substituting (7.4.26) into (7.4.16) leads a value of:

$$\beta = \frac{1}{120}. \tag{7.4.27}$$

Thus the longitudinal shear diffusion coefficient (7.4.17):

$$\kappa_L = \frac{1}{120}\frac{u_0^2 d^2}{\kappa}. \tag{7.4.28}$$

7.5. LONGITUDINAL DISPERSION IN A PIPE

As discussed in §7.2 and §7.4, the flow in a pipe is one example where all the conditions are satisfied to allow application of both the shear dispersion discussion and the application of the Lagrangian based kinematic dispersion arguments described in §7.4 as fluid particles wander across the pipe cross-section as they move downstream; in laminar pipe flow molecular diffusion diffuses molecules transversely and in turbulent pipe flow it is the turbulent eddy coefficient that disperse fluid transversely (similar to the schematic in Fig. 7.4.1).

Consider first longitudinal dispersion in laminar flow. The velocity distribution is repeated here for convenience:

$$\frac{v_1}{u_m} = \left(1 - \frac{r^2}{R^2}\right). \tag{7.5.1}$$

Simple integration shows that the discharge velocity u_d is connected to the maximum velocity $u_m = 2u_d$ so that:

$$(\bar{v}_1 - u_d) = u_m\left(\frac{1}{2} - \frac{r^2}{R^2}\right). \qquad (7.5.2)$$

In order to obtain an expression for the cross-sectional average concentration of a tracer we must proceed analogously to that in §7.4. The counterpart to (7.4.1) in cylindrical polar coordinates reads:

$$\frac{\partial C}{\partial t} + u_m\left(1 - \frac{r^2}{R^2}\right)\frac{\partial C}{\partial x_1} = D\left(\frac{\partial^2 C}{\partial r^2} + \frac{1}{r}\frac{\partial C}{\partial r} + \frac{\partial^2 C}{\partial x_1^2}\right). \qquad (7.5.3)$$

Again we neglect the last term of (7.5.3) and seek a solution of the form:

$$C(\xi, \zeta, \tau) = g(\xi, \tau) + h_d(\xi), \qquad (7.5.4)$$

and

$$\tau = t, \qquad (7.5.5)$$

$$\xi = \frac{r}{R}, \qquad (7.5.6)$$

$$\zeta = x_1 - u_d t. \qquad (7.5.7)$$

Following the same procedure as in §7.4 we get:

$$D_L = \frac{M}{A\dfrac{\partial g}{\partial \zeta}} = D_L = \frac{R^2 u_m^2}{192D}. \qquad (7.5.8)$$

Naturally, application of (7.5.8) are many and are not limited to laminar pipe flow. Arteries in our bodies are good examples of laminar flow. Consider the dispersion of oxygen in the blood stream ($D = 10^{-9}$ m^2 s^{-1}) as it flows from the lungs to your feet. Obviously in the actual body there are many bifurcations, bends and changes of cross-section, but for a first estimate we shall simplify the problem to that of a 1 mm diameter tube of length 3 m containing blood flowing with a discharge velocity of 0.01 m s^{-1}. Substituting these values into (7.5.8) leads to a value for D_L of 1.3×10^{-4} m^2 s^{-1}, which is larger than D by nearly a factor of 10^7.

We must, however, also check the applicability of this estimate from (7.4.21), i.e.

$$T_c = \frac{0.4R^2}{D} = 300 \text{ s} \tag{7.5.9}$$

which is somewhat longer than the travel time of 300 s, but not a great deal. This illustrates the restrictive nature of the shear dispersion mechanism when applied to laminar flow. Ignoring the initial set up time we can use the solution (7.3.25) to estimate the diffusion width at the end of the tube. It is customary to designate the width of the normal distribution by four times the standard deviation:

$$W\sqrt{4Dt} = 1.6\ m. \quad e^{-4} = 0.018$$

The result indicates that very considerable longitudinal dispersion takes place within the arteries of the body.

Let us now turn to turbulent flow. The analysis remains identical to that above, except that the velocity distribution is now described by the turbulent counter parts, §5.6 and §5.7. Further, instead of using the molecular diffusion coefficient it is the equivalent turbulent diffusion coefficient that is required.

The logarithmic velocity profiles are very difficult to integrate so it is more convenient to use the empirical distribution valid for turbulent flow in a smooth pipe:

$$\frac{u}{u_m} = \left(\frac{R-r}{R}\right)^{1/7}, \tag{7.5.10}$$

that has the property that:

$$u_d = \frac{49}{60}u_m. \tag{7.5.11}$$

The turbulent diffusion coefficient in a pipe is given by:

$$D = 0.06\ Ru_*. \tag{7.5.12}$$

Substituting (7.6.11) to (7.6.12) into the diffusion equation and proceeding as above leads to a longitudinal diffusion coefficient:

$$D_L = 10.1\ Ru_*. \tag{7.5.13}$$

As an example of the applicability of (7.5.13), consider a main pipeline to a remote township. Suppose the pipe diameter is 0.30 m, the discharge velocity is $0.2\ \mathrm{m\,s^{-1}}$ and the length of pipeline is 450 km. Can we use (7.5.13) to estimate the distribution of an accidentally introduced 1 kg slug of fluoride at the head works by the time it reaches the township?

First, we must calculate the shear velocity. Suppose the pipe is smooth. The Reynolds number:

$$\mathrm{Re} = 6 \times 10^4$$

leading to a friction factor from §5.6 $f = 0.02$ and $u_* = 10^{-2}\ \mathrm{m\,s^{-1}}$ and from (7.5.14):

$$D_L = 10^{-2}\ m^2\ s^{-1}$$

which means that the slug width W will be given by:

$$W = \frac{4(4D_L x)^{\frac{1}{2}}}{u_d} = 1131\ m.$$

The time to reach shear dispersion as captured by (7.5.14) is

$$T_c = \frac{04R^2}{D} = 66\ \mathrm{s}$$

a time the water has traveled 133 m. Thus in this case, our model is fully applicable.

REFERENCES

Fischer, H.B., List, E.J., Koh, C.Y., Imberger, J., Brooks, N.H., 1979. Mixing in Inland and Coastal Waters. Academic Press, 483 p.

Imberger, J., Ivey, G.N., 1993. Boundary mixing in stratified reservoirs. Journal of Fluid Mechanics 248, 477–491.

Taylor, G.I., 1921. Diffusion by continuous movements. Proceedings of London Mathematical Society 2, 196–211.

Yeates, P.S., Gomez-Giraldo, A., Imberger, J., 2012. Relationship between turbulent mixing and the gradient Richardson number in a thermally stratified lake. Journal of Environmental Fluid Mechanics (Submitted).

Yih, C.S., 1980. Dynamics of Nonhomogeneous Fluids. Macmillan, 306 p.

CHAPTER 8

Surface Waves

Contents

Surface gravity waves are a common feature in nature and we observe them whenever a water body is unconfined, such as in our swimming pool, the neighborhood river, estuaries, coastal seas and even the deep ocean. The objective of this chapter is to provide some introductory material on surface waves; long waves, linear waves and reflection and transmission of linear surface waves as an introduction to internal waves discussed in Chapter 09.

8.1. 2D LONG SURFACE WAVES

Consider a free surface undulation, $\eta(x_1, t)$, with a horizontal length scale λ in a shallow water body of depth h, as shown in Fig. 8.1.1, and suppose $\frac{h}{\lambda} < 1$. Under these conditions the pressure, $p(x_1, x_3, t)$, in the water will be hydrostatic:

$$p(x_1, x_3, t) = \rho_0 g(\eta(x_1, t) - x_3), \tag{8.1.1}$$

The momentum equation reduces to a simple balance of the net pressure force, increasing or decreasing the momentum in the control volume. Now

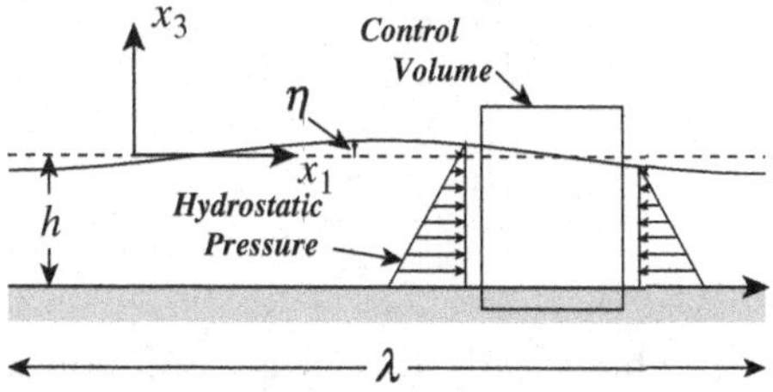

Figure 8.1.1 Schematic of long wave analysis control volume.

Environmental Fluid Dynamics
ISBN 978-0-12-088571-8, DOI: 10.1016/B978-0-12-088571-8.00008-5

the net pressure force F_p on the control volume, of width dx_1, is approximately given by:

$$dF_p \approx -\rho g h \frac{\partial \eta}{\partial x} dx. \tag{8.1.2}$$

The momentum conservation law applied to the control volume shown in Fig. 8.1.1, neglecting the advective momentum flux:

$$-\rho g h \frac{\partial \eta}{\partial x_1} dx = \rho h \frac{\partial v_1}{\partial t} dx_1. \tag{8.1.3}$$

The movement of the free surface must be such that the mass of water, in the control volume, is conserved and, since water is incompressible, volume is conserved:

$$h \frac{\partial v_1}{\partial x_1} = -\frac{\partial \eta}{\partial t}. \tag{8.1.4}$$

Combining (8.1.3) and (8.1.4) leads to a single equation for η:

$$\frac{\partial^2 \eta}{\partial t^2} - c^2 \frac{\partial^2 \eta}{\partial x_1^2} = 0, \tag{8.1.5}$$

where

$$c = \sqrt{gh}. \tag{8.1.6}$$

Equation (8.1.5) is fundamental for much of physics and is called the "wave equation". The basic property, important to us here, is that for any function f, the expression

$$\eta = f(x \pm ct), \tag{8.1.7}$$

is a solution of (8.1.5). To see this let us carry out the differentiation:

$$\frac{\partial^2 \eta}{\partial t^2} = c^2 \frac{\partial^2 f}{\partial \zeta_\pm^2}, \tag{8.1.8}$$

$$\frac{\partial^2 \eta}{\partial x^2} = \frac{\partial^2 f}{\partial \zeta_\pm^2}, \tag{8.1.9}$$

where $\zeta_\pm = x \pm ct$. Thus clearly (8.1.7) is a solution of (8.1.5) for any function f, provided the function has a second order derivative.

The simple interpretation of (8.1.7) is that along a trajectory

$$x = \zeta_{\pm} \pm ct, \tag{8.1.10}$$

in the (x,t) plane the solution remains constant for constant $\zeta_{\pm}$. In other words the solution consists of a simple translation of our function to the right $(-ve)$ or to the left $(+ve)$ with a speed c. These displacements are thus called waves and the interpretation of $\sqrt{gh}$ is the wave phase speed of long waves.

In §5.8, we introduced the concept of the Froude number:

$$Fr = \frac{U}{\sqrt{gh}}, \tag{8.1.13}$$

where U is the discharge velocity in the channel. Thus, the Froude number has the kinematic interpretation that it is the ratio of the discharge velocity to the wave speed. Provided that there is no interaction between the flow and the waves, then $Fr > 1$ means waves can only propagate downstream, but if $Fr < 1$ then waves can propagate both down and upstream. Hence we assign the word supercritical for $Fr > 1$ flows and subcritical for flows with $Fr < 1$. When $Fr = 1$ we say the flow is critical.

In the derivation of (8.1.5), two simplifying assumptions were made. First, the advective momentum flux was assumed to be small compared to the unsteady inertia:

$$\frac{v_1\dfrac{\partial v_1}{\partial x_1} + v_3\dfrac{\partial v_1}{\partial x_3}}{\dfrac{\partial v_1}{\partial t}} \sim \frac{v_1 T}{\lambda} = \frac{v_1 h}{\lambda c} \leq \frac{h\eta_{\max}}{\lambda^2}, \tag{8.1.14}$$

where $\eta_{\max}$ is the maximum value of the surface displacement and T is the period of the motion $\frac{h}{c}$. Second, we assumed the pressure was hydrostatic, requiring that:

$$\frac{\dfrac{\partial v_3}{\partial t}}{\rho_0 g} \sim \frac{h v_1}{\lambda \rho_0 g T} \leq \frac{h\eta_{\max}}{\lambda^2}. \tag{8.1.15}$$

8.2. LINEAR SURFACE WAVES

Consider once again the configuration shown in Fig. 8.1.1, but we now inquire whether we can relax the constraint (8.1.14). From §8.1, we saw that the velocity at the bottom is oscillatory under a surface gravity wave and from §3.6 we note that an oscillatory outer flow gives rise to a confined viscous boundary layer attached to the bottom. If this layer thickness is thin and remains confined close to the bottom, then we may assume that the outer flow, the region between the boundary layer and the free surface, will be irrotational as there is no source of vorticity. For such flows, we showed in §4.5, that there exists a velocity potential $\phi(x_1, x_3)$ that has the property that the velocity $\{v_i\}$ is given by:

$$v_i = \phi_{,i}. \tag{8.2.1}$$

Combining this with conservation of volume (incompressible flow) leads to:

$$v_{i,i} = \phi_{,ii} = 0, \tag{8.2.2}$$

or in cartesian coordinates

$$\frac{\partial^2 \phi}{\partial x_1^2} + \frac{\partial^2 \phi}{\partial x_3^2} = 0, \tag{8.2.3}$$

which is called the Laplace Equation.

Consider surface waves in a fluid of depth h moving over a flat horizontal bottom as shown in Fig. 8.1.1. The boundary condition to be satisfied at the bottom, at $x_3 = -h$, is zero vertical velocity (the outer problem) or in terms of the velocity potential:

$$\frac{\partial \phi}{\partial x_3} = 0 \quad x_3 = -h. \tag{8.2.4}$$

Before we can solve (8.2.2), we must determine the boundary conditions at the free surface. The first one of these is that the pressure on the free surface is zero (atmospheric pressure). If the wave amplitude is small, we can assume the pressure over a distance η, the height of the wave, is hydrostatic so that the condition of zero pressure at $x_3 = \eta$ may be applied at $x_3 = 0$ so that:

$$p(x_1, 0, t) = \rho g \eta. \tag{8.2.5}$$

Now if we neglect the non-linear acceleration in the momentum equation we may write:

$$\frac{\partial v_1}{\partial t} = -\frac{1}{\rho}\frac{\partial p}{\partial x_1} = -g\frac{\partial \eta}{\partial x_1}, \tag{8.2.6}$$

the same as (8.1.3). Substituting from (8.2.1) yields

$$\frac{\partial^2 \phi}{\partial t \partial x_1} = -g \frac{\partial \eta}{\partial x_1}. \tag{8.2.7}$$

Integrating once with respect to x_1 yields the required pressure boundary condition:

$$\frac{\partial \phi}{\partial t} = -g\eta; \quad x_3 = 0. \tag{8.2.8}$$

However, (8.2.8) contains two unknowns, ϕ and η, so we need one further equation at the boundary $x_3 = 0$. This is obtained by noting that the water surface moves with the same velocity as the fluid immediately below. Hence

$$v_3 = \frac{\partial \eta}{\partial t} = \frac{\partial \phi}{\partial x_3}; \quad x_3 = 0. \tag{8.2.9}$$

Combining (8.2.8) and (8.2.9) by eliminating η yields the equation:

$$g\frac{\partial \phi}{\partial x_3} + \frac{\partial^2 \phi}{\partial t^2} = 0; \quad x_3 = 0. \tag{8.2.10}$$

Thus the water motion is the solution to (8.2.2) subject to the boundary conditions (8.2.4) and (8.2.10). The difference between (8.2.2) and (8.1.8) is that here we have not assumed that the pressure is hydrostatic only that the advective non-linear terms are small in setting up the surface boundary condition.

Progressive Waves: Let us inquire whether a periodic solution of the form:

$$\phi(x_1, x_3, t) = \psi(x_1, x_3)e^{i\omega t}, \tag{8.2.11}$$

exists to this problem. Substituting (8.2.11) into (8.2.2) to (8.2.10) and (8.2.4):

$$g\frac{\partial \psi}{\partial x_3} - \omega^2 \psi = 0; \quad x_3 = 0, \tag{8.2.12}$$

$$\frac{\partial^2 \psi}{\partial x_1^2} + \frac{\partial^2 \psi}{\partial x_3^2} = 0; \quad -h < x_3 < 0, \tag{8.2.13}$$

$$\frac{\partial \psi}{\partial x_3} = 0; \quad x_3 = -h. \tag{8.2.14}$$

Let,

$$\psi(x_1, x_3) = h(x_1)f(x_3), \tag{8.2.15}$$

Substituting (8.2.15) into (8.2.12)–(8.2.14) leads to a separation of variables:

$$g\frac{\mathrm{d}f}{\mathrm{d}x_3} - \omega^2 f = 0; \quad x_3 = 0, \tag{8.2.16}$$

$$\frac{1}{h}\frac{\mathrm{d}^2 h}{\partial x_1^2} = -\frac{1}{f}\frac{\mathrm{d}^2 f}{\partial x_3^2}; \quad -h \le x_3 \le 0, \tag{8.2.17}$$

$$\frac{\partial f}{\partial x_3} = 0; \quad x_3 = -h. \tag{8.2.18}$$

From (8.2.17) it follows immediately that:

$$\frac{\mathrm{d}^2 h}{\partial x_1^2} + k^2 h = 0, \tag{8.2.19}$$

and

$$\frac{\mathrm{d}^2 f}{\partial x_3^2} - k^2 f = 0, \tag{8.2.20}$$

where k is a constant, yet to be determined. The solutions to (8.2.19) and (8.2.20) are

$$h = Ae^{ikx_1} + Be^{-ikx_1}, \tag{8.2.21}$$

and

$$f = C\cosh(kx_3) + D\sinh(kx_3), \tag{8.2.22}$$

where A, B, C and D are constants yet to be determined from compatibility with the boundary conditions.

Consider first the boundary condition at $x_3 = 0$. Substituting (8.2.22) into (8.2.16) and setting $x_3 = 0$ yields:

$$gDk - \omega^2 C = 0. \tag{8.2.23}$$

Similarly substituting (8.2.22) into (8.2.18):

$$-Ck\sinh(kh) + Dk\cosh(kh) = 0, \tag{8.2.24}$$

or

$$D = C\tanh(kh). \tag{8.2.25}$$

Substituting (8.2.25) into (8.2.23) leads to an equation for k in terms of the frequency ω:

$$gk\tanh(kh) = \omega^2. \tag{8.2.26}$$

Equation (8.2.26) is called the *dispersion relationship* because if we introduce the wave speed

$$c = \frac{\omega}{k}, \tag{8.2.27}$$

then (8.2.26) becomes

$$c^2 = \frac{g\tanh(kh)}{k}, \tag{8.2.28}$$

which implies that waves with different wave numbers k (different wavelengths) move with a different wave speed c. The relationship (8.2.28) is shown in Fig. 8.2.1. Given that $\tanh(kh) \to kh$ as $kh \to 0$ we see that in the limit of long waves ($kh \to 0$)

$$c = \sqrt{gh}, \tag{8.2.29}$$

which is the result derived in §8.1. Conversely, for very deep water ($kh \to \infty$) we get the result:

$$c = \sqrt{\frac{g}{k}}, \tag{8.2.30}$$

implying that, the shorter the waves, the slower they travel; waves in deep water are dispersive.

Substituting (8.2.21) and (8.2.22) into (8.2.15) and using the dispersion relation as well as (8.2.23) leads to an expression for the velocity potential for a progressive wave moving in the positive x_1 direction:

$$\varphi = \frac{ga}{\omega}\frac{\cosh k(x_3 + h)}{\cosh kh}\sin(kx_1 - \omega t), \tag{8.2.31}$$

where a is the wave amplitude.

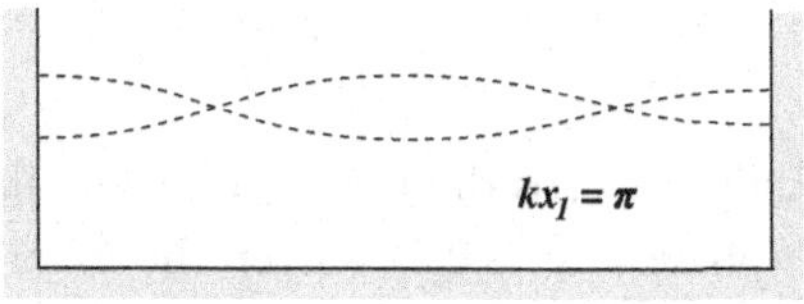

Figure 8.2.1 Example of a mode one standing wave.

The associated velocity fields are

$$v_1 = \frac{\partial \varphi}{\partial x_1} = \frac{gak}{\omega} \frac{\cosh k(x_3 + h)}{\cosh kh} \cos(kx_1 - \omega t), \tag{8.2.32}$$

$$v_3 = \frac{\partial \varphi}{\partial x_3} = \frac{gak}{\omega} \frac{\sinh k(x_3 + h)}{\cosh kh} \sin(kx_1 - \omega t), \tag{8.2.33}$$

$$\eta = -\frac{1}{g}\frac{\partial \varphi}{\partial t}\bigg|_{x_3=0} = a\cos(kx_1 - \omega t). \tag{8.2.34}$$

The system of equations that we have used to derive this solution are all linear, so we may combine solutions at will and the combination will again be a solution. An example of some importance is that of a standing wave, obtained by adding a left and right moving wave, of the type (8.2.33) and (8.2.34).

$$\eta = \frac{a}{2}\{\cos(kx_1 - \omega t) + \cos(kx_1 + \omega t)\}, \tag{8.2.35}$$

where we have chosen half the amplitude in order to make the final wave of amplitude a. Simple application of the cosine summation formula from trigonometry leads to the solution set:

$$\varphi = \frac{ga}{\omega} \frac{\cosh k(x_3 + h)}{\cosh kh} \sin(kx_1)\cos(\omega t), \tag{8.2.36}$$

$$v_1 = \frac{\partial \varphi}{\partial x_1} = \frac{gak}{\omega} \frac{\cosh k(x_3 + h)}{\cosh kh} \cos(kx_1)\cos(\omega t), \tag{8.2.37}$$

$$v_3 = \frac{\partial \varphi}{\partial x_3} = \frac{gak}{\omega} \frac{\sinh k(x_3 + h)}{\cosh kh} \sin(kx_1)\sin(\omega t), \tag{8.2.38}$$

$$\eta = \frac{a}{2}\{\cos(kx_1)\cos(\omega t)\}. \tag{8.2.39}$$

This is called a standing wave, as the water surface oscillates vertically with no translation of the phase (Fig. 8.2.1). Given the form of (8.2.35), we see that the horizontal velocity is such that it is zero whenever $kx_1 = n\pi$, so a standing wave, as the name implies, may be fitted into a rectangular basin as shown in Fig. 8.2.1.

8.3. REFLECTION AND TRANSMISSION OF SURFACE WAVES

Consider a simple linear plane progressive wave coming from $x_1 \to -\infty$ and moving toward $x_1 \to +\infty$. Suppose near $x_1 = 0$ there is a sill on the bottom, as shown in Fig. 8.3.1, causing both a reflection of the surface wave and transmission modification of the wave as it passes over the mound moving toward $x_1 \to +\infty$.

Let the bottom be described by the function:

$$x_3^{(b)} = -h + \Delta f(x_1), \tag{8.3.1}$$

where $f(x_1)$ is a dimensionless function describing the shape of the bottom, Δ is the amplitude of the bottom changes and h is the mean depth of the water domain. The solution to the problem when the bottom undulations are small was derived by Hurley and Imberger (1969).

Suppose we have a single wave coming from the left and impinging on the bottom undulation. This incoming wave may be written as (see §8.2)

$$\varphi(x_1, x_3, t) = A_i \frac{\cosh k(x_3 + h)}{\cosh kh} \sin(kx_1 - \omega t), \tag{8.3.2}$$

where k is the wave number and ω is the frequency of the incoming wave. As seen in §8.2, k and ω are connected through the dispersion relationship:

$$gk \tanh(kh) = \omega^2. \tag{8.3.3}$$

From §8.2 we see that, in order to obtain an expression for both the reflection and transmission properties, we must solve for the velocity potential ϕ that satisfies:

$$g\frac{\partial \varphi}{\partial x_3} + \frac{\partial^2 \varphi}{\partial t^2} = 0; \quad x_3 = 0, \tag{8.3.4}$$

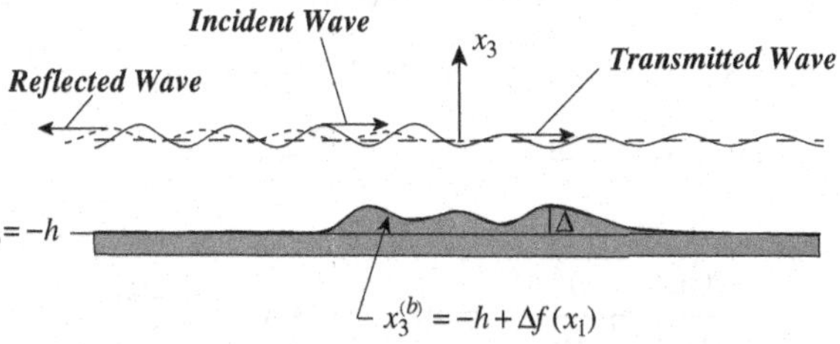

Figure 8.3.1 Schematic of wave reflection and transmission.

$$\frac{\partial^2\varphi}{\partial x_1^2}+\frac{\partial^2\varphi}{\partial x_3^2}=0; \quad 0>x_3>x_3^{(b)}, \tag{8.3.5}$$

$$\frac{\partial\varphi}{\partial x_3}=\Delta\frac{df}{dx_1}\frac{\partial\varphi}{\partial x_1}; \quad x_3=x_3^{(b)}, \tag{8.3.6}$$

where equation (8.3.6) comes directly from requiring that the velocity at the bottom is tangential to the bottom:

$$\frac{v_3}{v_1}=\Delta\frac{df}{dx_1}. \tag{8.3.7}$$

Now, before proceeding, it is convenient to introduce the following non-dimensional variables:

$$x_1^*=\frac{x_1}{h}; \quad x_3^*=\frac{x_3}{h}; \quad t^*=\omega t; \quad \varphi^*=\frac{\varphi}{A_i}, \tag{8.3.8}$$

so that (8.3.4)–(8.3.6) become

$$\frac{\partial^2\phi^*}{\partial^2 t^*}+\left(\frac{g}{h\omega^2}\right)\frac{\partial\phi^*}{\partial x_3^*}=0; \quad x_3^*=0, \tag{8.3.9}$$

$$\frac{\partial^2\varphi^*}{\partial x_1^{*2}}+\frac{\partial^2\varphi^*}{\partial x_3^{*2}}=0; \quad 0>x_3^*>x_3^{*(b)}, \tag{8.3.10}$$

$$\frac{\partial\varphi^*}{\partial x_3^*}=\varepsilon\frac{df}{dx_1^*}\frac{\partial\varphi^*}{\partial x_1^*}; \quad x_3^*=x_3^{*(b)}, \tag{8.3.11}$$

where $\varepsilon=\dfrac{\Delta}{h}$. Given that (8.3.2) is periodic with a frequency ω, the solution ϕ must also be periodic with a frequency ω and the non-dimensional variable, φ^* must thus be periodic with a period of unity. Hence we seek a solution of the form:

$$\varphi^*(x_1^*,x_3^*,t^*)=\psi(x_1^*,x_3^*)e^{-it^*}. \tag{8.3.12}$$

Substituting (8.3.12) into (8.3.9) to (8.3.11) yields

$$\psi-K\psi_{,3}=0; \quad K=\frac{g}{\omega^2 h}, \quad x_3^*=0, \tag{8.3.13}$$

$$\frac{\partial^2\psi^*}{\partial x_1^{*2}}+\frac{\partial^2\psi^*}{\partial x_3^{*2}}=0; \quad 0>x_3^*>x_3^{*(b)}, \tag{8.3.14}$$

$$\frac{\partial \psi}{\partial x_3^*} = \varepsilon \frac{df}{dx_1^*}\frac{\partial \psi}{\partial x_1^*}; \quad x_3 = x_3^{(b)}, \tag{8.3.15}$$

For clarity and ease of understanding the following perturbation analysis, we shall from now on drop the star superscript and also change to the index notation. The boundary condition (8.3.15) now needs to be reduced to a condition at $x_3 = -1$. This may be achieved by noting that, so $\psi_{,3} \cdot \psi_{,1}$ may be expanded in a Taylor series about $x_3 = -1$.

$$\psi(x_1, x_3^{(b)}) = \psi(x_1, -1) + \varepsilon f(x)\psi_{,3}(x_1, -1) + \frac{\varepsilon^2}{2!} f^2(x)\psi_{,33}(x_1, -1) + \ldots \tag{8.3.16}$$

and now seek a perturbation solution of the form:

$$\psi(x_1, x_3) = \psi^{(0)}(x_1, x_3) + \varepsilon\psi^{(1)}(x_1, x_3) + \frac{\varepsilon^2}{2}\psi^{(2)}(x_1, x_3) + \ldots \tag{8.3.17}$$

Substituting both (8.3.16) and (8.3.17) into equations (8.3.13)–(8.3.15) and equating equal powers of ε leads to:

$$\psi^{(i)}(x_1, 0) - K\psi_{,3}^{(i)}(x_1, 0) = 0, \tag{8.3.18}$$

$$\psi_{,11}^{(i)}(x_1, x_3) + \psi_{,33}^{(i)}(x_1, x_3) = 0, \tag{8.3.19}$$

$$\psi_{,3}^{(0)}(x_1, -1) = 0, \tag{8.3.20}$$

$$\psi_{,3}^{(1)}(x_1, -1) = (f\psi_{,3}^{(0)})_{,1}, \tag{8.3.21}$$

$$\psi_{,3}^{(2)}(x_1, -1) = \left(f\psi_{,1}^{(1)}\right)_{,1} + \left(\frac{f^2}{2}\psi_{,13}^{(0)}\right)_{,1}, \quad i = 1, 2, 3. \tag{8.3.22}$$

This set of equations is most easily solved by taking a Fourier Transform with respect to x_1. However, given that the function $\psi^{(i)}(x_1,x_3)$ does not go to zero at $x_1 = \pm\infty$, some care must be taken with the integration. We shall use the theory of generalized function described in Lighthill (1962).

Define the Fourier pair by

$$G(u) = \frac{1}{\sqrt{2\pi}} \int_{-\infty}^{+\infty} g(x_1) e^{-iux_1} \mathrm{d}x_1, \tag{8.3.23}$$

$$g(x_1) = \frac{1}{\sqrt{2\pi}} \int_{-\infty}^{+\infty} G(u) e^{iux_1} \mathrm{d}u, \tag{8.3.24}$$

Taking the Fourier Transform of equations (8.3.18)–(8.3.22) leads to:

$$\Psi^{(i)} - K\Psi,_{33}^{(i)} = 0; \quad x_3 = 0, \tag{8.3.25}$$

$$\Psi,_{33}^{(i)} - u^2\Psi^{(i)} = 0, \tag{8.3.26}$$

$$\Psi,_3^{(i)} = G^{(i)}(u) \quad x_3 = -1, \tag{8.3.27}$$

where $G^{(i)}(u)$ is the Fourier transform of the RHS of (8.3.22) for $i = 1, 2, 3, \ldots$

$$g^{(0)}(x_1) = 0, \tag{8.3.28}$$

$$g^{(1)}(x_1) = \left(f\psi,_1^{(0)}\right),_1, \tag{8.3.29}$$

$$g^{(2)}(x_1) = \left(f\psi,_1^{(1)}\right),_1 + \left(\frac{f^2}{2!}\psi,_{13}^{(0)}\right),_1. \tag{8.3.30}$$

Now the solution to (8.3.26) is

$$\Psi^{(i)}(u, x_3) = A^{(i)}(u)\cosh(ux_3) + B^{(i)}(u)\sinh(ux_3). \tag{8.3.31}$$

Substituting this solution into (8.3.25) reveals:

$$A^{(i)}(u) = KB^{(i)}(u), \tag{8.3.32}$$

so that (8.3.31) may be written as:

$$\Psi^{(i)}(u, x_3) = B^{(i)}(u)(Ku\cosh(ux_3) + \sinh(ux_3)). \tag{8.3.33}$$

This may now be substituted in the lower boundary condition (8.3.27):

$$B^{(i)}(u)(u\cosh(u) - Ku^2\sinh(u)) = H^{(i)}, \tag{8.3.34}$$

thus

$$B^{(i)}(u)\frac{G^{(i)}(u)}{u(\cosh(u)-Ku\sinh(u))}+C_1^{(i)}\delta(u-n_0)+C_2^{(i)}\delta(u+n_0), \tag{8.3.35}$$

where $\pm n_0$ are the symmetric roots to the equation

$$\cosh(u)-Ku\sinh(u) = 0, \tag{8.3.36}$$

which is (8.3.3) the dispersion equation. Equation (8.3.35) may now be substituted into (8.3.33) to yield the solution:

$$\begin{aligned}\Psi^{(i)}(u,x_3) &= \frac{ku\cosh(ux_3)+\sinh(ux_3)}{u(\cosh(u)-Ku\sinh(u))}G^{(i)}(u)+(Kn_0\cosh(n_0x_3)\\ &\quad +\sinh(n_0x_3))(C_1^{(i)}\delta(u-n_0)-C_2^{(i)}\delta(u+n_0)),\end{aligned} \tag{8.3.37}$$

where $\delta(u-n_0)$ is the Dirac Delta function.

The solution $\psi^{(i)}(x_1, x_3)$ may be found by inverting (8.3.37). From (8.3.24) it is seen that inversion may be achieved by taking the Fourier Transform of $\psi^{-(i)}(-u, x_3)$. Further, since we are only after the reflection and transmission properties it is sufficient to carry out an asymptotic Fourier Transform of $\psi^{-(i)}(-u, x_3)$ for x_1. This is achieved (see Lighthill, 1962) by extracting the singularities (poles) from (8.3.37) and transforming only these. The singularities in (8.3.37) are where the denominator is zero:

$$u(\cosh(u)-Ku\sinh(u)) = 0. \tag{8.3.38}$$

Designating the roots of (8.3.38) by $(0,\ n_0,\ +n_0)$, we see that n_0 is the solution to the dispersion equation (8.3.3). Since (8.3.38) is simple zeros at $u=\pm n_0$, we may write:

$$\begin{aligned}&\frac{ku\cosh(ux_3)+\sinh(ux_3)}{u(\cosh(u)-ku\sinh(u))}G^{(i)}(-u)\sim G^{(i)}(n_0)\\ &\quad\times\frac{kn_0\cosh(n_0x_3)+\sinh(n_0x_3)}{(n_0+\cosh(n_0)\sinh(n_0))}\frac{\sinh(n_0)}{u+n_0}\\ &\text{as } |u+n_0|\rightarrow 0\\ &G^{(i)}(-n_0)\frac{kn_0\cosh(n_0x_3)+\sinh(n_0x_3)}{(n_0+\cosh(n_0)\sinh(n_0))}\frac{\sinh(n_0)}{u-n_0}\\ &\text{as } |u-n_0|\rightarrow 0\end{aligned} \tag{8.3.39}$$

Hence,

$$\Psi^{(i)}(-u, x_3) = \frac{kn_0 \cosh(n_0 x_3) + \sinh(n_0 x_3)}{(n_0 + \cosh(n_0)\sinh(n_0))}$$

$$\times \sinh(n_0)\left[\left\{\frac{H^{(i)}(n_0)}{u + n_0} - \frac{H^{(i)}(n_0)}{u - n_0}\right\}\right.$$

$$+ \frac{n_0 + \cosh(n_0)\sinh(n_0)}{\sinh(n_0)} \tag{8.3.40}$$

$$\left.\times \left(C_2^{(i)}\delta(u + n_0) - C_1^{(i)}\delta(u - n_0)\right)\right] + F(u),$$

where $F(u)$ is an analytic function, the inverse of which does not contribute to the solution $\Psi^{(i)}(x_1, x_3)$ at $x_1 \to \pm\infty$. The inverse of eqn (8.3.40) for $x_1 \to \pm\infty$ may be written (see Lighthill, 1966) as

$$\Psi^{(i)}(-u, x_3) \sim i\left(\frac{\pi}{2}\right)^{1/2} \frac{\cosh(n_0(x_3 + 1))}{(n_0 + \cosh(n_0)\sinh(n_0))}\,[e^{in_0 x_1} G^{(i)} n_0(\operatorname{sgn} x_1 - 1)$$

$$- e^{-in_0 x_1} G^{(i)}(-n_0)(\operatorname{sgn} x_1 + 1)], \tag{8.3.41}$$

where the radiation conditions of $x_1 \to \pm\infty$ were used to evaluate $C_1^{(i)}$ and $C_2^{(i)}$.

$$C_1^{(i)} = \frac{-i\pi G^{(i)}(-n_0)\sinh(n_0)}{(n_0 + \cosh(n_0)\sinh(n_0))}, \tag{8.3.42}$$

$$C_2^{(i)} = \frac{-i\pi G^{(i)}(n_0)\sinh(n_0)}{(n_0 + \cosh(n_0)\sinh(n_0))}, \tag{8.3.43}$$

so that at $x_1 \to \pm\infty$ we have only the incoming wave.

The reflection coefficient α and the transmission coefficient β become

$$\alpha = \frac{i\sqrt{2\pi}}{n_0 + \cosh(n_0)\sinh(n_0)}\left\{\varepsilon H^{(1)}(n_0) + \varepsilon^2 H^{(2)}(n_0) + \ldots\right\}, \tag{8.3.44}$$

$$\beta = 1 + \frac{i\sqrt{2\pi}}{n_0 + \cosh(n_0)\sinh(n_0)}\left\{\varepsilon G^{(1)}(-n_0) + \varepsilon^2 G^{(2)}(-n_0) + \ldots\right\}, \tag{8.3.45}$$

where

$$g^{(0)}(x_1) = 0 \quad \text{so that } G^{(0)}(u) = 0, \tag{8.3.46}$$

and

$$g^{(1)}(x_1) = f\psi,^{(0)}_{11} + \frac{df}{dx_1}\psi,^{(0)}_{11}, \tag{8.3.47}$$

so that

$$G^{(1)}(u) = n_0 u F(u + n_0), \tag{8.3.48}$$

and

$$\alpha = \frac{i\sqrt{2\pi}}{n_0 + \cosh(n_0)\sinh(n_0)}\left\{\varepsilon n_0^2 F(2n_0) + O(\varepsilon^2)\right\}, \tag{8.3.49}$$

$$\beta = 1 + \frac{i\sqrt{2\pi}}{n_0 + \cosh(n_0)\sinh(n_0)}\left\{-\varepsilon n_0^2 F(0) + O(\varepsilon^2)\right\}, \tag{8.3.50}$$

where $F(0)$ is the area under the mound. We shall not take the analysis to higher orders; the reader is referred to Hurley and Imberger (1969).

Equations (8.3.50) and (8.3.51) offer a convenient tool to understand the reflection and transmission of surface waves from submerged obstacles. Consider the obstacle given by:

$$f(x) = \frac{1}{\sqrt{\pi}} \int_{\frac{x-l}{\tau}}^{\frac{x+l}{\tau}} e^{-\zeta^2} d\zeta, \tag{8.3.51}$$

that is shown schematically in Fig. 8.3.2 and represents a simple mound of length $O(2l)$ with transitions at each length $O(\tau)$.

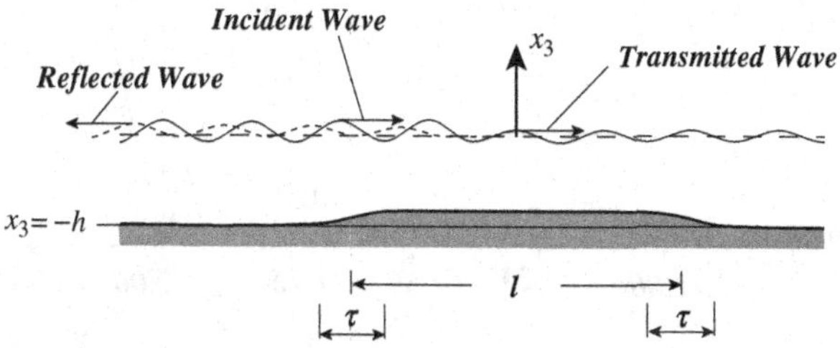

Figure 8.3.2 Reflection from a uniform, constant height sill.

The Fourier transform of (8.3.51) is given by:

$$F(u) = \sqrt{\frac{2}{\pi}}\frac{\sin(lu)}{u}e^{-\frac{\tau^2 u^2}{4}}, \tag{8.3.53}$$

so that

$$\alpha = \frac{i\varepsilon n_0 \sinh(2n_0 l)}{n_0 h + \cosh(n_0 h)\sinh(n_0 h)}e^{-\tau^2 v_0^2} + O(\varepsilon^2), \tag{8.3.54}$$

$$\beta = 1 + \frac{2il}{n_0 h + \cosh(n_0 h)\sinh(n_0 h)} + O(\varepsilon^2), \tag{8.3.55}$$

the i indicating a 90° phase shift and the wave number n_0 must satisfy the dispersion relationship:

$$gn_0 \tanh(n_0 h) = \omega^2$$

the amplitude $|\alpha|$ is shown in Fig. 8.3.3 where it is seen that $|\alpha|$ depends on $\sin(2n_0 l)$, indicating that the front and the rear of the mound cause an equal reflection, but with a different phase shift so that the reflections add constructively when $2n_0 l = \frac{(2k+1)\pi}{2}$, $k = 1,2,\ldots$, and destructively when $2n_0 l = 2k\pi$, $k = 1,2,\ldots$,. From Fig. 8.3.4 we see that, as $\frac{\tau}{\lambda}$ becomes large, $|\alpha|$ decreases exponentially and we approach what is called weak reflection; the incident wave can negotiate the sill without appreciable reflection.

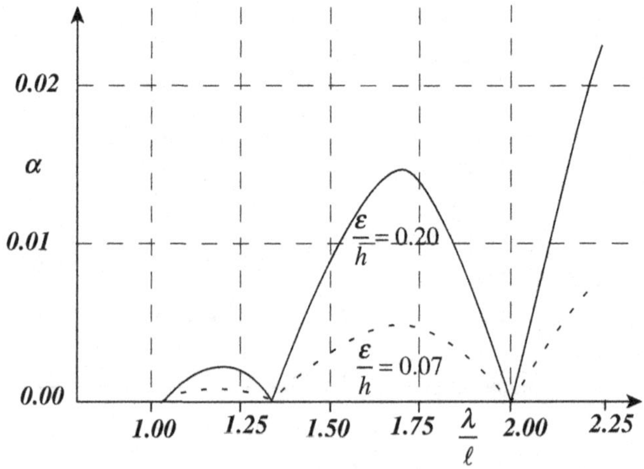

Figure 8.3.3 Reflection coefficient for a rectangular sill.

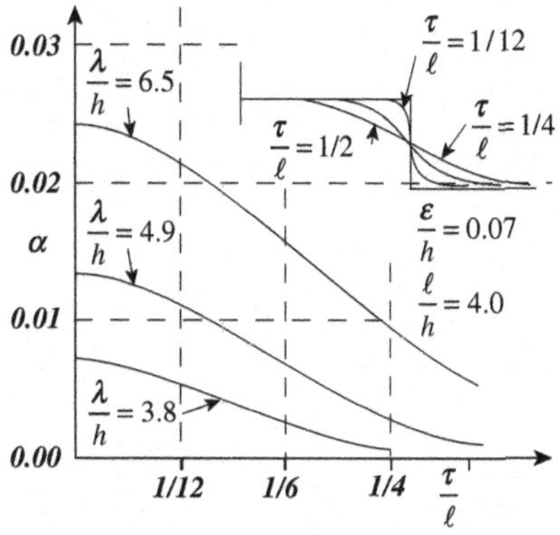

Figure 8.3.4 Effect of slope width on the reflection coefficient for a sill.

REFERENCES

Hurley, D.G., Imberger, J., 1969. Surface and internal waves in a liquid of variable depth. B. Aust. Math. Soc. 1, 29–46.

Lighthill, M.J., 1962. Fourier Analysis and Generalised Functions. Cambridge Univ. Press, pp. 79.

CHAPTER 9

Environmental Flows

Contents

In simple terms, the study of environmental flows involves the hydraulics of internal motions in a stratified fluid. Just as in hydraulics, surface waves are ubiquitous, stratified water bodies are normally filled with both travelling and standing internal wave motions, motions that distribute the energy introduced at the surface by wind throughout the lake. This is of enormous ecological importance, because the internal wave motions gently agitate the water body critical for primary production.

9.1. LONG INTERNAL WAVES: TWO-LAYER CASE

Consider the motion of an internal wave formed at the interface of a two-layer fluid with a free surface, shown in Fig. 9.1.1. We assumed that the interfacial and surface waves are much longer then the layers are deep, so that the pressure may be assumed to be hydrostatic.

Using Taylor's theorem, we can relate the interface displacement at section 2 to the displacement at section 1:

$$\zeta_2^{(2)} = \zeta_2^{(1)} + \frac{\partial \zeta_2}{\partial x_1} \mathrm{d}x_1, \tag{9.1.1}$$

Environmental Fluid Dynamics
ISBN 978-0-12-088571-8, DOI: 10.1016/B978-0-12-088571-8.00009-7

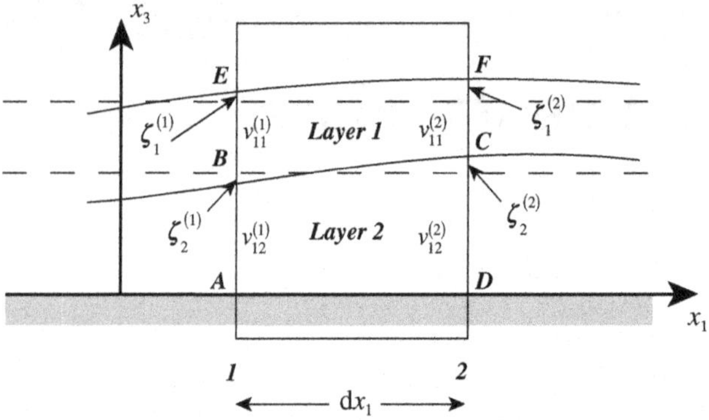

Figure 9.1.1 Schematic of two-layer structure. Notation: $v_{ij}^{(k)}$, subscript "i" refers to direction of velocity component, subscript "j" refers to layer number and superscript "(k)" refers to section number.

$$\zeta_1^{(2)} = \zeta_1^{(1)} + \frac{\partial \zeta_1}{\partial x_1} dx_1, \tag{9.1.2}$$

resulting in the following hydrostatic pressures:

$$p_B = \rho_1 g(h_1 + \zeta_1^{(1)} - \zeta_2^{(1)}), \tag{9.1.3}$$

$$p_C = \rho_1 g(h_1 + \zeta_1^{(2)} - \zeta_2^{(2)}), \tag{9.1.4}$$

Consider Layer 2:

Linear conservation of momentum:

$$F_{AB} + F_{BC} - F_{DC} = \rho_2 h_2 \frac{\partial v_{12}}{\partial t} dx_1, \tag{9.1.5}$$

where F_{AB} is the component of the pressure force acting in the horizontal direction on the interface AB, F_{BC} is the pressure force on face BC and F_{CD} is the pressure force on face DC.

Assuming the pressure is hydrostatic leads to:

$$\begin{aligned}
&(h_2 + \zeta_2^{(1)})\rho_1 g(h_1 + \zeta_1^{(1)} - \zeta_2^{(1)}) + \frac{1}{2}\rho_2 g(h_2 + \zeta_2^{(1)})^2 + \frac{1}{2}\rho_1 g(h_1 + \zeta_1^{(1)} \\
&\quad - \zeta_2^{(1)})\,(\zeta_2^{(2)} - \zeta_2^{(1)}) + \frac{1}{2}\rho_1 g(h_1 + \zeta_1^{(2)} - \zeta_2^{(2)})\,(\zeta_2^{(2)} - \zeta_2^{(1)}) \\
&\quad - \rho_1 g(h_1 + \zeta_1^{(2)} - \zeta_2^{(2)})\,(h_2 + \zeta_2^{(2)}) - \frac{1}{2}\rho_2 g(h_2 + \zeta_2^{(2)})^2 \\
&\quad = \rho_2 h_2 \frac{\partial v_{12}}{\partial t} dx_1.
\end{aligned} \tag{9.1.6}$$

Neglecting all second order terms in (9.1.6) yields:

$$\begin{aligned}
&\rho_1 g(h_1h_2 + \zeta_1^{(1)}h_2 - \zeta_2^{(1)}h_2 + \zeta_2^{(1)}h_1) + \frac{1}{2}\rho_2 g h_2^2 + \rho_2 g h_2 \zeta_2^{(1)} \\
&\quad + \rho_1 g(h_1\zeta_2^{(2)} - h_1\zeta_2^{(1)}) - \rho_1 g(h_1h_2 + \zeta_1^{(2)}h_2 - \zeta_2^{(2)}h_2 + \zeta_2^{(2)}h_1) \\
&\quad - \frac{1}{2}\rho_2 g h_2^2 - \rho_2 g h_2 \zeta_2^{(2)} = \rho_2 h_2 \frac{\partial v_{12}}{\partial t}\,dx_1,
\end{aligned} \tag{9.1.7}$$

which simplifies to:

$$-g'\frac{\partial \zeta_2}{\partial x_1} - \frac{\rho_1}{\rho_2}g\frac{\partial \zeta_1}{\partial x_1} = \frac{\partial v_{12}}{\partial t}, \tag{9.1.8}$$

where $\Delta\rho = \rho_2 - \rho_1$ and $g' = \dfrac{\Delta\rho}{\rho_2}g$.

Conservation of volume:

$$\frac{\partial}{\partial x_1}(h_2 v_{12}) = -\frac{\partial \zeta_2}{\partial t}. \tag{9.1.9}$$

Consider Layer 1:

Linear conservation momentum, using the same notation as above:

$$F_{BE} - F_{BC} - F_{FC} = \rho_1 h_1 \frac{\partial v_{11}}{\partial t}\,dx_1. \tag{9.1.10}$$

Again substituting for the pressure forces yields:

$$\begin{aligned}
&\frac{1}{2}\rho_1 g(h_1 + \zeta_1^{(1)} - \zeta_2^{(1)})^2 - \rho_1 g(h_1 + \zeta_1^{(1)} - \zeta_2^{(1)})\,(\zeta_2^{(2)} - \zeta_2^{(1)}) \\
&\quad - \frac{1}{2}\rho_1 g(h_1 + \zeta_1^{(2)} - \zeta_2^{(2)})^2 = \rho_1 h_1 \frac{\partial v_{11}}{\partial t}\,dx_1.
\end{aligned} \tag{9.1.11}$$

Simplifying and retaining only first order terms implies:

$$-g\frac{\partial \zeta_1}{\partial x_1} = \frac{\partial v_{11}}{\partial t}. \tag{9.1.12}$$

Conservation of volume:

$$\frac{\partial}{\partial x_1}(h_1 v_{11}) = \frac{\partial \zeta_2}{\partial t} - \frac{\partial \zeta_1}{\partial t}. \tag{9.1.13}$$

To find an equation for the interface displacement, ζ_2, we take the derivative with respect to x_3 of (9.1.8) and (9.1.12), substitute from the conservation of volume equations and subtract the resulting equations from each other:

$$\frac{\partial^2 \zeta_2}{\partial t^2} - c^2 \frac{\partial^2 \zeta_2}{\partial x_1^2} = 0, \tag{9.1.14}$$

$$c^2 = \frac{g' h_1 h_2 \rho_2}{\rho_1 h_2 + \rho_2 h_1}, \tag{9.1.15}$$

where we have assumed that $\zeta_1 \ll \zeta_2$.

Equation (9.9) is the wave equation identical to that for a long surface wave derived in §8.1 except now the wave speed is that of an internal wave (9.1.15).

9.2. INTERNAL WAVE MODES

In order to illustrate the concept of normal modes, consider the three-layer stratification shown in Fig. 9.2.1. Once again, as in §9.1, we make the assumption that the displacements, ζ_1, ζ_2, ζ_3 are small and the wavelength is long so that the pressure is hydrostatic and non-linear acceleration terms may be neglected. With these assumptions, consider the x_1 linear momentum equation for the control volumes shown in Fig. 9.2.1 with the convention that F is the net force on a surface.

Consider Layer 1:

$$F_{DC} - F_{EF} - F_{CF} = \rho_1 h_1 \frac{\partial v_{11}}{\partial t} dx_1, \tag{9.2.1}$$

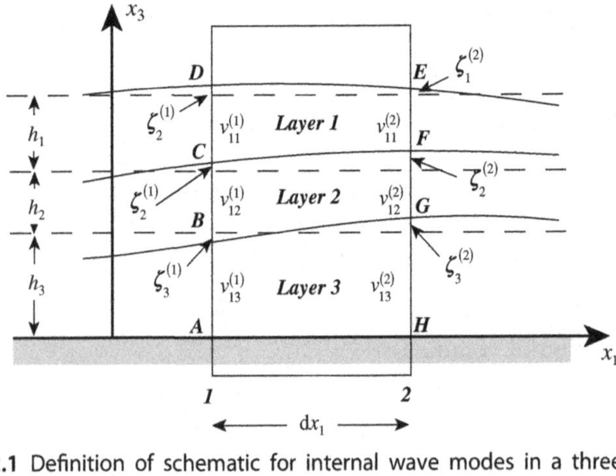

Figure 9.2.1 Definition of schematic for internal wave modes in a three-layer fluid configuration.

where

$$F_{DC} = \frac{1}{2}\rho_1 g h_1^2 + \rho_1 g h_1 (\zeta_1^{(1)} - \zeta_2^{(1)}), \tag{9.2.2}$$

$$F_{EF} = F_{DC} + \rho_1 g h_1 \mathrm{d}x_1 \left(\frac{\partial \zeta_1^{(1)}}{\partial x_1} - \frac{\partial \zeta_2^{(1)}}{\partial x_1} \right), \tag{9.2.3}$$

$$F_{CF} = \rho_1 g h_1 \frac{\partial \zeta_2}{\partial x_1} \mathrm{d}x_1. \tag{9.2.4}$$

Substituting (9.2.2), (9.2.3) and (9.2.4) into (9.2.1) yields:

$$\frac{\partial v_{11}}{\partial t} = -g \frac{\partial \zeta_1}{\partial x_1}. \tag{9.2.5}$$

Conservation of volume:

$$\frac{\partial v_{11}}{\partial x_1} = -\frac{1}{h_1}\frac{\partial \zeta_1}{\partial t} + \frac{1}{h_1}\frac{\partial \zeta_2}{\partial t}. \tag{9.2.6}$$

Consider Layer 2:

$$F_{BC} + F_{CF} - F_{FG} - F_{BG} = \rho_2 h_2 \frac{\partial v_{12}}{\partial t} \mathrm{d}x_1, \tag{9.2.7}$$

where

$$F_{CB} = \rho_1 g h_1 h_2 + \frac{1}{2}\rho_2 g h_2^2 + \rho_1 g h_1 \zeta_2 - \rho_1 g h_1 \zeta_3 + \rho_1 g h_2 \zeta_1$$
$$- \rho_2 g h_2 \zeta_3 + \Delta\rho_{21} g h_2 \zeta_2, \tag{9.2.8}$$

$$F_{FG} = F_{CB} - \rho_1 g h_1 \frac{\partial \zeta_2}{\partial x_1}\mathrm{d}x_1 + \rho_1 g h_1 \frac{\partial \zeta_3}{\partial x_1}\mathrm{d}x_1 - \rho_1 g h_2 \frac{\partial \zeta_1}{\partial x_1}\mathrm{d}x_1$$
$$+ \rho_2 g h_2 \frac{\partial \zeta_3}{\partial x_1}\mathrm{d}x_1 - \Delta\rho_{21} g h_2 \frac{\partial \zeta_2}{\partial x_1}\mathrm{d}x_1, \tag{9.2.9}$$

$$F_{CF} = \rho_1 g h_1 \frac{\partial \zeta_2}{\partial x_1}\mathrm{d}x_1, \tag{9.2.10}$$

$$F_{GB} = (\rho_1 g h_1 + \rho_2 g h_2) \frac{\partial \zeta_3}{\partial x_1}\mathrm{d}x_1, \tag{9.2.11}$$

where $\Delta\rho_{21} = \rho_2 - \rho_1$.

Substituting (9.2.7), (9.2.8) and (9.2.9) into (9.2.6) yields:

$$\frac{\partial v_{12}}{\partial t} = -g\frac{\partial \zeta_1}{\partial x_1} - \frac{\Delta\rho_{21}}{\rho_0}g\frac{\partial \zeta_2}{\partial x_1}, \tag{9.2.12}$$

where again we have assumed that ρ_2 is approximately equal to ρ_1.
Conservation of volume:

$$\frac{\partial v_{12}}{\partial x_1} = -\frac{1}{h_2}\frac{\partial \zeta_2}{\partial t} + \frac{1}{h_2}\frac{\partial \zeta_3}{\partial t}. \tag{9.2.13}$$

Consider Layer 3:

$$F_{AB} + F_{BG} - F_{GH} = \rho_3 h_3 \frac{\partial v_{13}}{\partial t}\mathrm{d}x_1, \tag{9.2.14}$$

where

$$\begin{aligned} F_{AB} &= \rho_1 g h_1 h_3 + \rho_2 g h_2 h_3 + \frac{1}{2}\rho_3 g h_3^2 + \rho_1 g h_1 \zeta_3 + \rho_1 g h_3 \zeta_1 \\ &\quad + \rho_2 g h_2 \zeta_3 + \Delta\rho_{21} g h_3 \zeta_2 + \Delta\rho_{32} g h_3 \zeta_3, \end{aligned} \tag{9.2.15}$$

$$F_{GB} = (\rho_1 g h_1 + \rho_2 g h_2)\frac{\partial \zeta_3}{\partial x_1}\mathrm{d}x_1, \tag{9.2.16}$$

$$\begin{aligned} F_{GH} &= F_{AB} + \rho_1 g h_1 \frac{\partial \zeta_3}{\partial x_1}\mathrm{d}x_1 + \rho_1 g h_3 \frac{\partial \zeta_1}{\partial x_1}\mathrm{d}x_1 + \rho_2 g h_2 \frac{\partial \zeta_3}{\partial x_1}\mathrm{d}x_1 \\ &\quad + \Delta\rho_{21} g h_3 \frac{\partial \zeta_2}{\partial x_1}\mathrm{d}x_1 + \Delta\rho_{32} g h_3 \frac{\partial \zeta_3}{\partial x_1}\mathrm{d}x_1. \end{aligned} \tag{9.2.17}$$

Substituting (9.2.13), (9.2.14) and (9.2.15) into (9.2.12) yields:

$$\frac{\partial v_{13}}{\partial t} = -g\frac{\partial \zeta_1}{\partial x_1} - \frac{\Delta\rho_{21}}{\rho_0}g\frac{\partial \zeta_2}{\partial x_1} - \frac{\Delta\rho_{32}}{\rho_0}g\frac{\partial \zeta_3}{\partial x_1}. \tag{9.2.18}$$

Conservation of volume:

$$\frac{\partial v_{13}}{\partial x_1} = -\frac{1}{h_3}\frac{\partial \zeta_3}{\partial t}. \tag{9.2.19}$$

Hence we have established six equations [(9.2.5), (9.2.11), (9.2.16)–(9.2.19)] for the six unknowns $(v_{1i}, \zeta_j;\ i = 1,2;\ j = 1,2,3)$. These equations are coupled so that the motion on one interface influences motion on another interface. However, a judicial combination of the motions v_{11}, v_{12} and v_{13} allows the formation of uncoupled set of equations. To facilitate the uncoupling, let us first write these equations in a more compact form:

$$\frac{\partial v_{1i}}{\partial t} = -gA_{ij}\frac{\partial \zeta_j}{\partial x_1}, \tag{9.2.20}$$

$$\frac{\partial v_{1i}}{\partial x_1} = -B_{ij}\frac{\partial \zeta_j}{\partial t}, \tag{9.2.21}$$

where

$$A_{ij} = \begin{pmatrix} A_{11} & A_{12} & A_{13} \\ A_{21} & A_{22} & A_{23} \\ A_{31} & A_{32} & A_{33} \end{pmatrix} = \begin{pmatrix} 1 & 0 & 0 \\ 1 & \varepsilon_{21} & 0 \\ 1 & \varepsilon_{21} & \varepsilon_{32} \end{pmatrix}, \tag{9.2.22}$$

$$B_{ij} = \begin{pmatrix} B_{11} & B_{12} & B_{13} \\ B_{21} & B_{22} & B_{23} \\ B_{31} & B_{32} & B_{33} \end{pmatrix} = \begin{pmatrix} 1/h_1 & -1/h_1 & 0 \\ 0 & 1/h_2 & -1/h_2 \\ 0 & 0 & 1/h_3 \end{pmatrix}, \tag{9.2.23}$$

where $\varepsilon_{ij} = \dfrac{\rho_i - \rho_j}{\rho_0}$.

Suppose that we define a set of coefficients a_{mi}, yet to be determined, so that:

$$Q_{1m} = a_{mi}v_{1i}, \tag{9.2.24}$$

$$\Pi_m = a_{mi}B_{ij}\zeta_j, \tag{9.2.25}$$

with the condition that,

$$a_{mi}(A_{ij} - \beta_{(m)}B_{ij}) = 0. \tag{9.2.26}$$

If a solution to equation (9.2.26) exists then (9.2.20) and (9.2.21) may be rewritten in the form:

$$\frac{\partial Q_{1m}}{\partial t} = -\beta_{(m)}g\frac{\partial \Pi_m}{\partial x_1}, \tag{9.2.27}$$

and

$$\frac{\partial Q_{1m}}{\partial x_1} = -\frac{\partial \Pi_m}{\partial t}, \tag{9.2.28}$$

where the bracket around the subscript means that there is no sum on the subscript m. Cross-differentiating (9.2.27) and (9.2.28) and eliminating Q_{1m} leads to the wave equation:

$$\frac{\partial^2 \Pi_m}{\partial t^2} - \beta_{(m)} g \frac{\partial^2 \Pi_m}{\partial x_1^2} = 0. \tag{9.2.29}$$

The solution pair (Q_{1m}, Π_m) are solutions to the wave equation which are called the normal modes of the system and they are uncoupled. However, it is important to firmly keep in mind that Π_m represents a linear combination of the original interfacial waves.

In books on linear algebra it is shown that (9.2.26) has a solution provided:

$$\left| A_{ij} - \beta_{(m)} B_{ij} \right| = 0. \tag{9.2.30}$$

Substituting from (9.2.22) and (9.2.23) into (9.2.30) leads to a cubic equation for $\beta_{(m)}$:

$$\beta_m^3 - H\beta_m^2 + \gamma H^2 \beta_m - \alpha H^3 = 0, \tag{9.2.31}$$

where H is the total depth ($H = h_1 + h_2 + h_3$),

$$\gamma = \frac{1}{H^2}\{\varepsilon_{21} h_1 h_2 + \varepsilon_{31} h_1 h_3 + \varepsilon_{32} h_2 h_3\} \tag{9.2.32}$$

and

$$\alpha = \frac{1}{H^3}\{\varepsilon_{21}\varepsilon_{32} h_1 h_2 h_3\}. \tag{9.2.33}$$

The values β_m are called the eigenvalues and from (9.2.29) we see that $(\beta_m g)^{1/2}$ is the phase speed of the normal mode m.

The solution of (9.2.31) is facilitated if we note:

$$\begin{aligned}\beta_m^3 - H\beta_m^2 + \gamma H^2 \beta_m - \alpha H^3 &= (\beta_m - \beta_1)(\beta_m - \beta_2)(\beta_m - \beta_3) \\ &\quad + \gamma(H\beta_m^2 - \alpha H^2 \beta_m),\end{aligned} \tag{9.2.34}$$

where

$$\beta_1 = H, \tag{9.2.35}$$

$$\beta_2 = \frac{H}{2}\left(\gamma + \left(\gamma^2 - 4\alpha\right)^{1/2}\right), \tag{9.2.36}$$

$$\beta_3 = \frac{H}{2}\left(\gamma - \left(\gamma^2 - 4\alpha\right)^{1/2}\right). \tag{9.2.37}$$

Now $\gamma = O(\varepsilon_{ij})$ and $\alpha = O(\varepsilon_{ij}^2)$, so the last term in (9.2.34) are small so that β_1, β_2 and β_3 are approximate roots to equation (9.2.31).

The eigen functions a_{mi} are obtained by solving equation (9.2.25):

$$a_{m1}\left(A_{11} - \beta_{(m)}B_{11}\right) + a_{m2}\left(A_{21} - \beta_{(m)}B_{21}\right) + a_{m3}\left(A_{31} - \beta_{(m)}B_{31}\right) = 0, \tag{9.2.38}$$

$$a_{m1}\left(A_{12} - \beta_{(m)}B_{12}\right) + a_{m2}\left(A_{22} - \beta_{(m)}B_{22}\right) + a_{m3}\left(A_{32} - \beta_{(m)}B_{32}\right) = 0, \tag{9.2.39}$$

$$a_{m1}\left(A_{13} - \beta_{(m)}B_{13}\right) + a_{m2}\left(A_{23} - \beta_{(m)}B_{23}\right) + a_{m3}\left(A_{33} - \beta_{(m)}B_{33}\right) = 0, \tag{9.2.40}$$

where $m = 1, 2, 3$ and no sum on m.

Once a_{mi} has been determined, (9.2.24) can be used to determine Q_{1m}. However, (9.2.29) or its equivalent for Q_{1m}, leads to a solution for Π_m and Q_{1m}, the normal mode composite displacement and velocity, and what is then required is the inverse a_{ij}^{-1}. Let b_{km} be a nine component quantity such that:

$$b_{km}a_{mi} = \delta_{ki}, \tag{9.2.41}$$

then from (9.2.24):

$$v_{1k} = b_{km}Q_{1m}. \tag{9.2.42}$$

The inverse b_{km} can be found by solving (9.2.42) for v_{1k} and in general:

$$b_{km} = \frac{1}{\left|a_{ij}\right|}\left\{\begin{matrix} \left(a_{22}a_{33} - a_{23}a_{32}\right) & \left(a_{12}a_{33} - a_{13}a_{32}\right) & \left(a_{12}a_{23} - a_{13}a_{22}\right) \\ \left(a_{21}a_{33} - a_{23}a_{31}\right) & \left(a_{11}a_{33} - a_{13}a_{31}\right) & \left(a_{11}a_{23} - a_{13}a_{21}\right) \\ \left(a_{21}a_{32} - a_{22}a_{31}\right) & \left(a_{11}a_{31} - a_{13}a_{31}\right) & \left(a_{11}a_{22} - a_{12}a_{21}\right) \end{matrix}\right\}, \tag{9.2.43}$$

where

$$\begin{aligned}\left|a_{ij}\right| &= a_{11}\left(a_{22}a_{33} - a_{23}a_{32}\right) - a_{12}\left(a_{21}a_{33} - a_{23}a_{31}\right) \\ &\quad + a_{13}\left(a_{21}a_{32} - a_{22}a_{31}\right),\end{aligned} \tag{9.2.44}$$

is the determinant of the matrix a_{ij}.

9.3. PLANE INTERNAL KELVIN WAVE

Consider the rotating counterpart to the wave motion discussed in §9.2; a two-layered fluid system in a rotating frame of reference as shown in Fig. 9.3.1. It is envisaged that a wave traveling in the x_1 direction, with velocities in the (x_1,x_3) plane will experience a Coriolis force due to the rotation Ω_3 (Fig. 9.3.1). We shall now inquire whether a solution exists to the equations of motion that remains planar, but in which the x_2 variation adjusts so that the x_2 hydrostatic pressure gradient exactly balances the x_2 Coriolis force.

Once again, we shall assume that the wavelength in the x_1 direction and the variation in the x_2 directions are long compared to the total depth H, so that the pressure in the fluid layers may be assumed to be hydrostatic and the velocity in each layer is uniform over the depth.

Consider Layer 2: The x_1 linear momentum equation is given by:

$$F_{AB} - F_{DC} + F_{BC} - \rho_2 h_2 \frac{\partial v_{12}}{\partial t} \mathrm{d}x_1 = 0. \tag{9.3.1}$$

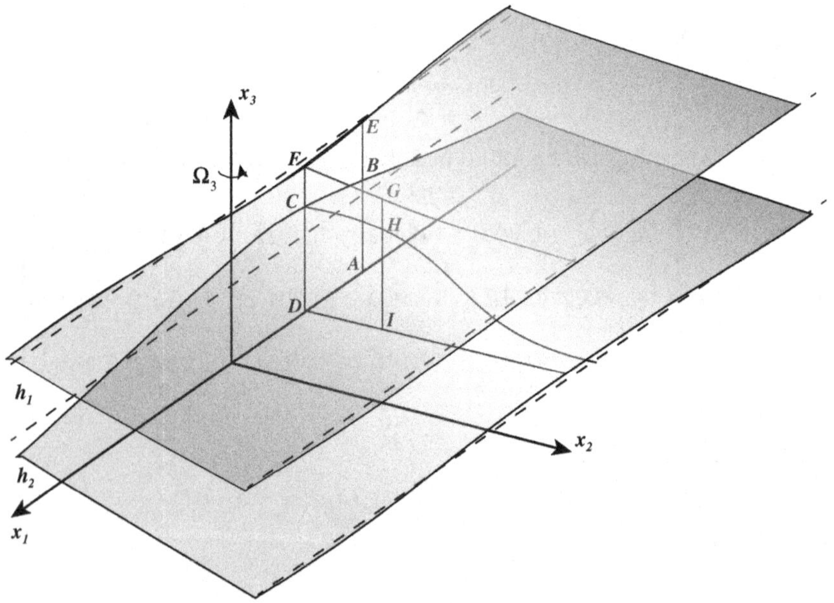

Figure 9.3.1 Schematic of a plane Kelvin wave.

Now it is assumed that both the displacement of the free surface and that of the interface are small so that:

$$
\begin{aligned}
F_{AB} - F_{DC} &= -\frac{\partial F_{AB}}{\partial x_1}\mathrm{d}x_1 \\
&= -\frac{\partial}{\partial x_1}\left\{\rho_1 g(h_1+\zeta_1-\zeta_2)\,(h_2+\zeta_2)+\frac{1}{2}\rho_2 g(h_2+\zeta_2)^2\right\}\mathrm{d}x_1 \\
&= -gh_2\Delta\rho\frac{\partial\zeta_2}{\partial x_1}\mathrm{d}x_1 - \rho_1 gh_2\frac{\partial\zeta_1}{\partial x_1}\mathrm{d}x_1 - \rho_1 gh_1\frac{\partial\zeta_2}{\partial x_1}\mathrm{d}x_1 + O(\zeta_2^2),
\end{aligned}
\tag{9.3.2}
$$

and

$$
F_{BC} = \rho_1 gh_1\frac{\partial\zeta_2}{\partial x_1}\mathrm{d}x_1 + O(\zeta_2^2). \tag{9.3.3}
$$

Substituting (9.3.3) and (9.3.2) into (9.3.1) yields the equation for conservation of momentum in the x_1 direction, with the assumption that $v_2 = 0$ in both layers 1 and 2.

$$
-g'\frac{\partial\zeta_2}{\partial x_1} - \frac{\rho_1}{\rho_2}g\frac{\partial\zeta_1}{\partial x_1} = \frac{\partial v_{12}}{\partial t}, \tag{9.3.4}
$$

that is the same as (9.1.15) and where $g' = \Delta\rho g/\rho_2$.

The x_2 linear momentum equation is simply a balance between hydrostatic pressure and the Coriolis force arising from the v_1 velocity:

$$
F_{CD} - F_{HI} + F_{CH} = 2\rho_2\Omega_3 v_{12}h_2\mathrm{d}x_2. \tag{9.3.5}
$$

Once again we may use Taylor's expansion to write:

$$
\begin{aligned}
F_{CD} - F_{HI} &= -\frac{\partial}{\partial x_2}\left\{\rho_1 g(h_1+\zeta_1-\zeta_2)\,(h_2+\zeta_2)+\frac{1}{2}\rho_2 g(h_2+\zeta_2)^2\right\}\mathrm{d}x_2 \\
&= -\rho_1 gh_1\frac{\partial\zeta_2}{\partial x_2}\mathrm{d}x_2 - \rho_1 gh_2\left(\frac{\partial\zeta_1}{\partial x_2}-\frac{\partial\zeta_2}{\partial x_2}\right)\mathrm{d}x_2 - \rho_2 gh_2\frac{\partial\zeta_2}{\partial x_2}\mathrm{d}x_2 + O(\zeta_2^2) \\
&= -\rho_1 gh_1\frac{\partial\zeta_2}{\partial x_2}\mathrm{d}x_2 - \rho_1 gh_2\frac{\partial\zeta_1}{\partial x_2}\mathrm{d}x_2 - \Delta\rho gh_2\frac{\partial\zeta_2}{\partial x_2}\mathrm{d}x_2 + O(\zeta_2^2),
\end{aligned}
\tag{9.3.6}
$$

and

$$
F_{CH} = \rho_1 gh_1\frac{\partial\zeta_2}{\partial x_2}\mathrm{d}x_2 + O(\zeta_2^2). \tag{9.3.7}
$$

Substituting (9.3.6) and (9.3.7) into (9.3.6) yields:

$$-g'\frac{\partial \zeta_2}{\partial x_2} - \frac{\rho_1}{\rho_2} g \frac{\partial \zeta_1}{\partial x_2} = 2\Omega_3 v_{12}. \tag{9.3.8}$$

Conservation of volume remains unchanged to that from §9.1:

$$h_2 \frac{\partial v_{12}}{\partial x_1} = -\frac{\partial \zeta_2}{\partial t}, \tag{9.3.9}$$

Consider Layer 1: The linear momentum equation in the x_1 direction in layer 1:

$$F_{BE} - F_{FC} - F_{BC} = \rho_1 h_1 \frac{\partial v_{11}}{\partial t} \mathrm{d}x_1, \tag{9.3.10}$$

so that:

$$-g\frac{\partial \zeta_1}{\partial x_1} = \frac{\partial v_{11}}{\partial t}. \tag{9.3.11}$$

Conservation of linear momentum equation in the x_2 direction:

$$F_{FC} - F_{GH} - F_{CH} = 2\rho_1 \Omega_3 v_{11} h_1 \mathrm{d}x_2, \tag{9.3.12}$$

where

$$\begin{aligned} F_{FC} - F_{GH} &= -\frac{\partial F_{FC}}{\partial x_2}\mathrm{d}x_2 = -\frac{\partial}{\partial x_2}\left\{\frac{1}{2}\rho_1 g(h_1 + \zeta_1 - \zeta_2)^2\right\}\mathrm{d}x_2 \\ &= \rho_1 g h_1 \left(\frac{\partial \zeta_2}{\partial x_2} - \frac{\partial \zeta_1}{\partial x_2}\right)\mathrm{d}x_2, \end{aligned} \tag{9.3.13}$$

and

$$F_{CH} = \rho_1 g h_1 \frac{\partial \zeta_2}{\partial x_2} \mathrm{d}x_2. \tag{9.3.14}$$

Substituting (9.3.13) and (9.3.14) into (9.3.12) leads to the force balance:

$$-g\frac{\partial \zeta_1}{\partial x_2} = 2\Omega_3 v_{11}. \tag{9.3.15}$$

Conservation of volume remains unchanged to that from §9.1:

$$h_1 \frac{\partial v_{11}}{\partial x_1} = \frac{\partial \zeta_2}{\partial t} - \frac{\partial \zeta_1}{\partial t}. \tag{9.3.17}$$

By combining (9.3.11) (9.3.15), (9.3.16) and (9.3.17) we may derive an equations for ζ_1 and ζ_2 with respect to the x_1 variation that is the same as (9.1.20), repeated here in the form:

$$\frac{\partial^2\zeta_1}{\partial t^2} - c_s^2\frac{\partial^2\zeta_1}{\partial x_1^2} = 0, \tag{9.3.18}$$

and

$$\frac{\partial^2\zeta_2}{\partial t^2} - c_i^2\frac{\partial^2\zeta_2}{\partial x_1^2} = 0, \tag{9.3.19}$$

where

$$c_s = \sqrt{gH} \quad \text{and} \quad c_i = \left(\frac{g' h_1 h_2 \rho_2}{\rho_1 h_2 + \rho_2 h_1}\right)^{\frac{1}{2}}. \tag{9.3.20}$$

However, now the solution must incorporate the x_2 variation:

$$\zeta_1 = g_s(x_2) f_s(x_1 - c_s t), \tag{9.3.21}$$

and

$$\zeta_2 = g_i(x_2) f_i(x_1 - c_i t), \tag{9.3.22}$$

where $f_s(\varphi_s)$ and $f_i(\varphi_i)$ are functions given by the initial displacement of the free surface and the interface and φ is the phase:

$$\varphi_s = x_1 - c_s t \quad \text{and} \quad \varphi_i = x_1 - c_i t. \tag{9.3.23}$$

As we will see, when we derive the functions $g_i(x_2)$ and $g_s(x_2)$ only waves moving in the positive x_1 directions are admitted when rotation is present. The functions $g_s(x_2)$ and $g_i(x_2)$, may be determined by substituting (9.3.21) and (9.3.22) into the x_2 momentum equations (9.3.9) and (9.3.15) to yield:

$$\frac{-\rho_1}{\rho_2} g f_s \frac{dg_s}{dx_2} - g' f_i \frac{dg_i}{dx_2} = \frac{2\Omega_3 c_i g_i f_i}{h_2}, \tag{9.3.24}$$

and

$$- g f_s \frac{dg_s}{dx_2} = \frac{2\Omega_3}{h_2}(c_s g_s f_s - c_i g_i f_i). \tag{9.3.25}$$

Rearranging (9.3.24) and (9.3.25) leads to the equations:

$$\frac{dg_i}{dx_2} + n_i g_i = \frac{n_i c_i c_s}{h_1 g' f_i} g_s, \tag{9.3.26}$$

and

$$\frac{dg_s}{dx_2} + n_s g_s = -\frac{c_i^2 f_i}{h_1 g f_s} \frac{dg_i}{dx_2}, \quad (9.3.27)$$

where

$$n_s = \frac{2\Omega_3}{c_s} \quad \text{and} \quad n_i = \frac{2\Omega_3}{c_i}. \quad (9.3.28)$$

The quantities n_s^{-1} and n_i^{-1} are called the Rossby radius of deformation for the surface and internal waves, respectively.

It is not difficult to show that, since $\zeta_1 \ll \zeta_2$ (see also §9.2), the right hand side of (9.3.26) and (9.3.27) are approximately zero (compared to the terms on the left hand side). Equations (9.3.26) and (9.3.27) thus reduce to the simple first order differential equations:

$$\frac{dg_i}{dx_2} + n_i g_i = 0, \quad (9.3.29)$$

and

$$\frac{dg_s}{dx_2} + n_s g_s = 0, \quad (9.3.30)$$

admitting the solution:

$$g_i = Ae^{-n_i x_2}, \quad (9.3.31)$$

and

$$g_s = Be^{-n_s x_2}, \quad (9.3.32)$$

where A and B are arbitrary coefficients that can also be incorporated into the functions f_i and f_s (equations (9.3.21) and (9.3.22)). Equations (9.3.31) and (9.3.32) clearly show that the surface and interface displacement decays exponentially with x_2 over a scale n_s^{-1} and n_i^{-1}, respectively. Hence the name (Rossby) radius of deformation. The reader may wish to show that, if in (9.3.21) and (9.3.22) it is assumed that the wave moves in the opposite direction, then (9.3.32) and (9.3.33) imply an exponential growth with x_2; hence the waves, just derived, called Kelvin waves, can only travel cyclonically, in the same sense as the rotation Ω_3. Kelvin waves are thus simply long waves, as in a non-rotating fluid system, with the Coriolis force, introduced by the rotation of the frame of reference, being balanced by a simple exponential variation of the pressure in the x_2 direction.

9.4. POINCARÉ WAVES IN A TWO-LAYER SYSTEM

We now turn our attention to an example of a non-planar motion in a rotating two-layer system. The conservation equations from §9.3 needs to be generalized to allow for non-zero velocities, v_{21} and v_{22}:

Consider Layer 2:

The x_1 linear momentum equation follows directly from (9.3.4):

$$-2\Omega_3 v_{22} - g'\frac{\partial \zeta_2}{\partial x_1} - \frac{\rho_1}{\rho_2} g \frac{\partial \zeta_1}{\partial x_1} = \frac{\partial v_{12}}{\partial t}, \tag{9.4.1}$$

the x_2 linear momentum equation follows from (9.3.9):

$$-2\Omega_3 v_{12} - g'\frac{\partial \zeta_2}{\partial x_2} - \frac{\rho_1}{\rho_2} g \frac{\partial \zeta_1}{\partial x_2} = \frac{\partial v_{22}}{\partial t}, \tag{9.4.2}$$

and conservation of volume follows from (9.3.16):

$$h_2 \frac{\partial v_{12}}{\partial x_1} + h_2 \frac{\partial v_{22}}{\partial x_2} = -\frac{\partial \zeta_2}{\partial t}. \tag{9.4.3}$$

Consider Layer 1:

Once again the x_1 momentum equation follows from a simple generalization of (9.3.11) and reads:

$$-2\Omega_3 v_{21} - g\frac{\partial \zeta_1}{\partial x_1} = \frac{\partial v_{11}}{\partial t}, \tag{9.4.4}$$

the x_2 momentum equation follows from (9.3.15):

$$-2\Omega_3 v_{11} - g\frac{\partial \zeta_1}{\partial x_2} = \frac{\partial v_{21}}{\partial t}, \tag{9.4.5}$$

and conservation of volume follows from (9.3.17):

$$h_1 \frac{\partial v_{11}}{\partial x_1} + h_1 \frac{\partial v_{21}}{\partial x_2} = \frac{\partial \zeta_2}{\partial t} - \frac{\partial \zeta_1}{\partial t}. \tag{9.4.6}$$

Equations (9.4.1)–(9.4.6) form a set of six equations for the six unknowns v_{11}, v_{12}, v_{21}, v_{22}, ζ_1, ζ_2. We may combine these equations to obtain a single equation for the interface displacement ζ_2. This is most easily achieved by subtracting the layer 2 equations from the layer 1 equation, and assuming that $\rho_1 \approx \rho_2$:

$$g'\frac{\partial \zeta_2}{\partial x_1} = \frac{\partial}{\partial t}\Delta v_1 + 2\Omega_3 \Delta v_2, \tag{9.4.7}$$

$$g'\frac{\partial \zeta_2}{\partial x_2} = \frac{\partial}{\partial t}\Delta v_2 + 2\Omega_3 \Delta v_1, \tag{9.4.8}$$

and

$$\frac{\partial}{\partial x_1}\Delta v_1 + \frac{\partial}{\partial x_2}\Delta v_2 = \frac{1}{h}\frac{\partial \zeta_2}{\partial t}, \tag{9.4.9}$$

where

$$\Delta v_1 = \frac{1}{\rho_2}(\rho_1 v_{11} - \rho_2 v_{12}), \tag{9.4.10}$$

$$\Delta v_2 = \frac{1}{\rho_2}(\rho_1 v_{21} - \rho_2 v_{22}), \tag{9.4.11}$$

and

$$h = \frac{h_1 h_2 \rho_2}{h_2 \rho_1 + h_1 \rho_2}. \tag{9.4.12}$$

and for (9.4.9) it was assumed that $\frac{\partial \zeta_1}{\partial t} << \frac{\partial \zeta_2}{\partial t}$.

A single equation for ζ_2 may be obtained by differentiating (9.4.7) with respect to x_1, and (9.4.8) with respect to x_2 and adding the resulting equations to yield:

$$\frac{\partial^2 \zeta_2}{\partial x_1^2} + \frac{\partial^2 \zeta_2}{\partial x_2^2} - \frac{1}{g'h}\frac{\partial^2 \zeta_2}{\partial t^2} = \frac{2\Omega_3}{g'}\left(\frac{\partial}{\partial x_2}\Delta v_1 + \frac{\partial}{\partial x_1}\Delta v_2\right). \tag{9.4.13}$$

Differentiating (9.4.13) with respect to t and then substituting for $\frac{\partial \Delta v_1}{\partial t}$ and $\frac{\partial \Delta v_2}{\partial t}$ from (9.4.7) and (9.4.8):

$$\frac{\partial^2 \zeta_2}{\partial x_1^2} + \frac{\partial^2 \zeta_2}{\partial x_2^2} - \frac{1}{c_i^2}\left(\frac{\partial^2 \zeta_2}{\partial t^2} + f^2 \zeta_2\right) = 0, \tag{9.4.14}$$

where

$$f = 2\Omega_3, \tag{9.4.15}$$

and

$$c_i = (g'h)^{1/2} = \left(\frac{h_1 h_2 \rho_2 g'}{h_2\rho_1 + h_1\rho_2}\right)^{1/2}, \tag{9.4.16}$$

is the phase speed in a non-rotating two-layer system.

We now seek a simple translation (progressive) wave solution of the form:

$$\zeta_2 = \tilde{\zeta}\cos\phi, \tag{9.4.17}$$

where the phase

$$\phi = k_1x_1 + k_2x_2 - \omega t, \tag{9.4.18}$$

and where $\tilde{\zeta}$, assumed to be real, is the wave amplitude. The angle given by θ between the phase lines and the x_1 axis is given by: $\tan\theta = -\dfrac{k_2}{k_1}$.

Substituting (9.4.17) into (9.4.14) shows that (9.4.17) is a solution provided ω satisfies the dispersion relationship:

$$\omega^2 = f^2 + c_i^2(k_1^2 + k_2^2), \tag{9.4.19}$$

indicating that the frequency ω of the waves lies between f (long waves; $k_1 \to 0$, $k_2 \to 0$) and $c_i(k_1^2 + k_2^2)^{1/2}$ (short waves, the non-rotating limit; $k_1 \to \infty$ and or $k_2 \to \infty$).

The corresponding velocity differences, Δv_1 and Δv_2, follow if we seek solutions of the form:

$$\Delta v_1 = A_1\cos\phi + B_1\sin\phi, \tag{9.4.20}$$

$$\Delta v_2 = A_2\cos\phi + B_2\sin\phi, \tag{9.4.21}$$

where A_1, A_2, B_1 and B_2 are constants to be determined. Substituting (9.4.20) and (9.4.21) into (9.4.7), (9.4.8) and (9.4.9) allows these constants to be determined so that:

$$\Delta v_1 = -\frac{\tilde{\zeta}^2}{hk_H^2}(k_1\omega\cos\phi + fk_2\sin\phi), \tag{9.4.22}$$

$$\Delta v_2 = -\frac{\tilde{\zeta}^2}{hk_H^2}(k_2\omega\cos\phi + fk_1\sin\phi), \tag{9.4.23}$$

where

$$k_H^2 = k_1^2 + k_2^2. \tag{9.4.24}$$

To illustrate the relationship between the interface displacement and the shear across the interface we suppose that $k_2 = 0$, so that $\theta = 0$ and the phase velocity is along the x_1 axis. For this case (9.4.22) and (9.4.23) reduce to:

$$\Delta v_1 = -\frac{\tilde{\zeta}}{h_1 k_1}\omega \cos\phi = -\Delta V_1 \cos\phi, \tag{9.4.25}$$

$$\Delta v_1 = +\frac{\tilde{\zeta}}{h_1 k_1} f \sin\phi = \Delta V_2 \sin\phi, \tag{9.4.26}$$

where ΔV_1 and ΔV_2 are the amplitudes of the shear. This solution is represented schematically in Fig. 9.4.1

From Fig. 9.4.1, we see that as the wave crest moves along the x_1 axis, the velocity difference rotates in the same direction as the rotation of the reference axes; such a rotation is called cyclonic; when $\omega > f$, the velocity difference is greatest when oriented in the direction of the phase velocity (x_1 is this example).

For the above example, the velocities components in the upper and lower layers are given by:

$$\zeta_2 = \tilde{\zeta}\cos\phi, \tag{9.4.27}$$

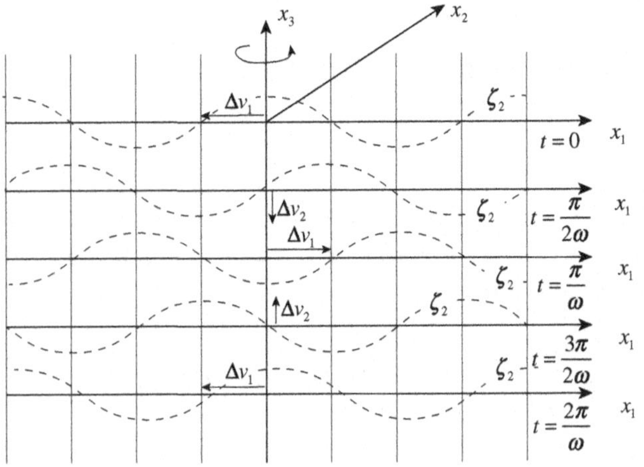

Figure 9.4.1 Interface displacement and associated shear across layer interface for a Poincare Wave moving in the x_1 direction.

$$v_{11} = -\frac{\tilde{\zeta}\omega}{k_1 h_1}\cos\phi, \qquad (9.4.28)$$

$$v_{12} = -\frac{\tilde{\zeta}\omega}{k_1 h_2}\cos\phi, \qquad (9.4.29)$$

$$\Delta v_1 = -\frac{\tilde{\zeta}\omega}{k_1 h}\cos\phi, \qquad (9.4.30)$$

$$v_{21} = \frac{\tilde{\zeta} f}{k_1 h_1}\sin\phi, \qquad (9.4.31)$$

$$v_{22} = -\frac{\tilde{\zeta} f}{k_1 h_2}\sin\phi, \qquad (9.4.32)$$

$$\Delta v_2 = \frac{\tilde{\zeta} f}{k_1 h}\sin\phi, \qquad (9.4.33)$$

where $\phi = k_1 x_1 - \omega t$.

As the governing equations have remained linear, as in the non-rotating case, it is possible to add solutions of the type (9.4.27)–(9.4.33) and obtain standing waves. By the way of example, consider the combination of a right moving wave as described by (9.4.27) and a left moving wave of the same amplitude:

$$\zeta_2 = \tilde{\zeta}\cos\phi^{+} + \tilde{\zeta}\cos\phi^{-}, \qquad (9.4.34)$$

where

$$\phi^{\pm} = k_1 x_1 \pm \omega t, \qquad (9.4.35)$$

Simple addition of the trigonometric functions leads to the expressions:

$$\zeta_2 = 2\tilde{\zeta}\cos(k_1 x_1)\cos(\omega t), \qquad (9.4.38)$$

$$v_{11} = -\frac{2\tilde{\zeta}\omega}{k_1 h_1}\sin(k_1 x_1)\sin(\omega t), \qquad (9.4.39)$$

$$v_{12} = \frac{2\tilde{\zeta}\omega}{k_1 h_2}\sin(k_1 x_1)\sin(\omega t), \qquad (9.4.40)$$

$$\Delta v_1 = -\frac{2\tilde{\zeta}\omega}{k_1 h}\sin(k_1 x_1)\sin(\omega t), \tag{9.4.41}$$

$$v_{21} = \frac{2\tilde{\zeta} f}{k_1 h_1}\sin(k_1 x_1)\cos(\omega t), \tag{9.4.42}$$

$$v_{22} = -\frac{2\tilde{\zeta} f}{k_1 h_2}\sin(k_1 x_1)\cos(\omega t), \tag{9.4.43}$$

$$\Delta v_2 = \frac{2\tilde{\zeta} f}{k_1 h}\sin(k_1 x_1)\cos(\omega t). \tag{9.4.44}$$

This combination has the property that v_{11} and v_{12}, the components in the x_1 direction, are zero at $x_1 = 0, \pi, 2\pi, 3\pi, \ldots$ and so represent standing waves in a duct bounded at walls between $x_1 = 0$ and any integer of π. It is noteworthy that the velocities are independent of x_2 and so the solution is uniform in the x_2 direction.

9.5. GEOSTROPHIC MOTION IN A TWO-LAYER SYSTEM

The equations developed in §9.3 may also be used to describe the concept of potential vorticity and geostrophic motion. These concepts, as we shall show here, are important when describing coastal upwelling. Consider again a simple two-layer system, with distinct upper and lower layers and variables as described in §9.1. Given the velocity in each layer is uniform with depth, the vorticity in each layer (see §2.12) will only have a vertical component:

$$\Gamma_{3i} = \frac{\partial v_{2i}}{\partial x_1} - \frac{\partial v_{1i}}{\partial x_2}, \tag{9.5.1}$$

where ζ_{3i} is the vertical component of vorticity in layer i. The difference in vorticity between the upper and lower layers:

$$\begin{aligned}\Delta\Gamma_3 &= \frac{\partial}{\partial x_1}(\Delta v_2) - \frac{\partial}{\partial x_2}(\Delta v_1) \\ &= \frac{\rho_1}{\rho_2}\left\{\frac{\partial}{\partial x_1}v_{21} - \frac{\partial}{\partial x_2}v_{11}\right\} - \left\{\frac{\partial}{\partial x_1}v_{22} - \frac{\partial}{\partial x_2}v_{12}\right\} \\ &= \left(\frac{\rho_1}{\rho_2}\zeta_{31} - \zeta_{32}\right).\end{aligned} \tag{9.5.2}$$

Differentiating (9.4.7) with respect to x_2 and (9.4.8) with respect x_1 subtracting the result leads to:

$$\frac{\partial}{\partial t}\left(\frac{\Delta\Gamma_3}{f} + \frac{\zeta_2}{h}\right) = 0. \tag{9.5.3}$$

The quantity

$$\Delta Q = \frac{\Delta\Gamma_3}{f} + \frac{\zeta_2}{h}, \tag{9.5.4}$$

is called the differential potential vorticity and equation (9.5.3) shows that the differential potential vorticity is constant as the motion proceeds. This result may be used to advantage when seeking steady state solutions to the equations established in §9.4.

For steady flow, (9.4.7), (9.4.8) and (9.4.9) become:

$$g'\frac{\partial\zeta_2}{\partial x_1} = -f\Delta v_2, \tag{9.5.5}$$

$$g'\frac{\partial\zeta_2}{\partial x_2} = f\Delta v_1, \tag{9.5.6}$$

$$\frac{\partial}{\partial x_2}(\Delta v_2) + \frac{\partial}{\partial x_2}(\Delta v_2) = 0. \tag{9.5.7}$$

Differentiating (9.5.5) with respect to x_1 and (9.5.6) with respect to x_2, adding the result and noting the definition of the potential vorticity (9.5.4):

$$-\left(\frac{\partial^2\zeta_2}{\partial x_1^2} + \frac{\partial^2\zeta_2}{\partial x_2^2}\right) + \frac{f^2}{g'h}\zeta_2 = \frac{f^2}{g'}Q(x_1, x_2, 0), \tag{9.5.8}$$

where $Q(x_1, x_2, 0)$ is the initial distribution of the differential potential vorticity.

Equation (9.5.8) describes the interface displacement due to a steady motion where the Coriolis force exactly balances the pressure force due to the interface displacement η and the free surface displacement ζ. By the way of example consider a coastal current such as the Leeuwin current flowing southward along the Western Australian coastline. If we idealize the coastline as a straight line oriented north south, the current as a surface layer of depth h_1 and a deep lower layer that we shall assume to be stationary. Further, we may seek a solution that is parallel to the coast as this coastline is nearly 3000 km long and the current is known to be quite narrow and

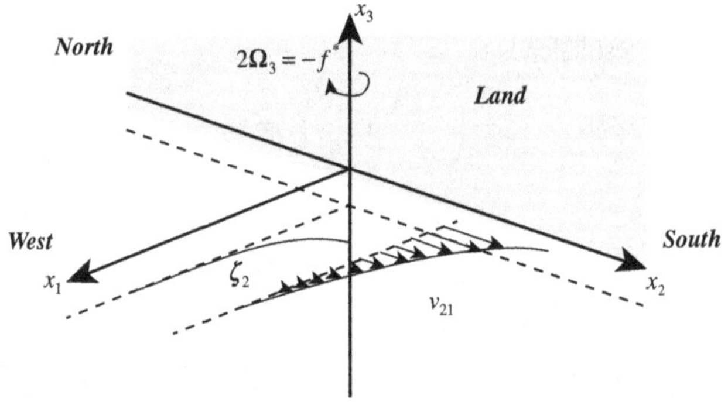

Figure 9.5.1 Schematic of a surface coastal current in the southern hemisphere.

shallow, flowing southward adjacent to the shoreline as shown schematically in Fig. 9.5.1. If we assume that the initial potential vorticity was zero, the onshore offshore component v_{1i} is zero (flow is parallel to the coastline) and the current is in the southern hemisphere so we let $f = -f^*$, then the solution to (9.5.8) is given by:

$$\zeta_2 = -Ae^{-\frac{x_1}{R}}; \quad R = \frac{(g'h)^{\frac{1}{2}}}{f^*}, \tag{9.5.10}$$

$$v_{21} = A\left(\frac{g'}{h}\right)^{\frac{1}{2}} e^{-\frac{x_1}{R}}; \quad A > 0, \tag{9.5.11}$$

where R is called the Rossby radius of deformation.

9.6. GEOSTROPHIC FLOW IN A CONTINUOUSLY STRATIFIED FLUID

In the previous sections (§9.1–§9.5) we assumed that the fluid column was made up of a two-layer stratification. Consider now the case when the fluid column is vertically stratified so that:

$$\rho(x_1, x_2, x_3, t) = \rho_e(x_1, x_2, x_3, t) + \rho'(x_1, x_2, x_3, t), \tag{9.6.1}$$

where ρ_e (x_1, x_2, x_3, t) is the background density structure and ρ' (x_1, x_2, x_3, t) is the density anomaly due to any unsteady motion. The coordinate system (x_1, x_2, x_3) is again taken as a right-handed Cartesian system with x_3 in the vertical direction. In such a system, the earth's rotation Ω_3, relative to a flat plane at particular latitude, is positive in the northern hemisphere and negative in the southern hemisphere. The linear equation of motion, assuming the fluid is incompressible and inviscid:

$$\rho_0 \frac{\partial v_1}{\partial t} - \rho_0 f v_2 = -\frac{\partial p}{\partial x_1}, \tag{9.6.2}$$

$$\rho_0 \frac{\partial v_2}{\partial t} + \rho_0 f v_1 = -\frac{\partial p}{\partial x_2}, \tag{9.6.3}$$

$$\rho_0 \frac{\partial v_3}{\partial t} = -\rho g - \rho_0 \frac{\partial p}{\partial x_3}, \tag{9.6.4}$$

where the Coriolis parameter:

$$f = 2\Omega_3 = 2\Omega_0 \sin \theta, \tag{9.6.5}$$

where Ω_0 is the earth's rotation and θ is the latitude.

The equations governing steady geostrophic motion are obtained by setting the time derivatives in (9.6.2)–(9.6.4) to zero:

$$\rho_0 f v_2 = -\frac{\partial p}{\partial x_1}, \tag{9.6.6}$$

$$\rho_0 f v_1 = -\frac{\partial p}{\partial x_2}, \tag{9.6.7}$$

$$0 = \rho_e g + \frac{\partial p}{\partial x_3}, \tag{9.6.8}$$

If we multiply (9.6.6) by v_1 and (9.6.7) by v_2 and subtract the resulting equations it follows immediately that:

$$v_1 \frac{\partial p}{\partial x_1} + v_2 \frac{\partial p}{\partial x_2} = 0, \tag{9.6.9}$$

which may be written as

$$v_i p_{,i} = 0; \quad i = 1, 2, \tag{9.6.10}$$

Equation (9.6.10) shows that for such flows the velocity vector is aligned with the lines of constant pressure; this is a well-known result and is familiar to anyone who has looked at weather maps in the newspaper.

To derive the variation of the velocity field with altitude, we differentiate both (9.6.6) and (9.6.7) with respect to x_3 and use (9.6.8) to eliminate the pressure. This yields the equations:

$$\frac{\partial v_2}{\partial x_3} = \frac{1}{f}\frac{\partial g_e}{\partial x_1}, \tag{9.6.11}$$

$$\frac{\partial v_1}{\partial x_3} = -\frac{1}{f}\frac{\partial g_e}{\partial x_2}, \tag{9.6.12}$$

where

$$g_e = \frac{g\rho_e}{\rho_0}, \tag{9.6.13}$$

is the modified acceleration due to gravity acting on the stratification. These equations are of great importance in both meteorology and oceanography as they allow the determination of the horizontal velocity field from a knowledge of the vertical variation of the density field (9.6.1) and the velocity at any one point from direct measurements. In meteorology, they are called the thermal wind equations.

9.7. WEDDERBURN AND LAKE NUMBERS

Consider the motion resulting from a surface wind stress being suddenly applied to a lake surface, idealized as a rectangular basin with a three-layer stratification as shown in Fig. 9.7.1. This example illustrated the role that end boundaries play in confining the motion in a finite domain.

The conservation of momentum (9.2.20) and volume (9.2.21), in a non-rotating frame of reference, applied to the three-layer stratification remain the same except now there is a wind stress exerting a surface stress on the upper layer (wind water shear velocity u_*) and the layers are confined by end boundaries:

$$\frac{\partial v_{1i}}{\partial t} = -gA_{ij}\frac{\partial \zeta_j}{\partial x_1} + \frac{u_*^2}{h_1}\delta_{1i}, \tag{9.7.1}$$

$$\frac{\partial v_{1i}}{\partial x_1} = -B_{ij}\frac{\partial \zeta_j}{\partial t}, \tag{9.7.2}$$

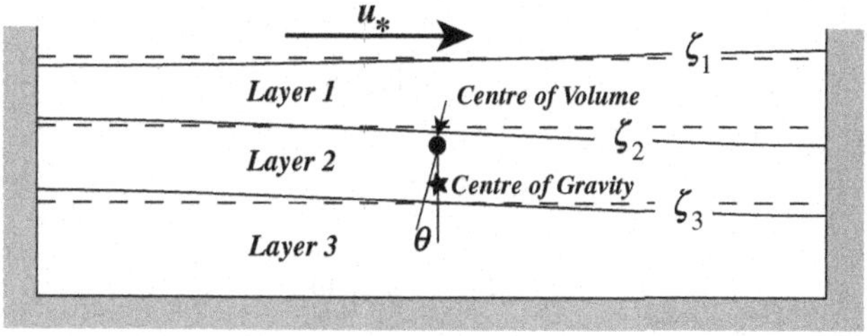

Figure 9.7.1 Schematic of a three layer fluid system in a rectangular container.

where the coefficients A_{ij} and B_{ij} are again given by:

$$A_{ij} = \begin{pmatrix} 1 & 0 & 0 \\ 1 & \varepsilon_{21} & 0 \\ 1 & \varepsilon_{21} & \varepsilon_{32} \end{pmatrix}, \tag{9.7.3}$$

$$B_{ij} = \begin{pmatrix} 1/h_1 & -1/h_1 & 0 \\ 0 & 1/h_2 & -1/h_2 \\ 0 & 0 & 1/h_3 \end{pmatrix}, \tag{9.7.4}$$

and the symbol is the Kroneker Delta:

$$\delta_{1i} \begin{cases} = 1 & i = 1 \\ = 0 & otherwise \end{cases}. \tag{9.7.5}$$

As before

$$\varepsilon_{ij} = \frac{\rho^i - \rho^j}{\rho_0} \quad \text{and} \quad g'_{ij} = g\varepsilon_{ij}. \tag{9.7.6}$$

Once again we seek a normal mode solutions by multiplying (9.7.1) and (9.7.2) by coefficients a_{mi} and summing over i, so that:

$$\frac{\partial Q_{1m}}{\partial t} = -\beta_{(m)} g \frac{\partial \Pi_m}{\partial x_1} + \frac{u_*^2}{h_1} a_{mi}, \tag{9.7.7}$$

and

$$\frac{\partial Q_{1m}}{\partial x_1} = -\frac{\partial \Pi_m}{\partial t}, \tag{9.7.8}$$

where

$$Q_{1m} = a_{mi}v_i, \tag{9.7.9}$$

$$\Pi_m = a_{mi}B_{ij}\zeta_j, \tag{9.7.10}$$

$$a_{mi}A_{ij}\zeta_j = \beta_{(m)}\Pi_m, \tag{9.7.11}$$

and the boundary conditions at $x_1 = 0$, L require $Q_{1m} = 0$.

Before we present the solution of (9.7.7) and (9.7.8) we shall introduce two very important non-dimensional numbers. First, consider a simple force balance of layer 1 once the wind stress has been applied and all motion has come to rest. If it is assumed that the surface, ζ_1 and interface, ζ_2 respond, but, ζ_3 remains horizontal, then the simple force balance (9.7.1), with $v_i = 0$, becomes

$$g\frac{\partial\zeta_1}{\partial x_1} \sim \frac{u_*^2}{h_1}, \tag{9.7.12}$$

and

$$g\frac{\partial\zeta_1}{\partial x_1} \sim g\varepsilon_{21}\frac{\partial\zeta_2}{\partial x_1}, \tag{9.7.13}$$

$$\text{with } \zeta_3 = 0. \tag{9.7.14}$$

Form this it follows that:

$$\frac{\zeta_2}{h_1} \sim \left(\frac{g'_{21}h_1^2}{u_*^2 L}\right)^{-1}. \tag{9.7.15}$$

The right hand side of (9.7.15) is defined as the Wedderburn number W:

$$W = \frac{g'_{21}h_1^2}{u_*^2 L}. \tag{9.7.16}$$

Hence if the wind is strong enough to yield a value of the Wedderburn number $W < 1$, we may expect $\zeta_2 > h_1$, implying that metalimnetic water will upwell to the surface.

The second non-dimensional number is called the Lake Number L_N and is defined as the ratio of the moment of the center of gravity about the center of volume, divided by the overturning moment about the center of volume due to the wind stress acting on the surface:

$$L_N = \frac{Mg(x_{3V} - x_{3g})\sin\theta}{\rho_1 u_*^2 L(H - x_{3V})}, \tag{9.7.17}$$

where x_{3V} is the height to the center of volume, x_{3g} is the height to the center of gravity, M is the total mass of water and θ is the average tilt angle (Fig. 9.7.1).

Now in general for an arbitrary shaped basin:

$$Mg(x_{3V} - x_{3g}) = \int_0^H g(x_{3V} - x_3)\rho(x_3)A(x_3)\mathrm{d}x_3 = g\rho_0 S, \tag{9.7.18}$$

and

$$S = \int_0^H \frac{\rho(x_3)}{\rho_0}(x_{3V} - x_3)A(x_3)\mathrm{d}x_3, \tag{9.7.19}$$

is called the water column stability. If the stabilizing moment is evaluated for an incipient upwelling inclination then:

$$\sin\theta = \frac{H - x_{3T}}{L}, \tag{9.7.20}$$

where x_{3T} is the height of the thermocline and L is the horizontal length scale. Substituting (9.7.20) and (9.7.18) into (9.7.17) yields:

$$L_N = \frac{gS}{v_*^2 A_0^{3/2}} \frac{\left(1 - \dfrac{x_{3T}}{H}\right)}{\left(1 - \dfrac{x_{3V}}{H}\right)}, \tag{9.7.21}$$

where it is assumed that $L = A_0^{1/2}$ where A_0 is the surface area, in the present case $A_0 = Lx_1$.

The stratification in most lakes may be approximated by a three-layer stratification as shown in Fig. 9.7.1 with the additional assumptions:

$$\left.\begin{array}{l} \varepsilon_{21} \ll \varepsilon_{32} \\ h_1 < h_2 < h_3 \\ \text{and } \varepsilon_{32} < \dfrac{h_1}{H} \\ \text{with } H = h_1 + h_2 + h_3 \end{array}\right\}. \tag{9.7.22}$$

For such a situation (9.7.21) reduces to:

$$L_N = \frac{g_{32} h_3 h_2^3}{v_*^2 LH}. \tag{9.7.23}$$

The assumption (9.7.22) may also be used to simplify the calculation discussed in §9.2:

$$\gamma = \frac{\varepsilon_{32} h_2 h_3}{H^2}, \tag{9.7.24}$$

$$\alpha = \frac{1}{H^3} \varepsilon_{21} \varepsilon_{32} h_1 h_2 h_3, \tag{9.7.25}$$

$$(\gamma^2 - 4\alpha)^{1/2} = \frac{\varepsilon_{32} h_2 h_3}{H^2} - \frac{2\varepsilon_{21} h_1}{H}, \tag{9.7.26}$$

so that:

$$\beta_1 = H, \tag{9.7.27}$$

$$\beta_2 = \frac{\varepsilon_{32} h_2 h_3}{H}, \tag{9.7.28}$$

$$\beta_3 = \varepsilon_{21} h_1, \tag{9.7.29}$$

and

$$a_{ij} = \begin{pmatrix} 1 & \frac{h_2}{h_1} & \frac{h_3}{h_1} \\ 1 & \frac{h_2}{h_1} & -\frac{h_2}{h_1} \\ 1 & -1 & \frac{\varepsilon_{21} h_1}{\varepsilon_{32} h_2} \end{pmatrix}. \tag{9.7.30}$$

The excursion at $x_1 = 0$ or L, in response to a wind stress, may be obtained by noting that at these extremes $v_i = 0$ so that (9.7.7) becomes

$$\frac{\partial}{\partial x_1}(a_{mi} B_{ij} \zeta_j) = \frac{u_*^2 a_{mi}}{h_1 g \beta_{(m)}}. \tag{9.7.31}$$

This implies:

$$a_{mi} B_{ij} \zeta_j \big|_{x_1 = 0, L} \sim \frac{u_*^2 a_{mi} L}{g h_1 \beta_{(m)}}. \tag{9.7.32}$$

Substituting from (9.7.4) and (9.7.30) and finding the inverse for ζ_i leads to:

$$\left.\frac{\zeta_1}{H}\right|_{x_1=0,L} \sim \frac{u_*^2 L}{gH^2} = Fr^2 \frac{L}{H}, \tag{9.7.33}$$

$$\left.\frac{\zeta_2}{h_1}\right|_{x_1=0,L} \sim \frac{u_*^2 L}{g\varepsilon_{21} h_1^2} = \frac{1}{W}, \tag{9.7.34}$$

$$\left.\frac{\zeta_3}{h_2}\right|_{x_1=0,L} \sim \frac{u_*^2 L}{g\varepsilon_{32} h_2^2} \sim \frac{1}{L_N}, \tag{9.7.35}$$

where

$$Fr = \left(\frac{u_*^2}{gH}\right)^{1/2}. \tag{9.7.36}$$

The dynamical significance of Fr, W and L_N is thus clear; the first interface displacement is sensitive to the Wedderburn number W and the lower interface is sensitive to the Lake Number L_N.

9.8. THREE-LAYER FORCED MOTIONS

In §9.2 and §9.7, it was shown that the equations of motion for the normal mode velocities Q_m and interfacial displacements Π_m is given by:

$$\frac{\partial Q_m}{\partial t} = -\beta_{(m)} g \frac{\partial \Pi_m}{\partial x_1} + \frac{u_*^2 a_{mi}}{h_1}, \tag{9.8.1}$$

$$\frac{\partial Q_m}{\partial x_1} = -\frac{\partial \Pi_m}{\partial t}, \tag{9.8.2}$$

where the definition of all the variables is given in §9.2 and §9.7.

Before proceeding to solve these equations it is convenient to non-dimensionalize all the variables as follows:

$$C_m^2 = \beta_{(m)} g, \tag{9.8.3}$$

$$\tau = \frac{C_m}{L} t, \tag{9.8.4}$$

$$\xi = \frac{x_1}{L}, \tag{9.8.5}$$

$$\theta_m = \frac{C_m}{FL} Q_m, \tag{9.8.6}$$

$$\Omega_m = \frac{C_m^2}{LF} \Pi_m, \tag{9.8.7}$$

$$F = \frac{u_*^2 a_{mi}}{h_1}. \tag{9.8.8}$$

Introducing (9.8.3)–(9.8.8) into (9.8.1) and (9.8.2) leads to equations for the non-dimensional velocities θ_m and interfacial displacements Ω_m:

$$\frac{\partial \theta_m}{\partial \tau} = -\frac{\partial \Omega_m}{\partial \xi} + 1, \tag{9.8.9}$$

$$\frac{\partial \theta_m}{\partial \xi} = -\frac{\partial \Omega_m}{\partial \tau}, \tag{9.8.10}$$

with the boundary conditions,

$$\xi = 0, 1; \quad \theta_m = 0, \tag{9.8.11}$$

and the initial condition,

$$\tau = 0; \quad \theta_m = \Omega_m = 0. \tag{9.8.12}$$

We shall solve these equations first via a series expansion and then by the method of characteristics. First seek a solution of the form:

$$\theta_m = \sum_{n=1}^{\infty} A_{mn} \sin \omega_{mn} \tau \sin k_{mn} \xi. \tag{9.8.13}$$

This satisfies the initial conditions (9.8.12) and the boundary conditions (9.8.11) provided:

$$k_{mn} = n\pi \quad n = 1, 2, \ldots. \tag{9.8.14}$$

The unit value on the right hand side of (9.8.9), arising from the wind forcing, may be expanded in terms of an odd Fourier series so that:

$$1 = \sum_{n=1}^{\infty} \frac{-4}{(2n-1)\pi} \sin(2n-1)\pi\xi. \tag{9.8.15}$$

Substituting (9.8.15) and (9.8.13) into (9.8.9) and (9.8.10) leads to the solution:

$$\theta_m = \sum_{n=1}^{\infty} \frac{-4}{\pi^2(2n-1)^2} \sin(2n-1)\pi\tau \sin(2n-1)\pi\xi, \qquad (9.8.16)$$

$$\Omega_m = \sum_{n=1}^{\infty} \frac{4}{\pi^2(2n-1)^2} \{\cos((2n-1)\pi\tau) - 1\} \cos(2n-1)\pi\xi. \quad (9.8.17)$$

The displacement function Ω_n is the same for all modes in terms of the non-dimensional variable Ω_m.

More insight into the nature of the motion may be obtained from what is called the method of characteristics or ray tracing. This method involves writing:

$$\zeta = \xi + \tau, \qquad (9.8.18)$$

$$\eta = \xi - \tau, \qquad (9.8.19)$$

so that (9.8.9) and (9.8.10) become

$$\frac{\partial}{\partial \zeta}(\theta_m + \Omega_m) - \frac{\partial}{\partial \eta}(\theta_m - \Omega_m) = 1, \qquad (9.8.20)$$

$$\frac{\partial}{\partial \zeta}(\theta_m + \Omega_m) + \frac{\partial}{\partial \eta}(\theta_m - \Omega_m) = 0, \qquad (9.8.21)$$

adding and subtracting (9.8.20) and (9.8.21) lead to the pair:

$$\frac{\partial}{\partial \zeta}(\theta_m + \Omega_m) = \frac{1}{2}, \qquad (9.8.22)$$

$$\frac{\partial}{\partial \eta}(\theta_m - \Omega_m) = -\frac{1}{2}. \qquad (9.8.23)$$

These equations may be integrated to yield:

$$\theta_m + \Omega_m = \frac{1}{2}\zeta + f(\eta), \qquad (9.8.24)$$

$$\theta_m - \Omega_m = -\frac{1}{2}\eta + g(\zeta). \qquad (9.8.25)$$

Thus if we move along the axis (or parallel to it, $\eta = \text{constant}$) by (9.8.24), $\theta_m + \Omega_m$ increase by half the amount ζ increases, and similarly $\theta_m - \Omega_m$ changes along a line $\zeta = \text{constant}$. The solution at any point (ξ, τ) can thus be traced back to the origin $\tau = 0$ where both θ_m and Ω_m are zero.

9.9. LINEAR WAVES IN A STRATIFIED FLUID

Suppose that a stratified fluid, otherwise stationary, is being traversed by small amplitude internal waves. We shall assume that the perturbations of the velocity, pressure and density fields are all small enough so that we can neglect the non-linear terms in the equations of motion. With this assumption, the linear conservation equations, (§2.11), assuming an incompressible inviscid, non-diffusive fluid, become

$$\rho_0 \frac{\partial v_1}{\partial t} = -\frac{\partial p}{\partial x_1}, \tag{9.9.1}$$

$$\rho_0 \frac{\partial v_3}{\partial t} = -\frac{\partial p}{\partial x_3} - \rho g, \tag{9.9.2}$$

$$\frac{\partial \rho}{\partial t} + v_1 \frac{\partial \rho}{\partial x_1} + v_3 \frac{\partial \rho}{\partial x_3} = 0, \tag{9.9.3}$$

$$v_2 = 0, \tag{9.9.4}$$

$$\frac{\partial v_1}{\partial x_1} + \frac{\partial v_3}{\partial x_3} = 0. \tag{9.9.5}$$

To clarify the role of the density field it is helpful to decompose the density into three components and to separate out the hydrostatic pressure field from the dynamic pressure variations, such that,

$$\rho = \rho_0 + \rho_e(x_3) + \rho'(x_1, x_3, t), \tag{9.9.6}$$

and

$$p = p_e(x_3) + p'(x_1, x_3, t), \tag{9.9.7}$$

where

$$\frac{\partial p_e}{\partial x_3} = -(\rho_0 + \rho_e(x_3))g, \tag{9.9.8}$$

and ρ' and p' are the fluctuations due to the internal wave motion, ρ_0 is the background density, $\rho_e(x_3)$ is the vertical variation of density in the absence of motion and p_e is the hydrostatic pressure due to the background stratification.

Substituting (9.9.6)–(9.9.8) into (9.9.1)–(9.9.5) and neglecting non-linear product terms leads to the equations for the velocity, pressure and density perturbations:

$$\frac{\partial v_1}{\partial x_1} + \frac{\partial v_3}{\partial x_3} = 0, \tag{9.9.9}$$

$$\rho_0 \frac{\partial v_1}{\partial t} = -\frac{\partial p'}{\partial x_1}, \tag{9.9.10}$$

$$\rho_0 \frac{\partial v_3}{\partial t} = -\rho' g - \frac{\partial p'}{\partial x_3}, \tag{9.9.11}$$

$$\frac{\partial \rho'}{\partial t} - \frac{v_3 \rho_0}{g} N^2(x_3) = 0, \tag{9.9.12}$$

where

$$N^2 = -\frac{g}{\rho_0} \frac{\partial \rho_0}{\partial x_3}, \tag{9.9.13}$$

is the buoyancy frequency squared.

Eliminating ρ', p' and v_1 leads to an equation for the vertical velocity v_3:

$$\frac{\partial^2}{\partial t^2}\left\{\frac{\partial^2 v_3}{\partial x_3^2} + \frac{\partial^2 v_3}{\partial x_1^2}\right\} + N^2(x_3)\frac{\partial^2 v_3}{\partial x_1^2} = 0. \tag{9.9.14}$$

To illustrate simple linear wave motions let us assume that:

$$N^2(x_3) = N^2 = \text{constant}, \tag{9.9.15}$$

then (9.9.14) admits the solution:

$$v_3 = A\cos(k_1 x_1 + k_3 x_3 - \omega t), \tag{9.9.16}$$

provided:

$$\omega = \frac{k_1}{(k_1^2 + k_3^2)^{1/2}} N = N\cos\theta. \tag{9.9.17}$$

Equation (9.9.16) represents a wave traveling in the positive x_1 and x_3 directions with phase lines (crest or troughs) given by:

$$k_1 x_1 + k_3 x_3 - \omega t = \phi, \tag{9.9.18}$$

where ϕ is the phase. This is shown in Fig. 9.9.1. The parameters k_1 and k_3 are called the horizontal and vertical wave numbers, connected to the horizontal (λ_1) and vertical (λ_3) wavelengths through the relations:

$$k_1 = \frac{2\pi}{\lambda_1}, \tag{9.9.19}$$

$$k_3 = \frac{2\pi}{\lambda_3}. \tag{9.9.20}$$

If there is also motion in the x_2 direction, then we also have a wave number k_2 and a relationship:

$$k_2 = \frac{2\pi}{\lambda_2}. \tag{9.9.21}$$

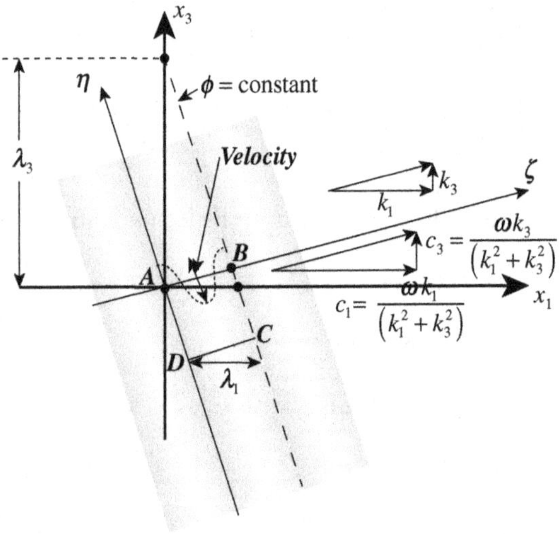

Figure 9.9.1 Schematic showing the directions of the wave fronts and the various wave properties defined in the text.

The phase velocity $\{c_i\}$ of the wave (9.9.16) is the speed with which a phase front moves and is directed perpendicular to the wave fronts:

$$\{c_i\} = \frac{\omega}{\left(k_1^2 + k_3^2\right)^{1/2}} \left\{ \frac{k_1}{\left(k_1^2 + k_3^2\right)^{1/2}} \underset{\sim}{\hat{i}_1} + \frac{k_3}{\left(k_1^2 + k_3^2\right)^{1/2}} \underset{\sim}{\hat{i}_3} \right\}. \tag{9.9.22}$$

where $\hat{i}_j$ are the unit vectors in the jth direction.

The solution (9.9.16) may be substituted into (9.9.5) to show:

$$k_1 v_1 + k_3 v_3 = 0. \tag{9.9.23}$$

This shows that the velocity vector $\{v_i\}$ is perpendicular to the wave number $\{k_i\}$. Internal waves in a stratified fluid are thus translational, with the velocity of the fluid parallel to the phase lines, $\phi = \text{constant}$ (Fig. 9.9.1).

It is convenient to introduce local coordinates:

$$\zeta = \frac{k_1}{k} x_1 + \frac{k_3}{k} x_3, \tag{9.9.25}$$

$$\eta = \frac{k_1}{k} x_1 + \frac{-k_3}{k} x_3, \tag{9.9.26}$$

where k is the magnitude of the wave number. In terms of these new variables the solution for the velocity may be written:

$$v_\eta = B \cos(k\zeta - \omega t), \tag{9.9.27}$$

$$v_\zeta = 0, \tag{9.9.28}$$

where $B = \dfrac{k}{k_1} A$ provides the link with the solution (9.9.16) and v_η is the velocity component in the η direction, and v_ζ is the velocity component in the ζ direction.

From (9.9.10) it follows that:

$$p' = \frac{\rho_0 B k_3 \omega}{k_1 k} \cos(k\zeta - \omega t), \tag{9.9.29}$$

and from (9.9.12) we see that:

$$\rho' = \frac{-\rho_0 N^2 B k_1}{g k \omega} \sin(k\zeta - \omega t). \tag{9.9.30}$$

The vertical displacement of the isopycnals ξ_3 may be obtained from the equality:

$$\frac{\partial \xi_3}{\partial t} = v_3, \tag{9.9.31}$$

yielding an expression for the vertical particle displacement given by:

$$\xi_3 = -\frac{Bk_1}{\omega k}\sin(k\zeta - \omega t). \tag{9.9.32}$$

Substituting (9.9.32) into (9.9.30) relates the density perturbation to the vertical displacement:

$$\rho' = \rho_0 \frac{N^2}{g}\xi_3. \tag{9.9.33}$$

The energetics of a control volume ABCD (Fig. 9.9.1) may now be used to introduce the important concept of group velocity. The width AB of the control volume is taken as one wavelength λ, and the length AD is unity.

The rate of working of the pressure field on the surface AB is given by:

$$\dot{W} = \int_0^\lambda v_\eta p' \, d\zeta. \tag{9.9.34}$$

Substituting for v_η from (9.9.27) and p' from (9.9.29) and carrying out the integration yields:

$$\dot{W} = \rho_0 \frac{\pi B^2 k_3 \omega}{k_1 k^2}. \tag{9.9.35}$$

By comparison, the total kinetic and potential energy within the control volume may be calculated from the expression:

$$KE + PE = \int_0^\lambda \left(\frac{1}{2}\rho_0 v_\eta^2 + \frac{1}{2}\rho' \xi_3 g\right) d\zeta. \tag{9.9.36}$$

Substituting from (9.9.27), (9.9.30) and (9.9.32) and carrying out the integration leads to the expression:

$$KE + PE = \frac{B^2 \rho_0 \pi}{k}, \tag{9.9.37}$$

where the kinetic energy contribution was exactly equal to the potential energy contribution, as in all simple translational wave systems.

Hence the rate of pressure working "fills" the control volume ABCD with potential and kinetic energy at a rate defined by a velocity:

$$c_g = \frac{\dot{W}}{KE + PE} = \frac{k_3 N}{k^2}, \tag{9.9.38}$$

a velocity in the direction of the wave phase lines perpendicular to the phase and fluid velocity vector. Given that $v_\varsigma = 0$, no work is done in this direction.

The velocity c_g is the velocity with which the pressure rate of working transmits kinetic and potential energy through the fluid; it is thus the velocity with which groups of waves travel and is therefore called the "group velocity". Given the direction of c_g we may write:

$$\{c_{gi}\} = \left\{\frac{k_3^2 N}{k^3}, -\frac{k_3 k_1 N}{k^3}\right\}, \tag{9.9.39}$$

as the pressure rate of working $\dot{W}$ is positive on the upward pointing surface AB (Fig. 9.9.1).

It is simple to show from the dispersion relationship (9.9.17) that:

$$c_{g1} = \frac{\partial \omega}{\partial k_1}, \tag{9.9.40}$$

$$c_{g3} = \frac{\partial \omega}{\partial k_3}. \tag{9.9.41}$$

9.10. INTERNAL WAVES AND INTERNAL MODES

In §9.9, we saw how the equations of motion admit a solution that represents a set of internal waves for the case when the buoyancy frequency N was a constant with height in the fluid. These waves represent phase planes that move perpendicular to the phase plane and have velocities that are in the direction of the phase planes. Given that equation (9.9.14) is linear, two solutions of the form (9.9.16) when combined will also be a solution:

$$v_3 = \frac{A}{2}\{\cos(k_1 x_1 + k_3 x_3 - \omega t) - \cos(k_1 x_1 - k_3 x_3 - \omega t)\}, \tag{9.10.1}$$

is also a solution of (9.9.14). Simple trigonometry allows us to rewrite (9.10.1) in the form:

$$v_3 = A \sin(k_1 x_1 - \omega t) \sin k_3 x_3. \tag{9.10.2}$$

This solution has the property that at $x_3 = \frac{\pi}{k_3}x$ the velocity $v_3 = 0$, representing an impermeable boundary. At the free surface, from §8.2:

$$g\frac{\partial \psi_3}{\partial x_3} + \frac{\partial^2 \psi_3}{\partial t^2} = 0. \tag{9.10.3}$$

However since g is large, it is sufficient to require

$$x_3 = h: \quad v_3 = 0, \tag{9.10.4}$$

so that provided the vertical wave number satisfies:

$$k_3 = \frac{n\pi}{h}, \tag{9.10.5}$$

where n is an arbitrary integer (9.10.2) is a solution for internal waves in a duct.

From conservation of volume (9.9.9) and (9.10.2) it follows that:

$$v_1 = -\frac{An\pi}{hk_1}\cos\frac{n\pi}{h}x_3 \cos(k_1x_1 - \omega t), \tag{9.10.6}$$

leading to the ratio

$$\frac{v_3}{v_1} = -\frac{hk_1}{n\pi}\tan\frac{n\pi x_3}{h}\tan(k_1x_1 - \omega t), \tag{9.10.7}$$

illustrating that the solutions (9.10.2) and (9.10.6) represent oscillating elliptical cells propagating in the x_1 direction as shown in Fig. 9.10.1.

The dispersion relationship (9.9.17) becomes

$$\omega = \frac{Nk_1}{\left(k_1^2 + \frac{n^2\pi^2}{h^2}\right)^{1/2}}, \tag{9.10.8}$$

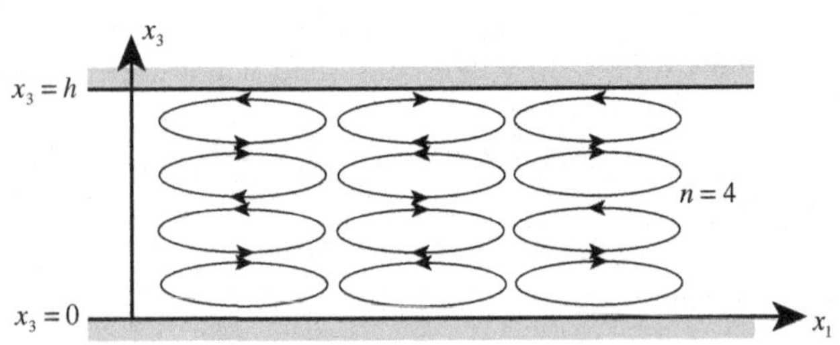

Figure 9.10.1 Propagating internal wave modes in a duct.

showing that the elongated cells (n large) have a small frequency, and thus a small phase speed along x_1 axis; the smaller the vertical scale (the larger n), the slower the cells move along the x_1 axis.

Two limiting cases arise when N^2 is not constant, but rather has a distribution typical of the stratification of a lake in mid-summer in a Mediterranean climate (see Fig. 9.10.2).

We seek a solution of the form of a propagating wave:

$$v_3 = A(x_3)\sin(k_1 x_1 - \omega t). \tag{9.10.9}$$

Substituting (9.10.10) into the equation of motion (9.9.14) leads to:

$$\frac{d^2A}{dx_3^2} - \left(\frac{N^2(x_3) - \omega^2}{\omega^2}\right)k_1^2 A = 0. \tag{9.10.10}$$

From (9.10.11) is clear that, when $N^2 \ll \omega^2$, then (9.10.11) reduces to:

$$\frac{d^2A}{dx_3^2} - k_1^2 A = 0, \tag{9.10.11}$$

showing that in regions where $N^2 \ll \omega^2$ the solution for A is in the form of an exponential

$$A \sim e^{-k_1(x_3 - h_2 - h_3)}, \tag{9.10.12}$$

above the upper $N^2 = \omega^2$ interface.

It is thus clear that the interfaces $N^2 = \omega^2$ have large vertical motions when the first few modes are excited, but the displacement progressively decreases with increasing vertical modal number; the first two modes are analogous to the two internal modes in a three-layer system discussed in §9.2.

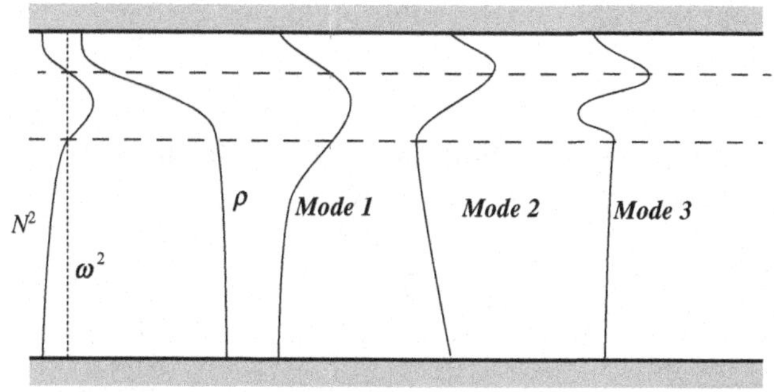

Figure 9.10.2 Internal wave vertical modal structure for a variable density profile.

On the other hand, higher modes have a decreasing vertical scale and quickly reach the stage where N is locally approximately constant over the vertical wavelength and the case where N^2 equal to a constant applies locally:

$$k_1 \sim \frac{n\pi\omega_0}{hN}, \tag{9.10.13}$$

indicating that, for a particular frequency ω_0, the horizontal wave number (wavelength) increases (decreases) with increasing (decreasing) wave modal number n and decreases (increases) with increasing maximum buoyancy frequency. Thus high vertical modal number waves have a longer horizontal wavelength, the sharper the density gradient for a particular frequency ω_0.

In §7.1, we used the gradient Richardson Number defined by:

$$Ri = \frac{N^2(x_3)}{\left(\frac{\partial v_1}{\partial x_3}\right)^2}, \tag{9.10.14}$$

to characterize the turbulence in the water column; of particular interest is the minimum value of this Richardson Number.

The Richardson Number resulting from internal waves thus follows directly by using (9.10.9) to evaluate (9.10.14):

$$Ri = \left[\frac{\omega^2}{(N^2 - \omega^2)}\right] \frac{N^2}{k_1^2 v_3^2(x_1, t')}, \tag{9.10.15}$$

Now N is maximum within the layer where $N^2 > \omega^2$ and if a_n is the amplitude of the vertical displacement associated with the nth vertical mode then:

$$v_3 \sim a_n\omega, \tag{9.10.16}$$

so that to a good approximation the minimum Richardson Number Ri_m associated with the nth mode is given by:

$$Ri_m = \frac{\omega^2}{N_m^2 k_1^2 a_n^2}, \tag{9.10.17}$$

where N_m is the maximum buoyancy frequency.

Internal waves can exist, with no rotation, for frequencies from zero to N_{max}, so that for a particular vertical displacement a_n the limit for Ri_m can be obtained from (9.10.17) for higher modes:

$$Ri_m \sim \frac{h^2}{n^2\pi^2 a_n^2} \sim \frac{\delta_n^2}{a_n^2}, \tag{9.10.18}$$

where δ_n is the vertical scale of the internal wave mode. This is independent of the maximum buoyancy frequency.

The minimum Ri_m for the first mode is more difficult to obtain because for the general case we do not know the value of k_1. However, a rough approximation can be obtained by assuming N is constant over h_2. With this assumption:

$$Ri_m \sim \frac{h_2^2}{a_n^2}, \tag{9.10.19}$$

similar in form to (9.10.18). These results are central to discussions on turbulent mixing in stratified lakes, estuaries and oceans.

9.11. INTERNAL WAVES IN A DOMAIN AT VARIABLE DEPTH

We shall now consider the propagation of an internal wave mode in a fluid of constant buoyancy frequency N, impinging on a section of the domain where the depth has small perturbations. This example is the counterpart to §8.3 where we discussed reflection of surface waves and illustrates many important features of internal waves such as a wave transmission, wave reflection and energy transfers to other modes. Suppose that the depth of domain containing a linearly stratified fluid is given by:

$$x_{3b} = h_b(x_1) = -h_0 + \varepsilon h_0 f(x_1), \tag{9.11.1}$$

where, as shown in Fig. 8.3.1, h_0 is the mean depth $f(x_1)$ is a function describing the depth variations and ε is a small parameter. It is assumed that at $x_1 = \pm\infty$, $f(x_1)$ is zero and the problem under consideration is that of a single mode, at a certain frequency ω, propagating in from $x_1 = -\infty$ so that from §9.10, the streamfunction ψ_i (x_1, x_3, t):

$$\psi_i(x_1, x_2, t) \sim a_i \sin(k_{1i}x_1 - \omega t)\sin k_3(x_3 - 1), \tag{9.11.2}$$

and

$$\omega = \frac{Nk_{1i}}{\left(k_{1i}^2 + k_{3i}^2\right)^{1/2}}; \quad k_{3i} = \frac{n_i^2 \pi^2}{h_0^2}, \tag{9.11.3}$$

where the subscript i designates the "incident" wave. The equation satisfied, generally by the solution streamfunction $\psi(x_1, x_3, t)$, for waves with a small amplitude, follows directly from §9.9:

$$\frac{\partial^2}{\partial t^2}\left(\frac{\partial^2 \psi}{\partial x_1^2} + \frac{\partial^2 \psi}{\partial x_3^2}\right) + N^2 \frac{\partial^2 \psi}{\partial x_1^2} = 0. \tag{9.11.4}$$

For simplicity of the calculation, we shall assume that the solution is periodic with frequency ω and so we seek a solution of the form:

$$\psi(x_1, x_3, t) = \phi(x_1, x_3) e^{i\omega t}, \tag{9.11.5}$$

where, in order to meet the radiation condition (9.11.2), we shall take the imaginary part of (9.11.5).

Substituting (9.11.5) into (9.11.4) yields:

$$\left(\frac{N^2}{\omega^2} - 1\right) \frac{\partial^2 \phi}{\partial x_1^2} - \frac{\partial^2 \phi}{\partial_3^2} = 0. \tag{9.11.6}$$

In §9.10, we saw that the boundary condition at the free surface $x_3 = 0$ may be approximated by a solid boundary and so,

$$\phi = 0 \tag{9.11.7}$$

and at $x_{3b} - h_0 + \varepsilon h_0 f(x_1)$ we also have

$$\phi = 0, \tag{9.11.8}$$

as we wish to have the bottom as a streamline.

Before proceeding, we non-dimensionalize the problem by introducing the non-dimensional coordinates:

$$x_1^* = \frac{x_1}{h_0}, \tag{9.11.9}$$

$$x_3^* = \frac{x_3}{h_0}, \tag{9.11.10}$$

$$\phi^* = \frac{\phi}{a_i}. \tag{9.11.11}$$

Introducing these into (9.11.6), and dropping the asterisk, yields the problem:

$$0 < x_3 < 1: \quad \left(\frac{N^2}{\omega^2} - 1\right)\frac{\partial^2 \phi}{\partial x_1^2} - \frac{\partial^2 \phi}{\partial x_3^2} = 0, \tag{9.11.12}$$

$$x_3 = 1 \quad \phi = 0, \tag{9.11.13}$$

$$x_3 = -1 + \varepsilon f(x_1) \quad \phi = 0, \tag{9.11.14}$$

and

$$x_1 \to -\infty \quad \phi \sim e^{-ik_{i1}x_1} \sin k_{3i}(x_3 - 1). \tag{9.11.15}$$

In order to solve this problem we reduce (9.11.14) to $x_3 = -1$ by noting:

$$\begin{aligned}\phi(x_1, -1 + \varepsilon f(x_1)) &= \phi(x_1, -1) + \varepsilon f \frac{\partial \phi}{\partial x_3}(x_1, -1) + \frac{(\varepsilon f)^2}{2!}\frac{\partial^2 \phi}{\partial x_3^2}(x_1, -1) \\ &\quad + \cdots + \frac{(\varepsilon f)^j}{j!}\frac{\partial^j \varphi}{\partial x_3^j}(x_1, -1),\end{aligned} \tag{9.11.16}$$

given that ε is a small parameter we may seek a solution of the form:

$$\phi(x_1, x_3) = \phi^{(o)}(x_1, x_3) + \varepsilon \phi^{(1)}(x_1, x_3) + \cdots \tag{9.11.17}$$

Substituting (9.11.17) into (9.11.12) and making use of (9.11.16) leads to a set of equations, one for each order of ε:

$$\left(\frac{N^2}{\omega^2} - 1\right)\frac{\partial^2 \phi^{(i)}}{\partial x_1^2} - \frac{\partial^2 \phi^{(i)}}{\partial x_3^2} = 0, \tag{9.11.18}$$

$$\phi^{(i)}(x_1, 0) = 0, \tag{9.11.19}$$

$$\phi^{(o)}(x_1, -1) = 0, \tag{9.11.20}$$

$$\phi^{(1)}(x_1, -1) = -f(x_1)\frac{\partial \varphi^{(1)}}{\partial x_3} - \frac{f^2(x_1)}{2!}\frac{\partial^2 \psi^{(o)}}{\partial x_3^2}, \tag{9.11.21}$$

and in general

$$\phi^{(i)}(x_1, -1) = -g^{(i)}(x_1), \tag{9.11.22}$$

where

$$g^{(i)}(x_1) = \sum_{j=1}^{i} \frac{f^j(x_1)}{j!} \frac{\partial^j}{\partial x_3^j} \phi^{(i-j)}(x_1, -1). \qquad (9.11.23)$$

We have reduced the original problem in a domain of variable depth with zero flow across the bottom boundary, to a problem in a uniform depth domain with source and sinks distributed along the bottom boundary; the sources and sinks being of such a strength so as to induce a flow that, when coupled with the flow of the incident wave (9.11.2), annuls the normal velocity across the variable depth boundary to the order under consideration.

We shall solve (9.11.18)–(9.11.23) by taking the Fourier Transform with respect to x_1:

$$\Phi(k_1) = \frac{1}{\sqrt{2\pi}} \int_{-\infty}^{\infty} \phi(x_1, x_3) e^{-ikx_1} \mathrm{d}x_1, \qquad (9.11.25)$$

where we have adopted the general convention of writing the Fourier Transform of a lower case variable as the upper case variable.

In terms of the Fourier Integral (9.11.18)–(9.11.23) become:

$$\frac{\partial^2 \Phi^{(i)}}{\partial x_3}(k, x_3) + \left(\frac{N^2}{\omega^2 - 1}\right) k^2 \Phi^{(i)}(k, x_3) = 0, \qquad (9.11.26)$$

$$\Phi(k, 0) = 0, \qquad (9.11.27)$$

$$\Phi^{(i)}(k, -1) = -G^{(i)}(k), \qquad (9.11.28)$$

$$\Phi^{(i)} \sim \delta(k + k_{i1}) \sin k_{31}(x_3 - 1); \quad x_1 \to \infty, \qquad (9.11.29)$$

The solution to (9.11.26) that also satisfied (9.11.29) is given by:

$$\Phi^{(i)}(k, x_3) = A^{(i)}(k) \sin\left\{\left(\frac{N^2}{\omega^2} - 1\right)^{1/2} k(x_3 - 1)\right\}. \qquad (9.11.30)$$

Now the boundary condition at $x_3 = -1$, (9.11.28), requires:

$$A^{(i)} \sin\left\{\left(\frac{N^2}{\omega^2} - 1\right)^{1/2} k\right\} = G^{(i)}(k). \qquad (9.11.31)$$

If we note that:

$$\sin\left(\frac{N^2}{\omega^2 - 1}\right)^{1/2} k = 0, \tag{9.11.32}$$

whenever

$$k = k_n : \quad k_n = \frac{n\pi}{\left(\frac{N^2}{\omega^2} - 1\right)^{1/2}}, \tag{9.11.33}$$

then (9.11.33) may be used to find $A^{(i)}$:

$$A^{(i)}(k) = \frac{G^{(i)}(k)}{\sin\left(\frac{N^2}{\omega^2} - 1\right)^{1/2} k} + \sum_{n=1}^{\infty} C_{1n}^{(i)}\delta(k - k_n) + C_{2n}^{(i)}\delta(k + k_n), \tag{9.11.34}$$

where $C_{1n}^{(i)}$ and $C_{2n}^{(i)}$ are arbitrary constants and $\delta(k \pm k_n)$ are delta functions. This result follows immediately by noting that the delta function has the property that:

$$k\delta(k) = 0. \tag{9.11.35}$$

Substituting (9.11.34) into (9.11.30) leads to the solution for the Fourier Transform of the solution:

$$\Phi^{(i)}(k, x_3) = \frac{G^{(i)}(k)\sin\left\{\left(\frac{N^2}{\omega^2 - 1}\right)^{1/2} k\,(x_3 - 1)\right\}}{\sin\left(\frac{N^2}{\omega^2} - 1\right)^{1/2} k}$$

$$+ \sum_{n=1}^{n=\infty} C_{1n}^{(i)}\sin\left\{\left(\frac{N^2}{\omega^2} - 1\right)^{1/2} k_n(x_1 - 1)\right\}\delta(k - k_n)$$

$$- \sum_{n=1}^{n=\infty} C_{2n}^{(i)}\sin\left\{\left(\frac{N^2}{\omega^2} - 1\right)^{1/2} k_n(x_3 - 1)\right\}\delta(k + k_n), \tag{9.11.36}$$

$$\Phi^{(i)}(k, x_3) = \sum_{n=1}^{\infty} \sin\left\{\left(\frac{N^2}{\omega^2} - 1\right)^{1/2} k_n(x_3 - 1)\right\}$$

$$\times \left\{ \frac{(-1)^n}{\left(\frac{N^2}{\omega^2} - 1\right)^{1/2}} \left(\frac{G^{(i)}(k_n)}{k - k_n} - \frac{G^{(i)}(-k_n)}{k + k_n} \right) \right.$$

$$\left. + C_{1n}^{(i)} \delta(k - k_n) - C_{2n}^{(i)} \delta(k + k_n) \right\} + \Phi_a^{(i)}(k, x_3),$$

(9.11.37)

where $\Phi_a^{(i)}(k, x_3)$ is an analytic function of k. From the theory of Fourier Transform it follows that $\Phi_a^{(i)}(k, x_3)$ does not contribute to the inverse at $x_1 \to \pm\infty$ and so in order to obtain an expression for the streamfunction away from the depth variations it is sufficient to take the inverse of the singularities in (9.11.37).

Following the discussion in §8.3, the general inverse of (9.11.37) becomes

$$\psi^{(i)}(x_1 x_3)_{x_1 \to \pm\infty} \sim \sum_{n=1}^{\infty} \frac{\sin\left\{\left(\frac{N^2}{\omega^2} - 1\right)^{1/2} k_n(x_3 - 1)\right\}}{\left(\frac{N^2}{\omega^2} - 1\right)^{1/2}} \tag{9.11.38}$$

$$\times \left\{ G^{(i)}(k_n) e^{-ik_n x_1} H(x_1) + G^{(i)}(-k_n) e^{ik_n x_1} H(-x_1) \right\},$$

where we have normalized the solution. Each one of the components of (9.11.38) represents a reflected wave component at $x_1 \to -\infty$ and a transmitted wave component of $x_1 \to +\infty$. Thus an incident wave gives rise to all admissible internal wave modes, with incident wave frequency, on impinging on a section of the domain where the depth varies.

We shall consider two examples to illustrate the usefulness of (9.11.38). First, suppose that the incident wave encounters a smooth mound of characteristic length of the form:

$$f(x) = e^{-\frac{x_1}{\ell^2}}. \tag{9.11.39}$$

Now

$$\frac{\partial\phi^{(0)}}{\partial x_3}(x_1, -1) = k_{3i}\, e^{-ik_{1i}x_1}, \tag{9.11.40}$$

so that

$$g^{(1)}(x_1) = -k_{3i}\, e^{-ik_{1i}x_1}\, e^{-\frac{x_1^2}{\ell^2}}, \tag{9.11.41}$$

and

$$G^{(1)}(k) = -k_{3i}\frac{\ell}{\sqrt{2}}\, e^{-\ell^2(k+k_{1i})^2}, \tag{9.11.42}$$

so that

$$\psi^{(1)}(x_1, x_3, t) \sim \sum_{n=1}^{\infty} \frac{-k_{3i}\ell e^{-\ell^2(k_n+k_{1i})^2}}{\sqrt{2}\left(\frac{N^2}{\omega^2}-1\right)^{1/2}}\, e^{i(k_n x_1+\omega t)} \times \sin\left\{\left(\frac{N^2}{\omega^2}-1\right)^{1/2} k_n(x_3-1)\right\}, \tag{9.11.43}$$

as $x_1 \to \infty$.

This represents a series of reflected waves the amplitudes of which are of order $e^{-\ell^2(k_n+k_{1i})^2}$, indicating that the longer the incident wavelength is relative to the obstacle length $\ell(\ell k_{1i} < 1)$ the larger will be the reflected wave amplitude. By contrast if the incident wavelength is short compared to $\ell(\ell k_{1i} > 1)$ then there will be negligible reflection, this is called weak reflection. The transition between strong and weak reflection; the transition between strong and weak reflection; the transition between strong and weak reflection is very abrupt.

The second example illustrates how multiple reflections may either add destructively or constructively depending on whether the spacing of the reflections leads to destructive or constructive addition of wave energy. To illustrate this consider a rectangular bottom hump given by:

$$f(x_1) = H(x_1 - \ell) - H(x_1 + \ell), \tag{9.11.44}$$

where $H(x_1)$ is the Heavyside step function. Combining (9.11.44) with (9.11.38) yields

$$g^{(1)}(x_1) = -k_{3i}\, e^{-ik_{1i}x_1}\{H(x_1 - \ell) - H(x_1 + \ell)\}, \tag{9.11.45}$$

so that

$$G^{(1)}(k_n) = \frac{-k_{31}}{\sqrt{2\pi}} \int_{-\ell}^{\ell} e^{-i(k_{1i}+k)x_1}\, dx_1$$
$$= -\left(\frac{2}{\pi}\right)^{1/2} \frac{k_{3i}\sin(k_{1i}+k_n)\ell}{(k_{1i}+k_n)}, \tag{9.11.46}$$

from which it follows immediately that as $x_1 \to -\infty$

$$\psi^{(1)}(x_1, x_3, t) \sim \sum_{n=1}^{\infty} -\left(\frac{2}{\pi}\right)^{1/2} k_{3i} \frac{\sin(k_{1i}+k_n)\ell}{(k_{1i}+k_n)\left(\frac{N^2}{\omega^2}-1\right)^{1/2}} e^{i(k_n x_1+\omega t)}$$
$$\times \sin\left\{\left(\frac{N^2}{\omega^2}-1\right)^{1/2} k_n(x_3-1)\right\}. \tag{9.11.47}$$

The sin $(k_{1i}+k_n)\ell$ dependence shows the interplay between reflection from the front of the obstacle ($x_1 = -\ell$) and equal reflection, but opposite sign, from the rear of the obstacle.

9.12. REFLECTION FROM A SLOPING WALL

In the previous section (§9.11) we showed how variations in the depth of a channel containing a linearly stratified fluid trigger a spectrum of reflected and transmitted internal modes in response to an incident internal mode impinging on the perturbations. We now investigate a single internal wave front, propagating along the group velocity ray, as shown in Fig. 9.12.1, impinging onto a sloping solid boundary and show that this gives rise to a single reflected wave train.

From §9.9, we may write the x_1 and x_3 velocity components (repeated here for convenience) and the streamfunction in the form:

$$v_1 = -B\sin\theta\cos(k_1x_1 + k_3x_3 - \omega t), \tag{9.12.1}$$

$$v_3 = B\cos\theta\cos(k_1x_1 + k_3x_3 - \omega t), \tag{9.12.2}$$

$$\psi = Bk^{-1}\cos(k_1x_1 + k_3x_3 - \omega t). \tag{9.12.3}$$

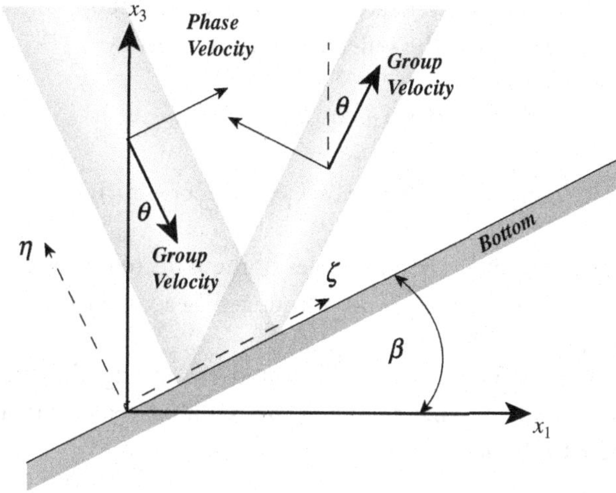

Figure 9.12.1 Schematic of internal wave reflection from a sloping wall in an infinite domain.

This represents a wave train with a phase velocity moving in the positive x_1 and x_3 direction (Fig. 9.12.1) and a group velocity given by:

$$c_{g1} = \frac{k_3 N \sin\theta}{k^2}, \tag{9.12.4}$$

$$c_{g3} = -\frac{k_3 N \cos\theta}{k^2}. \tag{9.12.5}$$

Energy thus propagates toward the sloping wall where it is reflected forming a standing wave. The reflected wave will be such that the velocity field from the combination of the incident and reflected field is such as to make the velocity, normal to the sloping wall, zero at all times.

Now since the reflected wave will have the same frequency ω as the incident wave and the magnitude of the angle θ is fixed by the frequency and the buoyancy frequency through the dispersion relationship:

$$\cos\theta = \frac{\omega}{N}, \tag{9.12.6}$$

so that the reflected wave will have a phase angle of $-\theta$ as shown in Fig. 9.12.1 and the group velocity (9.12.5) will be directed away from the sloping wall (energy being reflected) so that k_3' will be of opposite sign to k_3. We may thus assume that the reflected wave is given by:

$$v_1' = -B' \sin\theta \cos(k_1' x_1 - k_3' x_3 - \omega t), \tag{9.12.7}$$

$$v_3' = -B' \cos\theta \cos(k_1' x_1 - k_3' x_3 - \omega t), \tag{9.12.8}$$

where the primes donate the properties of the reflected wave train and the minus sign, in both v_1' and v_3', was introduced to account for the expected 180° phase shift on reflection.

The normal velocity v_η may be found by adding the incident and reflected wave velocities and resolving the resulting velocity in a direction perpendicular to the sloping wall:

$$v_\eta = -(v_1 + v_1')\sin\beta + (v_3 + v_3')\cos\beta, \tag{9.12.9}$$

where η is the coordinate normal to the slope and β is the slope of the bottom to the horizontal (Fig. 9.12.1). Substituting from (9.12.1), (9.12.2), (9.12.7) and (9.12.8) leads to the expression:

$$v_\eta = B\cos\phi \cos(\theta - \beta) - B' \cos\phi' \cos(\theta + \beta), \tag{9.12.10}$$

where

$$\phi = k_1 x_1 + k_3 x_3 - \omega t, \tag{9.12.11}$$

and

$$\phi' = k_1' x_1 - k_3' x_3 - \omega t. \tag{9.12.12}$$

Equation (9.12.10) represents the velocity normal to the wall and this must represent a standing wave with zero normal velocity at the wall, $\eta = 0$. This can only be achieved if:

$$B\cos(\theta - \beta) = B'\cos(\theta + \beta), \tag{9.12.13}$$

then

$$v_\eta = -2B\cos(\theta - \beta)\sin\left(\frac{\phi + \phi'}{2}\right)\sin\left(\frac{\phi - \phi'}{2}\right). \tag{9.12.14}$$

From (9.12.11) and (9.12.12) we may write:

$$\frac{\phi+\phi'}{2} = \left(\frac{k_1+k'}{2}\right)x_1 + \left(\frac{k_3-k_3'}{2}\right)x_3 - \omega t, \tag{9.12.15}$$

$$\frac{\phi-\phi'}{2} = \left(\frac{k_1-k_1'}{2}\right)x_1 + \left(\frac{k_3+k_3'}{2}\right)x_3. \tag{9.12.16}$$

Now let us introduce a change of coordinates from (x_1, x_3) to (η,ζ) as shown in Fig. 9.12.1 so that:

$$x_1 = \zeta\cos\beta - \eta\sin\beta, \tag{9.12.17}$$

$$x_3 = \zeta\sin\beta + \eta\cos\beta. \tag{9.12.18}$$

Substituting (9.12.17) and (9.12.18) into (9.12.15) and (9.12.16) yields:

$$\frac{\phi+\phi'}{2} = \left(\left(\frac{k_1+k_1'}{2}\right)\cos\beta + \left(\frac{k_3+k_3'}{2}\right)\sin\beta\right)\zeta - \left(\left(\frac{k_1+k_1'}{2}\right)\sin\beta - \left(\frac{k_3+k_3'}{2}\right)\cos\beta\right)\eta - \omega t, \tag{9.12.19}$$

$$\frac{\phi-\phi'}{2} = \left(\left(\frac{k_1-k_1'}{2}\right)\cos\beta + \left(\frac{k_3+k_3'}{2}\right)\sin\beta\right)\zeta - \left(\left(\frac{k_1-k_1'}{2}\right)\sin\beta + \left(\frac{k_3+k_3'}{2}\right)\cos\beta\right)\eta. \tag{9.12.20}$$

Now for (9.12.14) to be a standing wave $\left(\frac{\phi+\phi'}{2}\right)$ must be independent of ζ and η and dependent only on time and $\left(\frac{\phi+\phi'}{2}\right)$ must be independent of ζ. From this it follows that:

$$\left(k_1+k_1'\right)\cos\beta + \left(k_3+k_3'\right)\sin\beta = 0, \tag{9.12.21}$$

$$\left(k_1+k_1'\right)\sin\beta - \left(k_3+k_3'\right)\cos\beta = 0, \tag{9.12.22}$$

$$\left(k_1-k_1'\right)\cos\beta + \left(k_3+k_3'\right)\sin\beta = 0. \tag{9.12.23}$$

Now from the definition of θ it follows that:

$$\cos\theta = \frac{k_1}{k} = \frac{k_1'}{k'}, \qquad (9.12.24)$$

$$\sin\theta = \frac{k_3}{k} = -\frac{k_3'}{k'}. \qquad (9.12.25)$$

Substituting (9.12.24) and (9.12.25) into (9.12.21)–(9.12.23) and using some trigonometric identities:

$$k\cos(\theta+\beta) + k'\cos(\theta-\beta) = 0, \qquad (9.12.26)$$

$$k\sin(\beta-\theta) + k'\sin(\beta+\theta) = 0, \qquad (9.12.27)$$

$$k\cos(\theta-\beta) - k'\cos(\theta+\beta) = 0, \qquad (9.12.28)$$

thus

$$k_1' = k\frac{\cos(\beta-\theta)}{\cos(\theta+\beta)}\cos\theta, \qquad (9.12.29)$$

$$k_3' = k\frac{\cos(\beta-\theta)}{\cos(\theta+\beta)}\sin\theta. \qquad (9.12.30)$$

Now the energy flux, (9.9.38) associated with the reflected wave:

$$\dot{W}' = \frac{\rho_0\pi B'^2 k_3'\omega}{k_1' k'^2} = \frac{\rho_0\pi B^2}{k^2}\frac{k_3\omega}{k_1} = \dot{W}, \qquad (9.12.31)$$

which confirms that all the incident energy is reflected so that the energy reflection coefficient is unity.

By contrast the energy density, the energy flux per unit wavelength, of the reflected wave becomes

$$\left(\frac{KE+PE}{\lambda}\right)' = \frac{1}{2}\rho_0 B^2\frac{\cos^2(\beta-\theta)}{\cos^2(\beta+\theta)}. \qquad (9.12.33)$$

From this it is clear that when

$$\beta+\theta = \frac{\pi}{2}, \qquad (9.12.34)$$

the energy density in the reflected wave becomes infinite. This orientation of the incoming incident wave is referred to as the "critical" orientation.

$$\theta = \left(\frac{\pi}{2} - \beta\right). \tag{9.12.35}$$

For such incident waves, the reflected wave has an infinitesimal wavelength and the phase lines are parallel to the bottom slope.

REFERENCE

Rossby, C.-G., 1936. Dynamics of steady ocean currents in the light of experimental fluid dynamics. Pap. Phys. Oceanog. Meteor. 5, 1–43.

APPENDIX 1

Mathematical Preliminaries

The purpose of this appendix is to summarize the mathematical notation and most frequently used formulas used in the book. There is no attempt at completeness, this appendix is simply a convenient reference for the user of this book. In a small number of sections in the book, more advanced mathematical methods are used and for these references are given, the main ones being the use of generalized functions in §5 and §9.

Coordinates: Throughout the book use is made of the simple Cartesian coordinate system $\{x_1, x_2, x_3\}$. Many readers will be more familiar with the coordinate system $\{x, y, z\}$, but we prefer the former as it allows the use of the summation convention, that greatly simplifies the manipulation of the equations of motion.

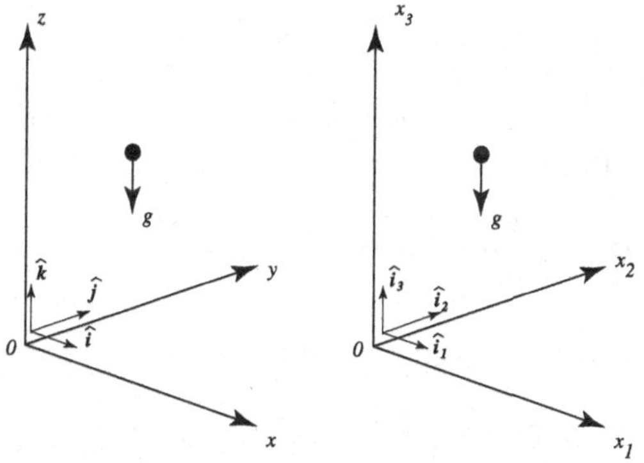

A vector a may be written as:

$$\boldsymbol{a} = (a_x, a_y, a_z) = (a_1, a_2, a_3) = a_1\hat{\boldsymbol{i}}_1 + a_2\hat{\boldsymbol{i}}_2 + a_3\hat{\boldsymbol{i}}_3 = a_k\hat{\boldsymbol{i}}_k. \quad \text{(A.1)}$$

The magnitude of a vector $\boldsymbol{a}$ is given by

$$a = (a_1^2 + a_2^2 + a_3^2)^{1/2} \quad \text{(A.2)}$$

Summation Notation

When a sub- or superscript is repeated we sum the quantities:

$$\sum_{i=1}^{3} a_i b_i = a_i b_i. \quad \text{(A.3)}$$

Differentiation Notation

$$()_{,k} = \frac{\partial}{\partial x_k} \qquad (A.4)$$

Permutation Symbols

$$\begin{aligned} \text{Let } e_{ijk} &= 1 \text{ if (ijk) is and even permutation of (123)} \\ &= -1 \text{ if (ijk) is and odd permutation of (123)} \\ &= 0 \text{ otherwise.} \end{aligned} \qquad (A.5)$$

$$\begin{aligned} \delta_{ij} &= 0 \quad i \neq j \\ &= 1 \quad i = j \end{aligned} \qquad (A.6)$$

This is called the Kronecker delta

Examples Of Notation Usage

(i) Inner product of two vectors

$$\boldsymbol{a} \cdot \boldsymbol{b} = ab\cos\boldsymbol{\theta} = a_i b_i \qquad (A.7)$$

(ii) Gradient of a scalar: If $\boldsymbol{\phi} = \boldsymbol{\phi}(x_1, x_2, x_3)$ a scalar point function of (x_1,x_2,x_3) then,

$$\nabla\boldsymbol{\phi} = \frac{\partial\boldsymbol{\phi}}{\partial x_1}\hat{\boldsymbol{i}}_1 + \frac{\partial\boldsymbol{\phi}}{\partial x_2}\hat{\boldsymbol{i}}_2 + \frac{\partial\boldsymbol{\phi}}{\partial x_3}\hat{\boldsymbol{i}}_3 = \frac{\partial\boldsymbol{\phi}}{\partial x_k}\hat{\boldsymbol{i}}_k = \boldsymbol{\phi}_{,k}\hat{\boldsymbol{i}}_k \qquad (A.8)$$

(iii) Inner product of a vector and the gradient of a scalar:

$$c = \boldsymbol{a} \cdot \nabla\boldsymbol{\phi} = a_1\frac{\partial\boldsymbol{\phi}}{\partial x_1} + a_2\frac{\partial\boldsymbol{\phi}}{\partial x_2} + a_1\frac{\partial\boldsymbol{\phi}}{\partial x_2} = a_k\boldsymbol{\phi}_{,k} \qquad (A.9)$$

(iv) Divergence of a vector:

$$div\ \boldsymbol{u} = \frac{\partial u_1}{\partial x_1} + \frac{\partial u_2}{\partial x_2} + \frac{\partial u_3}{\partial x_3} = u_{k,k} \qquad (A.10)$$

(v) Cross product of two vectors:

$$\boldsymbol{a} \times \boldsymbol{b} = \begin{vmatrix} \hat{\boldsymbol{i}}_1 & \hat{\boldsymbol{i}}_2 & \hat{\boldsymbol{i}}_3 \\ a_1 & a_2 & a_3 \\ b_1 & b_2 & b_3 \end{vmatrix}$$

$$= \hat{\boldsymbol{i}}_1(a_2b_3 - a_3b_2) - \hat{\boldsymbol{i}}_2(a_1b_3 - a_3b_1) + \hat{\boldsymbol{i}}_3(a_1b_2 - a_2b_1) = e_{ijk}a_jb_k\hat{\boldsymbol{i}}_i \qquad (A.11)$$

(vi) Curl of a vector:

$$\zeta = \text{curl}\,\boldsymbol{u} = e_{ijk}u_{k,j}\hat{\boldsymbol{i}}_i \tag{A.12}$$

(vii) Product of three vectors:

$$(\boldsymbol{a} \times \boldsymbol{b})\cdot\boldsymbol{c} = e_{ijk}a_ib_jc_k \tag{A.13}$$

(viii) Inner product of a vector and gradient of a vector:

$$v_\alpha v_{k,\alpha} = e_{k\beta\delta}\zeta_\beta v_\delta + \left(\frac{v_\beta v_\beta}{2}\right)_{,k}, \tag{A.14}$$

where we have used the identity: $e_{ijk}e_{k\alpha\beta} = \delta_{i\alpha}\delta_{i\beta} - \delta_{j\alpha}\delta_{i\beta}$.

Vectors and Tensors

Suppose we change coordinates from (x_1,x_2,x_3) to (y_1,y_2,y_3) by a rotation about the origin:

$$y_j = \ell_{ij}x_j; \quad \ell_{\alpha i}\ell_{\alpha\beta} = \boldsymbol{\delta}_{\alpha\beta} \tag{A.15}$$

A vector $\boldsymbol{a}$ is a quantity with three components (a_1,a_2,a_3) that transforms to $\boldsymbol{b}$ under a simple coordinate rotation, ℓ_{ij}, such that:

$$b_j = \ell_{ij}a_i \tag{A.16}$$

A second order tensor $\underline{\underline{A}}$ is a quantity with nine components that transforms, under a simple rotation, ℓ_{ij}, according to the rule:

$$\overline{A}_{pq} = \ell_{ip}\ell_{jq}A_{ij} \tag{A.17}$$

OR equivalently $\underline{\underline{A}}$ is a second order tensor if, when multiplied by an arbitrary vector $\boldsymbol{b}$ (inner product), we get another vector $\boldsymbol{c}$:

$$A_{ij}b_j = c_i \tag{A.18}$$

Properties of vectors and tensors

(i) $\underline{\underline{A}}$ is symmetric if $A_{ij} = A_{ji}$
(ii) $\underline{\underline{A}}$ is antisymmetric if $-A_{ij} = A_{ji}$
(iii) $\underline{\underline{A}}$ is isotropic if $\overline{A}_{ij} = A_{ij}$
(iv) the inner product of two vectors is a scalar $d = u_i\, v_i$
(v) the contraction of a tensor $\underline{\underline{A}}$ is A_{ii}
(vi) the inner product of two tensors A_{ij} and B_{ij}: $C_{ik} = A_{ij}\, B_{jk}$

Gauss Theorem

Let $\boldsymbol{v}$ be a vector field, V a fluid control volume, S the surface of the control volume and $\hat{\boldsymbol{n}}$ the unit normal to the surface, then:

$$\int_V \text{div}\, \boldsymbol{v} \mathrm{d}V = \int_S \boldsymbol{v} \cdot \hat{\boldsymbol{n}} \mathrm{d}V \tag{A.19}$$

Corollaries

$$1) \quad \int_V \nabla\phi \, \mathrm{d}V = \int_S \phi \hat{\boldsymbol{n}} \, \mathrm{d}S \tag{A.20}$$

$$2) \quad \int_V \text{curl}\, \boldsymbol{v} \mathrm{d}V = \int_S \hat{\boldsymbol{n}} \times \underset{\sim}{v} \, \mathrm{d}S \tag{A.21}$$

Stokes Theorem

Suppose C is a closed curve and S is the surface subtended by the curve C:

$$\int_C v \cdot \hat{\tau} \mathrm{d}C = \int_S \hat{\boldsymbol{n}} \cdot \text{curl}\, \boldsymbol{v} \mathrm{d}S \tag{A.22}$$

Leibnitz's rule (Transport Theorem)

One Dimensional:

$$\text{Suppose} \quad I(t) \int_{a(t)}^{b(t)} f(x, t) \mathrm{d}x \tag{A.23}$$

$$\text{then} \quad \frac{\mathrm{d}I}{\mathrm{d}t} = \int_{a(t)}^{b(t)} \frac{\partial f}{\partial t} \mathrm{d}x - f(a, t)\frac{\mathrm{d}a}{\mathrm{d}t} + f(b, t)\frac{\mathrm{d}b}{\mathrm{d}t} \tag{A.24}$$

Three Dimensional:

$$\frac{\mathrm{d}}{\mathrm{d}t} \int_{V(t)} f(x, t) \mathrm{d}V = \int_{V(t)} \frac{\partial f}{\partial t} \mathrm{d}V + \int_S f \, v_n \mathrm{d}S, \tag{A.25}$$

where v_n is the velocity component normal to the surface S:

$$F(x_1, x_2, x_3, t) = 0, \tag{A.26}$$

so that:

$$v_n = \hat{n}_1 \frac{dx_i}{dt} = \frac{\partial F/\partial_t dt}{|\nabla F|}. \tag{A.27}$$

Polar coordinates

Instead of Cartesian coordinates it is sometimes convenient to use spherical polar coordinates $\{r, \boldsymbol{\theta}, \boldsymbol{\phi}\}$, where r is the radial distance, $\boldsymbol{\theta}$ is the latitudinal and $\boldsymbol{\phi}$ is the longitude coordinate, then:

$$\text{div}\,\boldsymbol{u} = \frac{1}{r^2}\frac{\partial r^2 u_r}{\partial r} + \frac{1}{r\sin\boldsymbol{\theta}}\frac{\partial \sin\boldsymbol{\theta} u_{\boldsymbol{\theta}}}{\partial \boldsymbol{\theta}} + \frac{1}{r\sin\theta}\frac{\partial u_{\boldsymbol{\phi}}}{\partial \boldsymbol{\phi}} \tag{A.28}$$

$$\nabla^2\psi = \frac{1}{r^2}\frac{\partial}{\partial r}\left(r^2\frac{\partial \psi}{\partial r}\right) + \frac{1}{r^2\sin\boldsymbol{\theta}}\frac{\partial}{\partial \boldsymbol{\theta}}\left(\sin\boldsymbol{\theta}\frac{\partial \psi}{\partial \boldsymbol{\theta}}\right) + \frac{1}{r^2\sin^2\boldsymbol{\theta}}\frac{\partial^2 \psi}{\partial \boldsymbol{\phi}^2} \tag{A.29}$$

INDEX

Note: Page numbers with "f" denote figures; "t" denote tables.

H

I

J

K

L

M

R

S

T

U

V

W

Printed in the United States
By Bookmasters